CONE PENETRATION TESTING 2022

This abstracts volume (including full keynote and invited papers) contains the proceedings of the 5th International Symposium on Cone Penetration Testing (CPT'22), held in Bologna, Italy, 8-10 June 2022. More than 450 authors - academics, researchers, practitioners and manufacturers – contributed to the peer-reviewed papers included in this book, which contains three keynote lectures, four invited lectures and 170 abstracts. The contributions provide a full picture of the current knowledge and major trends in CPT research and development, with respect to innovations in instrumentation, latest advances in data interpretation, and emerging fields of CPT application. The paper topics encompass three well-established theme categories typically addressed in CPT events: Equipment and Procedures, Interpretation, Applications. Emphasis is placed on the use of non-deterministic approaches and innovative numerical strategies for CPT data interpretation, liquefaction studies, application of CPT to offshore engineering, comparative studies between CPT and other in-situ tests. Cone Penetration Testing 2022 contains a wealth of information that could be useful for researchers, practitioners and all those working in the broad and dynamic field of cone penetration testing.

PROCEEDINGS OF THE 5TH INTERNATIONAL SYMPOSIUM ON CONE PENETRATION TESTING (CPT'22), 8-10 JUNE 2022, BOLOGNA, ITALY

Cone Penetration Testing 2022

Editors

Guido Gottardi

Department of Civil, Chemical, Environmental and Materials Engineering, University of Bologna, Bologna, Italy

Laura Tonni

Department of Civil, Chemical, Environmental and Materials Engineering, University of Bologna, Bologna, Italy

CRC Press
Taylor & Francis Group
Boca Raton London New York

CRC Press is an imprint of the
Taylor & Francis Group, an **informa** business

A BALKEMA BOOK

Cover image: photo by Claudio Turci

First published 2022
by CRC Press/Balkema
Schipholweg 107C, 2316 XC Leiden, The Netherlands
e-mail: enquiries@taylorandfrancis.com
www.routledge.com – www.taylorandfrancis.com

CRC Press/Balkema is an imprint of the Taylor & Francis Group, an informa business

Library of Congress Cataloging-in-Publication Data

A catalog record has been requested for this book

ISBN: 978-1-032-35624-2 (hbk)
ISBN: 978-1-032-35870-3 (pbk)
ISBN: 978-1-003-32909-1 (ebk)
DOI: 10.1201/9781003329091

Cone Penetration Testing 2022 – Gottardi & Tonni (eds)
© 2022 the Editor(s), ISBN 978-1-032-35624-2

Table of contents

Session 3: Applications

© 2022 the Editor(s), ISBN 978-1-032-35624-2

Preface

This volume contains the contributions to the 5th International Symposium on Cone Penetration Testing, CPT'22, that took place in Bologna (Italy) from June 8th to 10th, 2022. The Symposium, built on the successful series of CPT events held in Delft (2018), Las Vegas (2014), Huntington Beach (2010) and Linköping (1995), was organized by the Italian Geotechnical Association (AGI) and the University of Bologna, under the auspices of the ISSMGE Technical Committee TC102.

Over the last decades, the use of the cone penetration testing for site investigation has steadily grown around the world, causing the CPT to be currently recognized as the major in situ testing technique in geotechnical engineering practice. Its widespread use has in turn stimulated the development of enhanced or new types of CPT equipment, aimed at both improving the accuracy of measurements and extending the field of application, along with the increasing demand for a sound understanding of penetration mechanisms and consistent interpretation methods, suitable for the characterization of a variety of soil deposits.

The CPT symposia have always provided a unique and fruitful forum for the exchange of new ideas and discussion on key issues within the largest gathering of world's experts, academics and non-academics, working in the broad and dynamic area of cone penetration testing. The objective of CPT'22 was therefore to foster a lively debate on current trends in cone penetration testing, among Researchers, Practitioners and Manufacturers, in a friendly and creative atmosphere for discussion and collaboration, as tradition of CPT events.

More than 450 Authors, coming from academic institutions, private companies and public bodies worldwide, contributed to the peer-reviewed papers included in this volume. The Symposium Organizing Committee was especially pleased with the unprecedented large number of contributions submitted, exceeding all expectations, and with the general high quality of the papers.

A total of 186 manuscripts were received and 170 of them were finally accepted for publication in the Symposium Proceedings. The papers were sorted into three well-established theme categories, according to the subject areas typically addressed in CPT events: Equipment and procedures, Interpretation, Applications. The volume also includes three outstanding keynote papers and four invited papers presented by renowned experts in specific thematic discussion sessions on selected key topics.

A significant effort was made to provide the Authors with a rigorous and fair review of the papers. The Editors are therefore very grateful to the almost one hundred Reviewers, for their generous and valuable work.

We gratefully acknowledge the support of the Italian Geotechnical Association, which organized the Symposium in cooperation with the University of Bologna. Special thanks are extended to the AGI Secretary, Mrs Susanna Antonielli, for her tireless work and patient assistance. We would also like to express our appreciation to all the Sponsors that helped us in making this conference a success.

Finally, many thanks to all Keynote Lecturers, Invited Speakers and Authors for their enthusiastic and proactive response to CPT'22, and for their contribution to this Proceedings volume.

We do hope you will find its content of valuable and long lasting use.

Guido Gottardi
Laura Tonni

Symposium organizers

Italian Geotechnical Association, AGI

Alma Mater Studiorum – Università di Bologna

Under the auspices of

ISSMGE TC 102 "In-Situ Testing"

© 2022 the Editor(s), ISBN 978-1-032-35624-2

Committees

ORGANIZING COMMITTEE

Laura **Tonni** – *Chair, Università di Bologna, Italy*
Sebastiano **Rampello** – *Co-Chair, AGI Delegate, Sapienza Università di Roma, Italy*
Claudio **Soccodato** – *AGI General Secretary*
Susanna **Antonielli** – *AGI Secretary*
Sara **Amoroso** – *Università di Chieti-Pescara, Italy*
Giovanni **Biondi** – *Università di Messina, Italy*
Giovanni **Bosco** – *Università dell'Aquila, Italy*
Simonetta **Cola** – *Università di Padova, Italy*
Andrea **Dominijanni** – *Politecnico di Torino, Italy*
Laura **Govoni** – *Università di Bologna, Italy*
Carmine Gerardo **Gragnano** – *Università di Bologna, Italy*
Chiara **Iodice** – *Università della Campania Luigi Vanvitelli, Italy*
Maria Clorinda **Mandaglio** – *Università di Salerno, Italy*
Michela **Marchi** – *Università di Bologna, Italy*
Luca **Masini** – *Sapienza Università di Roma, Italy*
Claudia **Meisina** – *Università di Pavia, Italy*
Lorella **Montrasio** – *Università di Parma, Italy*
Irene **Rocchi** –*Technical University of Denmark, Denmark*
Gianpiero **Russo** – *Università di Napoli Federico II, Italy*
Ermanno **Pagani** – *Pagani Geotechnical Equipment, Italy*
Daniele **Spizzichino** - *ISPRA Istituto Superiore per la Protezione e la Ricerca Ambientale, Italy*
Stefano **Stacul** – *Università di Pisa, Italy*
Marco **Uzielli** – *Università di Firenze, Italy*
Maurizio **Ziccarelli** – *Università di Palermo, Italy*

Cone Penetration Testing 2022 – Gottardi & Tonni (eds)
© 2022 the Editor(s), ISBN 978-1-032-35624-2

International advisory board

Reviewers

Glenda **Abate** – *Università di Catania, Italy*
Sara **Amoroso** – *Università di Chieti-Pescara, Italy*
Marcos **Arroyo** – *Universitat Politècnica de Catalunya, Spain*
Dusan **Berisavljevic** – *University of Belgrade, Serbia*
Silvia **Bersan** – *Crux Engineering BV, The Netherlands*
Ilaria **Bertolini** – *Università di Bologna, Italy*
Davide **Besenzon** – *Escuela Superior Politécnica del Litoral, Ecuador*
Katia V. **Bicalho** – *Federal University of Espirito Santo, Brazil*
Giovanni **Biondi** – *Università di Messina, Italy*
Giovanni **Bosco** – *Università di Messina, Italy*
Lorenzo **Brezzi** – *Università di Padova, Italy*
Vincenzo **Butticé** – *Università di Palermo, Italy*
Francesco **Calvetti** – *Politecnico di Milano, Italy*
Orazio **Casablanca** – *Università di Messina, Italy*
Zijun **Cao** – *Wuhan University, China*
Francesca **Casini** – *Università degli Studi di Roma Tor Vergata, Italy*
Francesca **Ceccato** – *Università di Padova, Italy*
Manuela **Cecconi** – *Università di Perugia, Italy*
Jared A. **Charles** – *University of Southampton, United Kingdom*
Anna **Chiaradonna** – *Università dell'Aquila, Italy*
Andrea **Ciancimino** – *Politecnico di Torino, Italy*
Matteo O. **Ciantia** – *University of Dundee, United Kingdom*
Simonetta **Cola** – *Università di Padova, Italy*
Giorgia **Dalla Santa** – *Università di Padova, Italy*
Andrea **Diambra** – *University of Bristol, United Kingdom*
Francesco **Di Buccio** – *Università di Chieti-Pescara, Italy*
Bruno **Di Buò** – *Tampere University, Finland*
Giuseppe **Di Filippo** – *Università di Messina, Italy*
Marco **D'Ignazio** – *University of Tampere, Finland*
Elena **Dodaro** – *Università di Bologna, Italy*
James **Doherty** – *University of Western Australia, Australia*
Settimio **Ferlisi** – *Università di Salerno, Italy*
Alessio **Ferrari** – *Università di Palermo, Italy*
Cristiana **Ferreira** – *Faculdade de Engenharia da Universidade do Porto, Portugal*
Aligi **Foglia** – *Fraunhofer IWES, Germany*
Fabio **Gabrieli** – *Università di Padova, Italy*
Andrea **Galli** – *Politecnico di Milano, Italy*
Domenico **Gaudio** – *Sapienza, Università di Roma, Italy*
Heraldo L. **Giacheti** – *São Paulo State University (Unesp), Brazil*

Domenico **Gioffré** – *Università di Pavia, Italy*
Daniela **Giretti** – *Università di Bergamo, Italy*
Guido **Gottardi** – *Università di Bologna, Italy*
Laura **Govoni** – *Università di Bologna, Italy*
Jürgen **Grabe** – *Hamburg University of Technology, Germany*
Carmine Gerardo **Gragnano** – *Università di Bologna, Italy*
Paolo Maria **Guarino** – *ISPRA Istituto Superiore per la Protezione e la Ricerca Ambientale, Italy*
Michael **Hicks** – *Delft University of Technology, The Netherlands*
Chiara **Iodice** – *Università degli Studi della Campania Luigi Vanvitelli, Italy*
Maria **Iovino** – *Università degli studi di Napoli "Parthenope", Italy*
Stefan **Kreiter** – *University of Bremen, Germany*
Tim **Länsivaara** – *Tampere University, Finland*
Valentina **Lentini** – *Università degli studi di Enna "Kore", Italy*
Jean-Sébastien **L'Heureux** – *Norwegian Geotechnical Institute, Norway*
Stefania **Lirer** – *Università degli Studi Guglielmo Marconi, Italy*
Diego **Lo Presti** – *Università di Pisa, Italy*
Tom **Lunne** – *Norwegian Geotechnical Institute, Norway*
Claudia **Madiai** – *Università di Firenze, Italy*
Rosa Maria Stefania **Maiorano** – *Università degli studi di Napoli "Parthenope", Italy*
Maria Clorinda **Mandaglio** – *Università di Salerno, Italy*
Alessandro **Mandolini** – *Università degli Studi della Campania Luigi Vanvitelli, Italy*
Diego **Marchetti** – *Studio Prof. Marchetti, Italy*
Michela **Marchi** – *Università di Bologna, Italy*
Luca **Masini** – *Sapienza Università di Roma, Italy*
Rossella **Massimino** – *Università di Catania, Italy*
Paul W. **Mayne** – *Georgia Institute of Technology, United States*
Claudia **Meisina** – *Università di Pavia, Italy*
Lucia **Mele** – *Università di Napoli Federico II, Italy*
Fausto Andrés **Molina-Gómez** – *University of Porto, Portugal*
Paola **Monaco** – *Università dell'Aquila, Italy*
Lorella **Montrasio** – *Università di Parma, Italy*
Diane **Moug** – *Portland State University, United States*
Zhongkun **Ouyang** – *University of New Haven, United States*
Priscilla **Paniagua** – *Norwegian Geotechnical Institute, Norway*
Paul José **Pinedo Vilcahuamán** – *Universitat Politècnica de Catalunya, Spain*
Nick **Ramsey** – *Fugro Australia, Australia*
Sara **Rios** – *University of Porto, Portugal*
Peter K. **Robertson** – *PK Robertson Inc & Gregg Drilling Inc, United States*
Irene **Rocchi** – *Technical University of Denmark, Denmark*
Fabio **Rollo** – *Sapienza Università di Roma, Italy*
Alessandra **Rosati** – *Sapienza Università di Roma, Italy*
Marco **Rosone** – *Università di Palermo, Italy*
Giacomo **Russo** – *Università di Napoli Federico II, Italy*
Gianpiero **Russo** – *Università di Napoli Federico II, Italy*
Fernando **Schnaid** – *Universidade Federal do Rio Grande do Sul, Brazil*

James **Schneider** – *US Army Corps of Engineers, United States*
Britta **Søndergaard** – *COWI, Denmark*
Daniele **Spizzichino** – *ISPRA Istituto Superiore per la Protezione e la Ricerca Ambientale, Italy*
Nunziante **Squeglia** – *Università di Pisa, Italy*
Stefano **Stacul** – *Università di Pisa, Italy*
Francine Chantal **Tchamaleu Pangop** – *Università di Padova, Italy*
Laura **Tonni** – *Università di Bologna, Italy*
Giuseppe **Tropeano** – *Università di Cagliari, Italy*
Marco **Uzielli** – *Università di Firenze, Italy*
Phil **Watson** – *UWA, University of Western Australia, Australia*
Varvara **Zania** – *Technical University of Denmark, Denmark*
Ningning **Zhang** – *RWTH Aachen University, Germany*
Maurizio **Ziccarelli** – *Università di Palermo, Italy*
Paolo **Zimmaro** – *Università della Calabria, Italy*

Sponsors

Platinum Sponsors

Gold sponsors

Exhibitors in Outdoor Spaces

Silver Sponsors

Bronze Sponsors

Keynote lectures

Cone Penetration Testing 2022 – Gottardi & Tonni (eds)
© 2022 the Author(s), ISBN 978-1-032-35624-2

New CPT methods for evaluation of the axial capacity of driven piles

B.M. Lehane & E.J. Bittar
University of Western Australia, Australia

S. Lacasse, Z. Liu & F. Nadim
Norwegian Geotechnical Institute, Norway

ABSTRACT: High costs associated with offshore foundation installations have provided strong impetus to the offshore energy sector in the search for more reliable design methods. This paper provides a summary of an Industry sponsored project that led to the development of new CPT-based design methods for the evaluation of the axial capacity of driven piles. Particular attention was given to the need for the new methods to be applicable to large diameter offshore piles given that many existing methods are derived by calibration with capacities measured in static pile load tests on smaller diameter onshore piles. The basic mechanisms supporting the general format of the expressions proposed for shaft friction and end bearing in sands and clays are described. It is shown how the new expressions, which are calibrated against a database of the most reliable load tests reported in the literature, lead to better predictions of capacity compared to other methods and can also satisfactorily predict the capacity of piles driven into deposits comprising interbedded layers of sand, silt and clay. Recommendations for the prediction of pile displacements at working loads using CPT data are also presented.

1 INTRODUCTION

The popularity of the CPT and the similarity between the mode of penetration of a cone and a driven pile have provided strong motivation in the search for direct correlations between the CPT q_t value and axial pile capacity. The best known of initial correlations proposed for pile shaft and base resistance in a range of soil types were developed about 40 years ago and include methods recommended by de Beer (1972), Schmertmann (1978), de Ruiter & Beringen (1979), Zhou et al. (1982), Bustamante & Gianeselli (1982) and Van Impe (1986).

These and other methods have since been assessed using specific databases of static load tests by workers such as Briaud & Tucker (1988), Niazi & Mayne (2013), Eslami et al. (2014), Hu et al. (2012) and Amirmojahedi & Abu-Farsakh (2019). These studies show that there is no consensus of opinion regarding the relative reliability of the methods with some methods providing good predictions for a given database but poor predictions for other databases. This lack of consensus is partly due to significant inconsistencies between the respective databases which arise due to (i) different definitions of ultimate axial capacity, (ii) contradictory interpretations of load test data, (iii) uncertainty in CPT q_t values at test pile locations, (iv) inclusion or exclusion of layered/mixed soil deposits and different pile types and (v) separate assumptions relating to the effects of time, loading direction, re-testing and loading rate. In addition, the formulations of most methods do not explicitly incorporate many characteristics of driven pile behaviour that have been well proven in experimental research programmes (discussed below).

The Offshore Energy sector recognized the advantages of a CPT-based method for axial capacity assessment but also understood the need to quantify the reliability of such a method using a database that had the backing of much of the profession. A Joint Industry Project (JIP) was set up in 2014 to achieve this objective. The JIP was managed by the Norwegian Geotechnical Institute (NGI) with support from the University of Western Australia (UWA) and was sponsored by Equinor, Ørsted, Lundin, ONGC, Petrobras and DNV GL. A 'Team of Experts' worked over a period of 3 years with guidance from the JIP steering committee to assess the suitability of about 600 pile load tests for inclusion in the database. A good CPT coverage close to pile tests was required and only first-time load tests on driven piles that were statically loaded to a displacement of 10% of the pile diameter were included. The final tests selected make up, what is referred to as, the 'Unified database' and comprises 71 test piles in silica sand and 49 pile tests in clay. Full details of the 'Unified database' and the steps followed in its compilation are provided in Lehane et al. (2017).

A subsequent phase of the JIP began in 2017 (sponsored by Equinor, Lundin Norway, Ørsted, ONGC, BP, Total, ExxonMobil, EnBW, EDF,

Aramco, SSER and HDEC) with the aim of developing new 'Unified' methods for the prediction of the axial capacity of driven piles in sand and clay. This initiative was motivated by the desire of Industry to replace the 4 CPT-based methods for driven piles in sand included in the API (2011) recommendations with a single method. The term 'unified' is employed as the method was developed with input from the proponents of the 4 most current CPT-based sand methods used offshore. Although no CPT method for driven piles in clay and silt are currently provided in API/ISO documentation, the inclusion of such a method was seen to be an important step forward to reduce dependence on laboratory measurements of undrained strength required by the existing 'alpha' method.

The JIP was concluded successfully in 2021 and the methods developed will be incorporated in the next versions of the ISO-19901-4 and API recommendations.

This paper provides an overview of the new CPT methods with a focus on justification of the basis of the formulations employed. A full description of the methods is provided in Lehane et al. (2020, 2022) and in Nadim et al. (2021). The 'sand' and 'clay' formulations were calibrated using the Unified database and their application to interbedded deposits comprising sand, silt and clay layers is examined in this paper. Recommendations are also provided to assist prediction of the displacement of driven piles at working loads using CPT data.

2 CPT METHODS FOR CAPACITY OF DRIVEN PILES IN SAND

2.1 *Approaches for evaluation of shaft friction*

Traditional CPT methods relate the local ultimate friction (τ_f) directly to the cone resistance (q_t) via a single factor, β:

$$\tau_f = q_t / \beta \quad (1)$$

Bustamante & Gianeselli (1982), for example, recommend β values for driven steel tubular piles of 120 and 200 in loose and dense sands respectively and propose limiting τ_f values in each of these deposits to respective maximum values of 35 kPa and 120 kPa. These and other similar recommendations reported in the literature ignore the following well-known characteristics of driven piles in sand observed from instrumented pile tests (Lehane et al. 2020):

i. Ultimate shaft friction (τ_f) varies is direct proportional to the tangent of the interface friction angle between the sand and pile shaft (tan δ) i.e. shaft friction is governed by Coulomb's law.
ii. τ_f developed in any soil horizon reduces with the distance of that soil horizon from the pile tip (*h*); this arises largely due to the progressive cycling of sand at any particular level as installation progresses.
iii. Smaller diameter piles generate larger frictions due to constrained dilation under shear at the interface.
iv. Open-ended pipe piles generate lower shaft frictions than closed-ended piles due to lower levels of displacement imparted to the sand mass.
v. Ageing effects lead to increases in shaft friction with time after pile installation.

These characteristics are not incorporated in Eqn. 1 and prompted the development of a new generation of CPT methods included in API (2011), namely those referred to as Fugro-05 (Kolk et al. 2005), ICP-05 (Jardine et al. 2005), NGI-05 (Clausen et al. 2005) and UWA-05 (Lehane et al. 2005). While all of these newer methods do not include all of the characteristics listed above, each incorporates a direct proportional relationship between τ_f and q_c and a reduction of τ_f with distance from the pile tip (*h*). These dependencies were first clearly revealed in experiments conducted in 1989 with the Imperial College instrumented pile in medium dense sand at Labenne, France (Lehane 1992). Figure 1 plots bounds to a large number of equalised radial stress measurements (σ'_{rc}) obtained in these experiments and displays an obvious trend for the σ'_{rc}/q_c ratio to be constant in a given soil horizon and to reduce with increasing distance of that horizon above the pile tip (noting that there is a direct relationship between τ_f and σ'_{rc} via Coulomb's law).

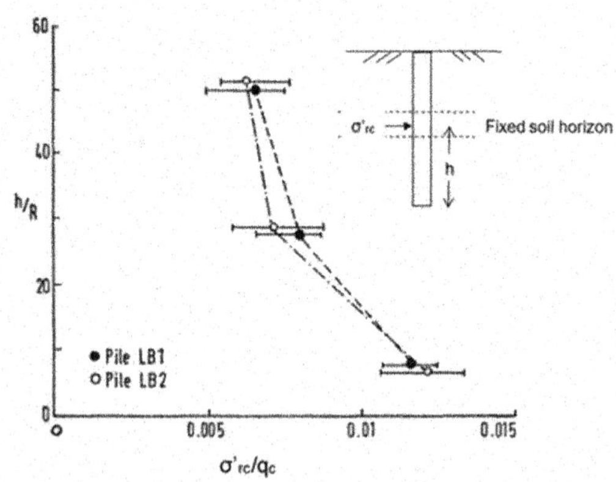

Figure 1. Dependence of equalised radial effective stress on q_c and normalised distance from the pile tip (*h/R*) at Labenne, France (Lehane 1992) [R=D/2 and τ_f varies directly with σ'_{rc}].

The trend shown on Figure 1, which has since been confirmed by Chow (1997) and others, is the key element of the shaft friction formulations of the

UWA-05, NGI-05, ICP-05 and Fugro-05 methods. These shaft frictions correspond to frictions that can be developed in the 2 to 4 week period following driving as the methods' calibrations employed static load test data recorded in the same period.

2.2 *Approaches for evaluation of base resistance*

Given the comparable modes of penetration, all historical CPT methods for closed-ended driven piles in sand relate the ultimate end bearing directly to the q_c value in the vicinity of the pile tip. However, the ultimate end bearing stress, defined at a displacement of 0.1D ($q_{b0.1}$), is typically only about 50% of q_c for a closed-ended piles driven in homogeneous sand deposits because a displacement of the pile tip of order of 1D is required to reach steady state penetration conditions (as exist during cone penetration). In addition to partial mobilisation at a displacement of 0.1D, Van Mierlo & Koppejan (1952) recognised the importance of the scale difference between a pile and a penetrometer. Recognition of this effect subsequently led to a variety of proposals for q_c averaging techniques, the most popular of which relates $q_{b0.1}$ to a simple average value of q_c values in a zone extending 1.5D above and 1.5D below a pile tip.

Compared with closed-ended piles, pipe piles induce lower levels of displacement (or disturbance) to the sand near their bases during installation. The degree of partial plugging (and hence disturbance) reduces as the pile diameter increases. Consequently, as shown by Gavin & Lehane (2003), the end bearing of large diameter piles under static loading (when the plug remains stationary) reduces to that of a bored pile for which $q_{b0.1}$ is typically about 15% of the q_c value (Lehane & Randolph 2002). This diameter dependence is incorporated in a number of different ways in the Fugro-05, ICP-05, NGI-05 and UWA-05 correlations for $q_{b0.1}$. No set-up of end bearing resistance has been observed for driven piles in sand.

2.3 *Performance of CPT methods in API (2011)*

The Fugro-05, ICP-05, NGI-05 and UWA-05 CPT methods included in API (2011) were calibrated using databases compiled specifically for the development of these methods. It is therefore of interest to examine their performance against the 'Unified database', which is fully supported by all researchers involved in the four methods. The findings from this exercise are summarised in Table 1, which presents statistics for ratios of measured to calculated capacities (Q_m/Q_c) for the four methods as well as for the earth pressure theory method in API RP2A (2011) and the new 'Unified method' (discussed below).

The results in Table 1 show that, while the earth pressure approach in API over-predicts pile capacities by an average of 66%, the mean Q_m/Q_c ratio of measured to calculated capacity for the four API CPT methods is close to unity. The spread in predictions for the earth pressure approach, as measured by the CoV for Q_m/Q_c, is also higher than for the CPT methods and is indicative of a significantly lower level of reliability. The UWA-05 and ICP-05 methods are the best performing methods with the lowest CoVs.

Table 1. Mean (μ) and coefficient of variation (CoV) of Q_m/Q_c for the Unified database of driven piles in silica sand.

	All open & closed-ended piles (total capacity) 71 Piles	
Method	μ	CoV
API (2011), K tan δ (β) approach	1.66	0.56
Fugro-05	0.99	0.40
ICP-05	1.04	0.27
NGI-05	0.99	0.34
UWA-05	1.06	0.26
Unified Method	1.05	0.24

3 UNIFIED CPT METHOD FOR DRIVEN PILES IN SAND

While the UWA-05 and ICP-05 methods performed relatively well against the 'Unified database', a primary aim of the JIP was to develop a unified and un-affiliated method that had the support of those involved in the derivation of the API β method and the four API CPT methods. The first point of agreement was that, in line with the findings of Lehane et al. (1993), and others, the following expression should form the basis of the correlation for peak shaft (τ_f):

$$\tau_f=(\sigma'_{rc}+\Delta\sigma'_{rd})\tan\delta_f \quad (2)$$

where σ'_{rc} is the stationary (equalised) radial effective stress, $\Delta\sigma'_{rd}$ is the increase in radial effective stress during pile loading (attributed to dilation) and δ_f is the constant volume sand-pile interface friction angle. On review of the state-of-the-art, it was agreed that the Unified method should incorporate considerations described in the following.

3.1 *Interface friction angle (δ_f)*

In the absence of site-specific tests to measure ultimate interface friction angles (δ_f), the ICP-05 and UWA-05 methods propose the variation with the mean effective particle size (d_{50}) indicated by the curve shown in Figure 2. This variation was deduced from direct shear box tests on steel interfaces with a roughness typical of industrial piles but that did not include pre-shearing to large relative displacements. Yang et al.

(2010), and others, have since shown that crushing of sand at the pile tip and subsequent shearing during installation reduces the grading to that of a fine sand. Interface shear angles measured in the Bishop ring shear apparatus in tests that induced a large level of pre-shearing are plotted on Figure 2 confirmed the relatively low sensitivity of δ_f to the initial d_{50} value. Similar tests reported by Liu et al. (2019) also show that δ_f has virtually no dependence on the (non-plastic) fines content of typical siliceous sands and on the normal stress level. It was therefore concluded that, in the absence of site specific ring shear interface tests, adoption of a constant δ_f value of 29° is a reasonable assumption for all piles in the load test database.

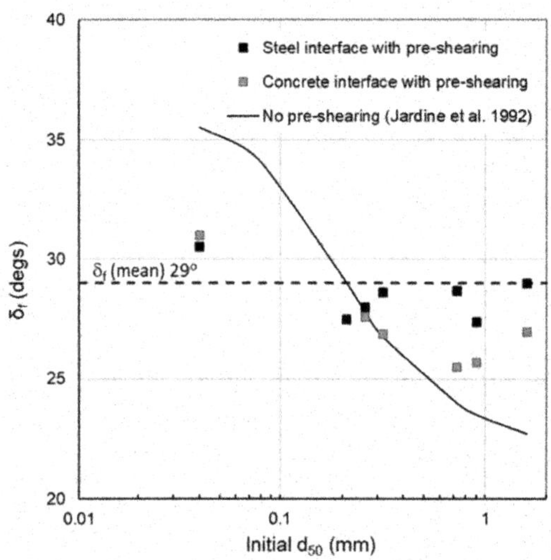

Figure 2. Interface friction angle and d_{50} for roughness typical of industrial piles (centreline average roughness of 5 to 20μm) (Barmpopoulos & Ho, 2009); Ho et al., 2011).

3.2 *Increase in radial stress ($\Delta\sigma'_{rd}$) during pile loading*

The restraint to dilation at the pile shaft during pile loading provided by the surrounding sand leads to an increase in radial stress on the pile shaft ($\Delta\sigma'_{rd}$) and hence to the peak shaft friction. This increase can be assessed from cavity expansion (CE) theory, where G is the operational shear modulus of the sand mass, y is the dilation of the sand at the shaft interface and $y/2D$ is the cavity strain:

$$\Delta\sigma'_{rd} = 4Gy/D \tag{3}$$

The UWA-05 and ICP-05 methods assume that the cavity strain is small enough for conditions to be fully elastic and hence the methods equate G with the small-strain elastic value (G_0). However, data from constant normal stiffness direct shear interface tests and tests on centrifuge piles with a range of diameters presented in Lehane et al. (2005) show that the cavity strains can be relatively large and that the operational G value is less than G_0 for typical pile diameters in the 'Unified database'. The following revised approximate expression was deduced using $\Delta\sigma'_{rd}$ measurements on jacked piles and parallel numerical analyses (Lehane et al. 2020):

$$\Delta\sigma'_{rd} = \left(\frac{q_c}{10}\right)\left(\frac{q_c}{\sigma'_v}\right)^{-0.33}\left(\frac{d_{CPT}}{D}\right) \tag{4}$$

with d_{CPT} =35.7mm, which is the usual CPT diameter.

Most piles in the Unified database have a diameter (D) between 350mm and 800mm. Equation (4) predicts that the increase in peak friction due to dilation (=$\Delta\sigma'_{rd}$ $\tan\delta_f$) for a 20m long pile in medium dense sand is about 35% for D=350mm but only 10% for D=800mm. The relative influence of dilation is clearly an important consideration when extrapolating from the smaller diameter piles in the database to larger diameter offshore piles. The relative influence of dilation is also greatest in looser sands and for longer piles.

3.3 *Allowance for partial plugging*

Many of the pipe piles in the Unified database (with D= 550 ± 250mm) experienced partial plugging during pile driving. The additional shaft and base capacity that such plugging induces (e.g. see Gavin & Lehane 2003) needs to be accounted for correctly in the database analysis to ensure safe extrapolation to the capacity of full scale offshore piles.

White et al. (2005) used a cavity expansion analogy to deduce that the equalised lateral effective stress acting on the pile shaft (σ'_{rc}) varies with the effective area ratio (A_{re}) raised to a power of between about 0.3 and 0.45 while Xu et al. (2008) present experimental data showing how $q_{b0.1}$ varies in proportion to A_{re}.

A_{re} provides a measure of the level of soil displacement in any given soil horizon, and is defined as:

$$A_{re} = 1 - IFR(D_i/D)^2 \tag{5}$$

where IFR is ratio of the change in plug length to change in pile embedment and D_i is the internal pile diameter. A_{re} varies from unity for a closed-ended or fully plugged pile to a value of $4t/D$ for a large diameter coring pile (where t is the pile wall thickness). As the IFR is not measured routinely, the analysis of the unified database substituted the plug length ratio (PLR) for IFR, noting that the PLR is the average IFR during pile driving and equal to the ratio of the final plug length to the embedded pile length.

The PLR is primarily a function of the internal pile diameter (D_i) (e.g. see Gudavalli et al. 2013) and the following approximate expression was derived based on available records (d_{CPT} =35.7mm):

$$PLR \approx IFR = \tanh\left[0.3\left(\frac{D_i}{d_{CPT}}\right)^{0.5}\right] \quad (6)$$

The database analysis examined the influence of various exponents of A_{re} in the search for a best-fit expression for τ_f.

3.4 *Time effects*

The shaft capacity of driven piles in silica sand increases with time over a period of least one year (e.g. Chow et al. 1998, Jardine et al. 2006, Karlsrud et al. 2014, Gavin et al. 2015). Such increases are not exhibited by bored piles and may be viewed as a recovery process following the 'trauma' of driven pile installation (Lim and Lehane 2014, Anusic et al. 2019). The new CPT method is calibrated using the Unified database comprising static load tests with a median equalisation period (or set-up time) of about two weeks. It is therefore likely to underestimate long term capacities and over-estimate short term capacities (including driving resistance).

3.5 *Formulation for σ'_{rc}*

The observed reduction in the equalised radial effective stress (σ'_{rc}) with the distance from the pile tip (h) or normalised distance from the pile tip (h/D) can be described as a power law relationship (as in the ICP-05 and UWA-05 methods) or as an exponential relationship (proposed by Randolph et al. 1994 and Salgado et al. 2011). Lehane et al. (2020) show that the bias in the database, with respect to diameter, of the ratio of measured to calculated capacities (Q_m/Q_c) can be removed when a h/D term is used instead of 'h' (as proposed by Alm & Hamre 2001). Lehane et al. (2020) also show that while the power law and exponential variations with h/D lead to very similar statistics for the Q_m/Q_c ratios, the exponential relationship tended to predict marginally larger capacities for piles with L/D<20 compared with the power law form. The power law form was retained as many piles used in the offshore wind Industry have lower L/D values.

3.6 *Tension compression ratio*

The database piles clearly showed that shaft friction measured in tension are, on average, about 75% of the shaft friction of compression piles. Limited data exist in relation to differences in the distribution of τ_f along test piles and, in the absence of other information, the global (best-fit) tension to compression ratio (f_L) of 0.75 was employed in the new method.

3.7 *Formulation for end bearing*

The Unified method relates the ultimate end bearing ($q_{b0.1}$) with the effective area ratio (A_{re}) for pipe piles in line with observations of Gavin & Lehane (2003) and as adopted in the UWA-05 method. The end bearing formulation of ICP-05 incorporates this dependency indirectly by allowing $q_{b0.1}$ for pipe piles to reduce with pile diameter.

The Unified method relates $q_{b0.1}$ with q_p, where q_p is the end bearing resistance expected for an 'imaginary cone' that has the same diameter as the pile being considered (or equivalent diameter for a pipe pile= $(A_{re}/\pi)^{0.5}$). The value of q_p is determined using a component of a 'thin-layer' procedure described by Boulanger and DeJong (2018) and its application in piling calculations is explained by Bittar et al. (2020). Although q_p is a rational and objective way of obtaining an averaged cone resistance near a pile base of given diameter, its determination requires use of software (freely downloadable from *https://faculty.engineering.ucdavis.edu/boulanger/research-interests/*) which is not appealing to many practitioners. The new ISO-19901-4 recommendations therefore suggest taking q_p as the average q_c value in the zone extending 1.5D above and below the pile tip, unless conditions at the pile tip are highly variable. Separate studies show that q_p for the Unified database piles is, on average, 20% higher than q_{Dutch}, where q_{Dutch} is the CPT resistance determined using the 'Dutch' averaging technique (Van Mierlo & Koppejan 1952)

3.8 *Formulation of Unified CPT method in sand and predictive performance*

The final formulations decided upon for the Unified CPT method for piles in sand (Zone 6 of the soil behaviour type chart) are provided in Table 2. When these are applied to the Unified database, the average ratio of measured to calculated capacities (Q_m/Q_c) is 1.05 while the CoV for Q_m/Q_c is 0.24. These statistics are a marginal improvement on those for the UWA-05 and ICP-05 methods (see Table 1). The method is, however, considered to be a significant step forward as it replaces the four API CPT methods with full support of the authors for these methods and it incorporates state-of-the-art understanding of the mechanisms controlling the axial capacity of driven piles in sand. It should be noted that the method potentially under-estimates axial capacity in gravelly sands (Zone 7 of the SBT chart), where the presence of gravels leads to higher average q_c values. The method is only applicable to piles driven in a conventional manner and should not be used for jacked piles or piles installed by vibration.

Table 2. General formulations for Unified method in sands, silts and clays *(see https://pile-capacity-uwa.com/).*

$Q_{shaft} = \pi D \int_0^L \tau_f \, dz$
$Q_{base} = q_{b0.1} (\pi D^2/4)$

Capacity estimate for piles for flexible piles in strain-softening clays requires load-transfer analysis using load transfer curves given in API (2011)

Sands: Zone 6 of SBT chart (I_c< 2.1)

$\tau_f = f_L (\sigma'_{rc} + \Delta\sigma'_{rd}) \tan 29°$
$q_{b0.1} = [0.12 + 0.38A_{re}]q_p$; for plugged base (expected when L/D>5)
$q_{b0.1} = A_{re}\, q_p$; for unplugged base
$\sigma'_{rc} = (q_c/44)\, A_{re}^{0.3}\, [\mathrm{Max}[1, h/D]]^{-0.4}$
$\Delta\sigma'_{rd} = (q_c/10)\, (q_c/\sigma'_v)^{-0.33}\, (d_{CPT}/D)$
$A_{re} = 1 - PLR\, (D_i/D)^2$
$PLR \approx \tanh\, [0.3\, (D_i/d_{CPT})^{0.5}]$
f_L= 0.75 in tension, 1.0 in compression
q_p can be taken as the average q_c within a zone 1.5D above and below the pile tip or determined using the procedure described Boulanger & De Jong (12018) and Bittar et al. (2020)

Clays: Zones 1,2,3 & 4 of SBT chart

$\tau_f = 0.07\, F_{st}\, q_t\, [\mathrm{Max}[1, h/D^*]]^{-0.25}$
$q_{b0.1} = [0.2 + 0.6A_{re}]q_p$

F_{st} =1 for clays with I_{z1}>0, in Zones 2, 3 and 4 on the SBT Chart ($I_c \geq 2.6$)
F_{st} =0.5±0.2 clays with I_{z1}<0, in Zone 1 on the SBT Chart
$D^* = (D^2 - D_i^2)^{0.5}$ for an open-ended pile and D^*=D for a closed-ended pile
$I_{z1} = Q_{tn} - 12exp^{(-1.4F_r)}$
q_p =average q_t value in zone between the pile tip and 1D below the pile tip (closed-ended/ plugged pile) or average q_t to a depth of 20*t* below the pile tip (large diameter, unplugged pile)

Silts: Zone 5 of the SBT chart (2.1< I_c <2.6)

Apply equations as for Zone 6 using corrected q_c value determined as:
$q_c = [3.93\, I_c^2 - 14.78\, I_c + 14.78]\, q_t$

Notation

h	*L* – *z*, where *L* is pile embedment length and *z* is the depth)
D	Pile outer diameter
D_i	Internal diameter of a pipe pile
f_s	Cone sleeve friction
F_r	Friction ratio (expressed as a percentage)=$f_s/(q_t-\sigma_{v0})$
q_t	Total (corrected) CPT end resistance (=q_c in sands)
Q_{tn}	Normalised cone resistance $[(q_t-\sigma_{v0})/p_a]/(p_a/\sigma'_{v0})^n$; p_a=ref. stress=100 kPa
I_c	Consistency index, function of Q_{tn} and F_r; see Robertson (2009)
n	Stress exponent for Q_{tn}, function of I_c and σ'_{v0}; see Robertson (2009)
t	Pile wall thickness
SBT	Soil Behaviour type; see Robertson (2009)
σ_{v0}	Total vertical stress
σ'_{v0}	Effective vertical stress

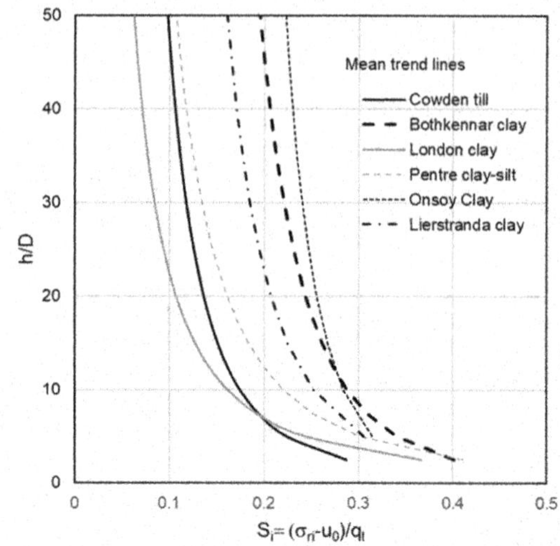

Figure 3. Mean recorded variations of normalised radial total stresses (S_i) with h/D measured during installation of closed-ended instrumented piles.

4 CPT METHODS FOR CAPACITY OF DRIVEN PILES IN CLAY

4.1 *Shaft friction*

Relationships between shaft friction and the measured and corrected cone resistances (q_c and q_t) for driven piles in clay have been proposed for many years e.g. Bustamante & Gianeselli (1982), Almeida et al. (1996), Lehane et al. (2000, 2013), Eslami & Fellenius (1997) and Niazi & Mayne (2016). The newly proposed Unified CPT method for clays builds on these methods but is based primarily on observations made in high quality instrumented pile experiments conducted by a number of institutions, most notably at the Norwegian Geotechnical Institute and Imperial College London.

The basis of the correlation between shaft friction and the (corrected) CPT end resistance (q_t) for the new Unified method is the observation that, in any particular clay, the radial total stress developed during installation (σ_{ri}) on the shaft of a closed-ended displacement pile varies directly with q_t and reduces with the normalised distance from the pile tip (h/D). Examples of observed trends are shown on Figure 3, which plots the mean measured variation of the normalised radial total stress in six different clays against h/D, where the normalised radial total stress (S_i) is defined as follows and u_0 is the hydrostatic or ambient pore pressure:

$$S_i = (\sigma_{ri} - u_0)/q_t \tag{7}$$

The mean trend lines indicated on Figure 3 have a typical standard deviation of 25% and are evidently dependent on the clay type. Following installation, instrumentation on driven piles shows that radial total stresses fall as excess pore pressure dissipate while radial effective stresses (σ'_r) increase. σ'_r reaches a fully equalised value (σ'_{rc}) after equalisation of radial total stresses and full pore pressure dissipation. When a pile is loaded to failure after equalisation, radial effective stresses reduce attaining a value of σ'_{rf} at peak local shear stress (τ_f).

These stages in the life of a driven pile are incorporated in the following expression for τ_f, which is based on Coulomb's friction law:

$$\tau_f = \sigma'_{rf}\tan\delta = q_t S_i (S_c/S_i)\ f_l \tan\delta_f \qquad (8a)$$

where

$$S_c = (\sigma_{rc} - u_0)/q_t = \sigma'_{rc}/q_t \qquad (8b)$$

$$f_l = \sigma'_{rf}/\sigma'_{rc} \qquad (8c)$$

Equation (8) illustrates that the relationship between τ_f and q_t depends on the degree of relaxation of radial total stresses during equalisation (expressed by S_c/S_i), the relative change in σ'_r during load testing (f_l) and the interface friction angle between the pile and clay (δ_f). Numerical simulations such as those of Whittle & Baligh (1988), as well as experiments, indicate that S_c/S_i ratios are lower in sensitive and low OCR clays where installation causes a significant degree of remoulding and reduction in effective stress. The S_c/S_i values measured experimentally range from 0.15 in (highly sensitive) Lierstranda clay to 1.0 in the heavily overconsolidated London clay. Measured values of f_l are in the range of 0.8 to 1.0 in all clays but average at about 0.6 in Lierstranda clay. The operational coefficient of friction ($\tan\delta_f$) is strongly affected by the clay mineralogy, amongst other factors, and for the six clays considered in Figure 3 varies by more than a factor of 2 (from a $\tan\delta_f$ value of 0.23 in London Clay to 0.55 in Bothkennar clay).

The compounding effects of differences in respective clays of the S_c/S_i, f_l and $\tan\delta_f$ values as well as the S_i relationship with h/D (shown on Figure 3) reveal a complex relationship between τ_f and q_t. Nevertheless, when ratios of τ_f/q_t with h/D measured in instrumented pile tests are plotted, as shown in Figure 4a, it is evident that a number of compensatory factors lead to a relatively consistent relationship between τ_f/q_t and h/D in all of clays apart from Lierstranda, where the product of S_c/S_i and f_l is, on average, about four times less than the other clays. Best-fit mean trend lines for the instrumented data included in Figure 4a are presented on Figure 4b and, as shown later, are consistent with best-fit expressions for the capacities of the piles in the Unified database.

4.2 *End bearing*

The contribution of end bearing to the capacity of driven piles in clay is relatively small and therefore

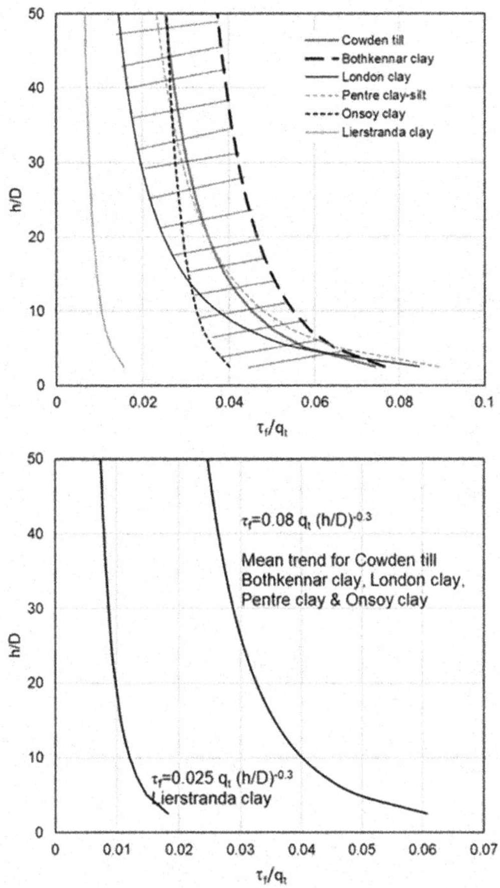

Figure 4. (a) Average measured variations of τ_f/q_t with h/D (showing envelope for all clays except Lierstranda clay), (b) best-fit mean trend lines.

has not been the subject of extensive research. The assumption of a simple direct relationship between $q_{b0.1}$ and the average q_t value near the pile tip is considered adequate for practical purposes. On the basis of a review of existing correlations, Lehane et al. (2022a) propose the following relationships for typical onshore driven piles (D <0.75m)

$$q_{b0.1} = 0.8 q_t (\text{closed-ended pile}) \qquad (9a)$$

$$q_{b0.1} = 0.4 q_t (\text{open-ended pile}) \qquad (9b)$$

5 UNIFIED CPT METHOD DRIVEN PILES IN CLAY

The development of the new Unified CPT method for clays recognised the compensatory effects of factors evident on Figure 5a but investigated potential systematic dependence on (i) clay plasticity index, (ii) soil behaviour type (SBT), (iii) CPT sleeve friction (f_s) and

(iv) vertical effective stress level (σ'_v). The calibration of the method with the Unified database also included consideration of the aspects detailed below.

5.1 Dependence of τ_f on loading direction

Compression and tension tests on piles with identical configurations in seven different clays in the Unified database indicate that, in general, there is not a clear dependence of ultimate shaft friction on the load direction. A load direction factor (f_L) of unity was found to provide a best fit to the full Unified database.

5.2 *Friction on open-ended piles*

Instrumented pile test data reported by Doherty & Gavin (2011) show that lower levels of soil displacement during installation of open-ended piles lead to lower installation radial total stresses (σ_{ri}) and lower ultimate shaft frictions after equalisation. There is, however, a considerable shortage of such data in different clays and, in the absence of such data, the regression analyses conducted for the Unified method assumed the following potential dependencies (where Equation 10a is comparable to the Unified CPT sand method and Equation 10b is a similar format to the ICP-05 method for sands):

$$\tau_f = f(A_{re}, h/D) \tag{10a}$$

$$\tau_f = f(h/D*); D* = (D^2 - D_i^2)^{0.5} \tag{10b}$$

Lehane et al. (2017) show that Equations 5 & 6, which were derived from data for piles in sand, may also be used to approximate A_{re} values for pipe piles in clays.

5.3 *Progressive failure*

A number of piles in the clay database exhibit post-peak softening of shaft shear stress. The calibration of the Unified method assumed that peak shear stresses corresponded to the measured peak axial capacity or, if such a peak was not observed, to the axial resistance at a displacement of 0.1D. This approach is considered reasonable for the majority of the database (which comprised relatively rigid piles) but is moderately conservative for a few very long piles in clays where strain softening can be significant.

5.4 *Silt layers*

Occasional silt layers with a soil behaviour type (SBT) index, I_c, in the range 2.05 to 2.6 occur within the clay strata of the database. For these deposits, the calibration process employed the Unified sand method adopting the equivalent clean sand q_t value ($q_{t,sand}$) for the silt. This was derived using the following relationship which is equivalent to the proposal of Robertson & Wride (1998) but adapted to a simplified format and modified to give a correction factor of unity at $I_c = 2.05$:

$$q_{t,sand} = q_c = \left[3.93 I_c^2 - 14.78 I_c + 14.78\right] q_t \text{ for } 2.05 < I_c < 2.6 \tag{11}$$

5.5 *End bearing of large diameter pipe piles*

Eqn. 9 was employed in the calibration of the relatively small diameter piles in the Unified database. For large diameter offshore piles, the following expression is considered more appropriate; this was based on findings of Doherty and Gavin (2011) from measurements of twin-walled instrumented piles during installation.

$$q_{b0.1} = [0.2 + 0.6 A_{re}]\, q_t \tag{12}$$

where q_t is the average corrected cone resistance value in the zone extending from the pile tip to a depth of $20t$ below the pile tip (t = pile wall thickness)

5.6 *Formulation of Unified CPT method in clay and predictive performance*

The formulations for the Unified clay method are provided in Table 2. These were established following various optimisation analyses which revealed that, as inferred from instrumentation pile test data, τ_f is primarily correlated to the CPT q_t value and the normalised distance from the pile tip (h/D*). The analyses also showed that the shaft friction generated in high sensitivity clays (within Zone 1 of the SBT chart) is typically 50% of the shaft friction in other clays (although there is significant variability). The formulations developed are remarkably similar to those deduced independently from instrumented pile test data (compare formulae in Figure 5 with those in Table 2). Surprisingly, the fit to the capacities of the database piles was not improved by consideration of additional factors such as cone sleeve friction, plasticity index and overconsolidation ratio (inferred by examination of a q_{net}/σ'_{v0} term).

The predictive performance for the Unified clay database of the new clay method was compared by Lehane et al. (2022a) with a number of other current methods. This comparison is summarised in Table 3, which also lists the main input parameters in the τ_f correlation for each method (noting that differences in the end bearing formulations had little impact on calculated capacities). Evidently the CoV of Q_m/Q_c for the new method is a good improvement on existing methods and is significantly lower than the corresponding CoV for the existing '$\alpha - s_u$' method recommended by API/ISO. It should be noted, however, that part of the reason for the substantial reduction in the CoV with respect to other methods arises due to application of a separate equation for Zone 1 (sensitive) clays.

Table 3. Comparison of predictive performance for driven piles in clay of the Unified method with other methods (in terms of mean and CoV of measured to calculated capacity ratios).

Method	Parameters controlling τ_f	Mean Q_m/Q_c	CoV for Q_m/Q_c
API (2011)	s_u & s_u/σ'_v	1.05	0.43
Fugro-96 (Kolk & van der Velde 1996)	s_u, h/D & s_u/σ'_v	1.04	0.35
ICP-05 (Jardine et al. 2005)	OCR, σ'_v, h/D*, δ & S_t	1.12	0.55
NGI-05 (Karlsrud et al. 2005)	s_u, s_u/σ'_v, I_p & σ'_v	1.1	0.36
UWA-13 (Lehane et al. 2013)	q_t & h/D*	1.12	0.33
Fugro-10 (Van Dijk & Kolk 2010)	q_{net}, h & q_{net}/σ'_v	0.98	0.37
Unified method (Lehane et al. 2022a)	q_t, h/D* & S_t	0.99	0.23

The Unified method in clay is calibrated on a relatively small number of pile tests (49) and, as may be inferred from the instrumented pile test data, its relatively good predictive performance arises partly because of compensating factors. For example, interface friction angles (δ) for clays in the Unified database range from 12° to 30°, implying a range in capacities of about 2.5 if τ_f varied only with tan δ. However, encouragingly, Lehane et al. (2022a) examined the performance of the new method against a different 'Test database' comprising 24 pile tests in clay and found a mean Q_m/Q_c ratio of 1.09 (i.e. a 9% over-prediction, on average) and a CoV for Q_m/Q_c of 0.22; this CoV is a substantial improvement on the CoV of 0.43 determined for the API $\alpha - s_u$ method (Table 3).

6 APPLICATION TO LAYERED DEPOSITS

The Unified CPT methods for sand and clay were calibrated against test data for piles that were driven into predominately sand or clay deposits. However, many deposits encountered in practice contain sand, silt and clay layers and it is therefore important to assess the reliability of the methods at such sites. Bittar et al. (2022) conducted such a study involving 23 load tests on piles driven into mixed stratigraphies. Ten of these load tests were on piles with diameters between 0.9m and 2m in diameter which is considerably larger than the mean D value of about 0.4m for the piles in the Unified database. An average of 55% of the layers in these case histories were coarse grained and two cases involved sensitive zone 1 soils.

Sample predictions for two driven steel pipe pile case histories examined are presented in Figure 6a for a tension test in Oakland California on a 13.3m long, 610mm pile and in Figure 6b for a compression test in Minnevika, Norway on a 40m long, 405mm diameter pile. As seen with reference to the plotted I_c profiles in these figures, a 5.5m thick layer of Zone 3 clay is present along the central portion of the pile shaft whereas the stratigraphy at Minnevika comprises alternating 1.5m to 2m thick layers of sand and clay. The calculated τ_f distributions were determined in a simple spreadsheet using the I_c dependent expressions, as summarized in Table 2; calculations can also be performed using UWA freeware at *https://pile-capacity-uwa.com/*. These τ_f distributions lead to ratios of measured to calculated capacities (Q_m/Q_c) of 1.15 at Oakland and 1.16 at Minnevika (ignoring potential strain softening).

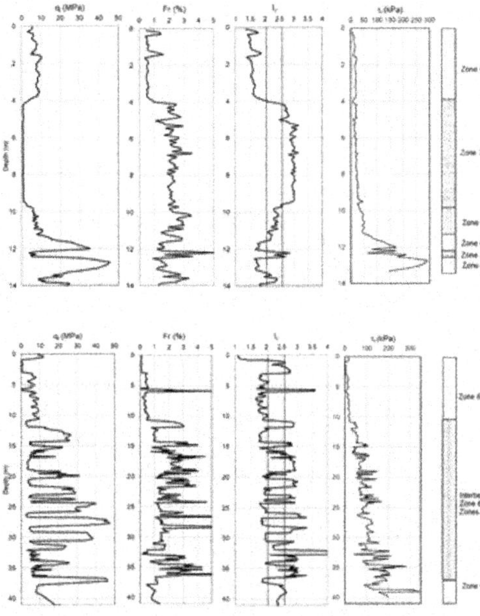

Figure 5. Stratigraphy and predicted τ_f distribution for (a) tension test on 610-mm diameter, 13.3m long pipe pile in Oakland and (b) compression tests on 405-mm diameter, 40m long pipe pile at Minneveka.

Bittar et al. (2022) show that the mean and coefficient of variation of the Q_m/Q_c values for the 23 load test database were 1.02 and 0.17 respectively i.e. the spread in predictions is less than that of the Unified databases for sand and clay sites. It can therefore be concluded that the Unified methods can be applied with the same level of confidence to layered stratigraphies as to 'single soil type' deposits. However well instrumented test piles that accurately measure the distribution of shaft friction in layered stratigraphies are needed for verification of the method. There is also a great shortage of skin friction data for piles installed in in Zone 5 (silt) of the SBT chart. Such data would help to resolve the discontinuity in the expressions for τ_f at the boundary between Zone 4 and 5 (at I_c=2.6), where τ_f calculated using the clay expression is typically double that calculated using the clean sand correction approach for silt.

7 LOAD DISPLACEMENT RESPONSE USING THE UNIFIED CPT METHOD

The load-displacement response of piles is normally predicted in a load transfer analysis where the shear stress-displacement (τ-w) springs (also called t-z springs) at various levels along the pile shaft are scaled in proportion to the ultimate shaft friction (τ_f) and the base spring (q_b-w_b) is scaled with the ultimate end bearing ($q_{b0.1}$). The application of the new Unified CPT method to this approach is examined in the following.

7.1 *Piles in sand*

Lehane et al. (2020b) examined the accuracy of pile displacement predictions for driven piles in the Unified sand pile database sand using the τ-w and $q_b - w_b$ load transfer relationships recommended in API (2011). These relationships are provided in tabular form in API (2011) but may be expressed as follows, where w_f is the displacement to peak shear stress (τ_f) and is assigned a mean value of 0.01D in API (2011):

$$\frac{\tau}{\tau_f} = 2\frac{w}{w_f}\left[1 - \frac{w}{2w_f}\right] \tag{13}$$

$$\frac{w_b}{D} = 0.01\left[\frac{q_b/q_{b0.1}}{1 - 0.9\,(q_b/q_{b0.1})}\right] \tag{14}$$

Although Equation (14) is not a perfect match to the API tabulated values when $q_b/q_{b0.1}$>0.5, this is not a concern as settlement predictions are required for the serviceability limit state when applied loads are rarely greater than 70% of the pile capacity.

API (2011) recommends adoption of a mean w_f/D value of 0.01. However the analyses showed that use of a constant value did not capture the tendency for softer pile responses in tension compared with compression and for stiffer responses in denser sands and longer piles. On examination of the trends shown by the piles in the Unified database, Lehane et al. (2020b) proposed the following expression for w_f/D:

$$\frac{w_f}{D} = \frac{q_c^{0.5}\sigma'^{0.25}_v}{Ap_a^{0.75}} \tag{15}$$

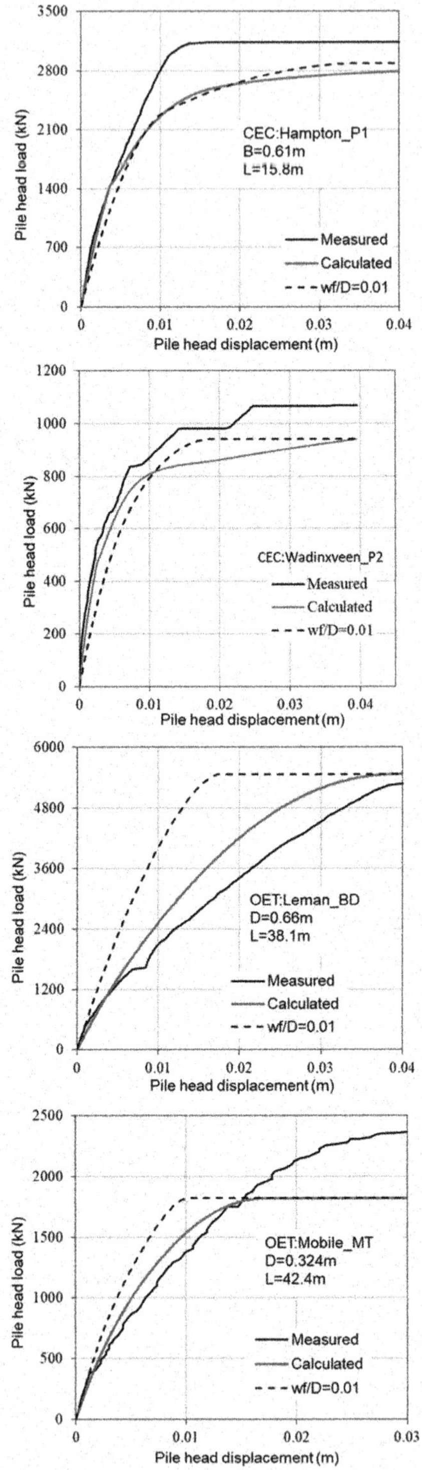

Figure 6. Comparison of measured pile load-displacement response with the calculated response using the Unified method with (i) w_fD given by Equation 15 and (ii) w_f/D=0.01. The symbols CEC and OET denote closed ended piles tested in compression and open-ended piles tested in tension respectively.

where the constant A is 1250 in compression and 625 in tension.

Comparisons of measured and calculated load-displacement responses are plotted for typical cases from the Unified database on Figure 7, where calculations were performed using Equations 13 and 14 combined with the Unified method to determine τ_f and $q_{b0.1}$. Figure 7 presents calculated responses for a constant w_f/D value of 0.01 and for w_f/D values at various levels along the pile shaft determined using Equation 15. The improved fit obtained using Equation 15 for tension piles and for longer piles is evident on Figure 7. Lehane et al. (2020b) show that the error in calculated displacements at 50% of the ultimate capacity is typically about 0.002D and is less than 0.005D. Equation 15 will be included in the next version of the API/ISO recommendations.

7.2 *Piles in clay*

Lehane & Bittar (2022) repeated the exercise described above for the Unified database of piles in clay, for which API (2011) recommends the same format of load transfer curves apart from the allowance for post-peak softening to a shear stress of (0.8 ±0.1) τ_f at a displacement of 0.02D. It was found that a combination of the Unified CPT method in clay (Table 2) with these load transfer curves provides good predictions of the displacements of both the tension and compression piles in the database. The standard deviation of the difference between measured and calculated displacements was 0.002D.

8 CONCLUSIONS

There are clear advantages to CPT-based design approaches for the assessment of the axial capacity of driven piles. The CPT-based methods have the full support of the geotechnical community as they remove subjectivity associated with parameter assessment, provide essentially continuous profiles of soil behaviour and allow direct rapid and automated calculation of capacity at any CPT location.

There has been a concerted effort over the past decade to first quantify and then increase the reliability of design methods for piles. This was made possible by the thorough, detailed and careful compilation of a database of static pile load tests (the Unified database) that had the support of the proponents of recent existing CPT methods and the support of a large and active group of energy companies from around the world.

The creation of the Unified database highlighted the level of uncertainty associated with design methods in common use and prompted the development of the new non-affiliated 'Unified CPT method' described in this paper. The new method captures key characteristics of driven piles, as observed in high quality instrumented pile test programmes completed over the past 30 years. The method is shown to have higher reliability than existing approaches.

As for most empirical methods, further developments and improvements are still to come. Such improvements include the need to better quantify the effects of time in various soil types and how the resistance to pile driving as well as the long term capacity can be assessed with more confidence. More static load tests on large diameter instrumented piles are clearly required while greater understanding of factors controlling the influence of the installation mode and friction in silt is needed.

It is hoped that the 'Unified database' can be expanded on an ongoing basis with each additional pile test included satisfying the strict selection criteria applied in the database development to date.

ACKNOWLEDGEMENTS

The authors gratefully acknowledge the funding and support provided under a Joint Industry Project (JIP) funded by Aramco, Equinor, Lundin, Ørsted, ONGC, BP, Total, ExxonMobil, EnBW, EDF and SSER. The significant contribution of the team of experts to the JIP, namely Prof. Richard Jardine, Dr. Philippe Jeanjean, Mr. Bas van Dijk, Dr. Mike Rattley and Mr. Pasquale Carotenuto is much appreciated.

REFERENCES

Alm T. and Hamre L. (2001). Soil model for pile driveability predictions based on CPT interpretations. Proc. 15th Int. Conf. on Soil Mechanics and Geotechnical Eng., 1297–1302.

Almeida M.S., Danziger F.A., and Lunne, T. (1996). Use of the piezocone test to predict the axial capacity of driven and jacked piles in clay. *Canadian Geotechnical J.*, 33 (1), 23–41.

Amirmojahedi M. and Abu-Farsakh M. (2019). Evaluation of 18 direct CPT methods for estimating the ultimate pile capacity of driven piles. Transport Research Record, 2673(9), 127–141.

Anusic I., Lehane B.M., Eiksund G., and Liingaard M. (2019). Evaluation of installation effects on the set-up of field displacement piles in sand. Canadian Geotechnical Journal, 56(4), 461–472.

API (2011). ANSI/API RP 2GEO: Geotechnical and Foundation Design Considerations. ISO 19901-4:2003 (Modified), Petroleum and natural gas industries-Specific requirements for offshore structures, Part 4-Geotechnical and foundation design considerations. 1st edition. Washington, DC: API Publishing Services.

Barmpopoulos I. and Ho T. (2009). The large displacement shear characteristics of granular media against concrete and steel interfaces. Proc Research Symp. Characterization & Behaviour of Interfaces, Atlanta, Georgia, USA Behaviour of Interfaces, 16–23.

Bittar E., Lehane B.M., Boulanger R.W., and DeJong J. (2020). CPT filter to estimate end bearing of closed-ended driven piles in layered sands. Proc. Int.

Symp. Frontiers in Offshore Geotechnics, Austin, Texas, 520–528.

Bittar E., Lehane B.M., Nadim F., Liu Z. and Lacasse S. (2022). Assessment of the unified CPT Design Method for Axial Capacity of Driven Piles in Interlayered Deposits. Canadian Goetechnical Journal (submitted).

Boulanger, R.W.W., and DeJong, J.T.T. (2018). Inverse filtering procedure to correct cone penetration data for thin-layer and transition effects. Proc 4th Int Symp. on Cone Penetration Testing (CPT'18), 21-22 June, 2018, Delft, The Netherlands, 25–44.

Briaud J.-L. and Tucker. L.M. (1988). Measured and Predicted Axial Response of 98 Piles. Journal of Geotechnical Engineering, 114(9), 984–1001.

Bustamante, M. and Gianeselli, L. (1982). Pile Bearing Capacity Prediction by Means of Static Penetrometer CPT. Proc. 2nd European Symposium on Penetration Testing, Amsterdam, 493–500.

Chow F. (1997). Investigations into the behaviour of displacement piles for offshore foundations. PhD thesis, Imperial College London (University of London).

Chow, F.C., Jardine, R.J., Nauroy, J.F., and Brucy, F. (1998). Effects of Time on Capacity of Pipe Piles in Dense Marine Sand. *J. Geotech. & Geoenv. Engineering*, 124, 254–264.

Clausen C. J. F., Aas P. M. & Karlsrud, K. (2005). Bearing capacity of driven piles in sand, the NGI approach. *Proc. 1st Int. Symp. on Frontiers in Offshore* Geotechnics *(ISFOG)*. Balkema, 574–580.

De Ruiter J. and Beringen F.L. (1979). Pile foundations for large North Sea structures. *Marine Geotechnology*, 5(3), 267–314.

Doherty, P. and Gavin, K. (2011). Shaft Capacity of Open-Ended Piles in Clay. J. Geotech. & Geoenv. Engrg, 137(11): 1090–1102.

Eslami, A. and Fellenius, B.H. (1997). Pile capacity by direct CPT and CPTu methods applied to 102 case histories. *Canadian Geotechnical J.*, 34(6), 886–904.

Eslami A., Tajvidi I. and Karimpour-Fard M. (2014). Efficiency of Methods for Determining Pile Axial Capacity – Applied to 70 Case Histories in Persian Gulf Northern Shore. International Journal of Civil Engineering, 12(1), 45–54.

Hu Z., McVay M., Bloomquist D., Horhota D. and P. Lai. P. (2012) New Ultimate Pile Capacity Prediction Method Based on Cone Penetration Test (CPT). *Canadian Geotechnical Journal*, 49(8), 961–967.

Gavin K., Jardine R., Karlsrud K., and Lehane B.M. (2015). The effects of pile ageing on the shaft capacity of offshore piles in sand. Proc. 3rd Int. Symp. Frontiers in Offshore Geotechnics, 1, 129–151.

Gavin K.G. and Lehane B.M. (2003). The shaft capacity of pipe piles in sand. Canadian Geotechnical Journal, 40 (1), 36–45.

Gudavalli S.R., Safaqah O. and Seo H. (2013). Effect of soil plugging on axial capacity of open-ended pipe piles in sand. Proc. 18th Int. Conf. Soil Mech. and Geotech. Eng., Paris, 1487–1490.

Ho T.Y.K., Jardine R.J. and Anh-Minh N. (2011). Large-displacement interface shear between steel and granular media. *Geotechnique*, 61(3): 221–234.

Jardine R.J., Lehane B.M. and Everton S.J. (1993). Friction coefficients for piles in sands and silts. In Offshore site investigation and foundation behaviour. Springer. 661–677.

Jardine, R. J., Chow, F. C., Overy, R. & Standing, J. R. (2005). *ICP design methods for driven piles in sands and clays*. Thomas Telford.

Jardine, R. Standing, J. and Chow, F, (2006). Some observations of the effects of time on the capacity of piles driven in sand. Géotechnique, 56 (4), 227–244.

Karlsrud, K., Clausen, C.J.F. and Aas, P.M. (2005). Bearing capacity of driven piles in clay, the NGI approach. Proc. Int. Symp. on Frontiers Offshore Geotechnics, Perth. pp. 775–782.

Karlsrud, K., Jensen, T. G., Wensaas Lied, E. K., Nowacki, F., and Simonsen, A. S. (2014) Significant ageing effects for axially loaded piles in sand and clay verified by new field load tests. Proc. Offshore Technology Conference, Houston, USA, OTC-25197-MS. doi:10.4043/25197-MS.

Kolk H. and van der Velde E. (1996). A Reliable Method to Determine Friction Capacity of Piles Driven into Clays. In Proc. Offshore Technology Conference (OTC), Houston, Texas, Pub No. 413532.

Kolk H.J., Baaijens A.E., and Vergobbi, P. (2005). Results of axial load tests on pipe piles in very dense sands: The EURIPIDES JIP. Proc. of the Int, Symp. Frontiers in Offshore Geomechanics, ISFOG. Taylor & Francis, London. pp. 661–667.

Lehane B.M. (1992). Experimental investigations of pile behaviour using instrumented field piles. Imperial College London (University of London).

Lehane B.M., Chow F.C., McCabe B.M. and Jardine R.J. (2000). Relationships between shaft capacity of driven piles and CPT end resistance. Geotech. Engrg., ICE, 143, 93–101.

Lehane B.M., Gaudin C., and Schneider J.A. (2005). Scale effects on tension capacity for rough piles buried in dense sand. Géotechnique, 55(10), 709–720.

Lehane B.M., Jardine R.J., Bond A.J. and Frank, R. (1993). Mechanisms of shaft friction in sand from instrumented pile tests. Journal of Geotechnical Engineering, ASCE,119(1), 19–35.

Lehane B., Li L. and Bittar E. (2020b). CPT-based load-transfer formulations for driven piles in sand. *Geotechnique Letters*, 10, 568–574.

Lehane B., Li L. and Bittar E. (2022). CPT-based load-transfer formulations for driven piles in clay (in preparation).

Lehane, B.M., Li, Y. and Williams, R. (2013). Shaft Capacity of Displacement Piles in Clay Using the Cone Penetration Test. J. Geotech. & Geoenv. Engrg., ASCE, 139(2):253–266.

Lehane B.M. and Randolph M.F. (2002). Evaluation of a Minimum Base Resistance for Driven Pipe Piles in Siliceous Sand. J. Geotech. and Geoenviron. Engineering, ASCE,128(3), 198–205.

Lehane, B.M., Liu, Z., Bittar, E., Nadim, F., Lacasse, S., Jardine, R., Carotenuto, P., Rattley, M., Jeanjean, P., Gavin, K., Gilbert, R., Bergan-haavik, J. and Morgan, N. (2020a). A new CPT-based axial pile capacity design method for driven piles in sand. Proc. 5th Int. Symp. Frontiers Offshore Geotechnics, ISFOG-21, 462–477. DFI publications, New Jersey.

Lehane B.M., Liu Z., Bittar E., Nadim F., Lacasse S., Bozorgzadeh N., Jardine E., Ballard J-C, Carotenuto P., Gavin K., Gilbert R., Bergan-Haavik J., Jeanjean P. and Morgan N. (2022). CPT-based axial capacity design method for driven piles in clay. J. Geotech. & Geoenv. Engineering, ASCE (under review)

Lehane, B.M., Lim, J.K., Carotenuto, P., Nadim, F., Lacasse, S., Jardine, R.J. and van Dijk, B.F.J. (2017). Characteristics of Unified Databases for Driven piles. Proc. 8th International Conf. Offshore investigation and Geotechnics: Smarter solutions for offshore developments, Society for Underwater Technology, 1, 162–194.

Lim J.K. and Lehane B.M. (2014). Characterisation of the effects of time on the shaft friction of displacement piles in sand. *Géotechnique*, 64(6), 476–485.

Liu T.F., Quinteros V.S., Jardine R.J, Carraro J. A. H., and Robinson J. (2019). A Unified database of ring shear steel-interface tests on sandy-silty soils. Proc. XVII European Conf. Soil Mech. and Geotechnical Engg. Reykjavik, Iceland. doi: 10.32075/17ECSMGE-2019-268.

Nadim F., Lehane B.M., Liu Z., Bittar E., Bozorgzadeh (2021). New CPT-based method for calculation of axial pile capacity in sand and in clay- Summary and Guidelines. Doc No. 20180089-03-R, Norwegian Geotechnical Institute, Oslo, Norway.

Niazi F. S. and Mayne P.W. (2013). Cone Penetration Test Based Direct Methods for Evaluating Static Axial Capacity of Single Piles. Geotechnical and Geological Engineering, 31(4), 979–1009.

Niazi, F.S. and Mayne P.W. (2016). CPTu-based enhanced UniCone method for pile capacity. Engineering Geology, 212: 21–34. Elsevier B.V.

Randolph M.F., Dolwin J., and Beck R. (1994). Design of driven piles in sand. *Geotechnique*, 44(3), 427–448.

Robertson, P.K. (2009). Interpretation of cone penetration tests - A unified approach. *Canadian Geotechnical J.*, 46(11):1337–1355.

Robertson P.K., and Wride C.E. (1998). Evaluating cyclic liquefaction potential using the cone penetration test. *Canadian Geotechnical Journal*, 35(3), 442–459.

Salgado R., Woo S. I., and Kim D. (2011). Development of load and resistance factor design for ultimate and serviceability limit states of transportation structure foundations. Rep. No. FHWA/IN/JTRP- 2011/03, Indiana Dept. of Transportation and Purdue Univ., West Lafayette, US.

Schmertmann J.H. (1978). Guidelines for cone penetration test, performance and design. Report FHWA-TS-78-209, Washington, 145pp.

Van Dijk B.F.J. and Kolk H.J. (2011). CPT-based design method for axial capacity of offshore piles in clays. In Proc. of the Int. Symposium on Frontiers in Offshore Geotechnics II. Taylor & Francis Group, London. 555–560.

Van Impe W.F. (1986). Evaluation of deformation and bearing capacity parameters of foundations from static CPT-results. Proc. 4th Int. Geotechnical on Field Instrumentation and in-situ measurements, Singapore, 51–70.

Van Mierlo, W.C. and Koppejan, A.W. 1952, 'Lengte en draagvermogen van heipalen', Bouw, January.

White D., Schneider J., and Lehane B.M. (2005). The influence of effective area ratio on shaft friction of displacement piles in sand. Proc. 1st Int. Symp. Frontiers in Offshore Geotechnics, Perth, Australia. pp. 741–747.

Whittle A.J. and Baligh M.M. (1988). A Model for Predicting the Performance of TLP Piles in Clays. Final Report Phase III to sponsors, Dept. of Civil Engineering, Massachusetts Institute of Technology.

Xu X., Schneider J.A., and Lehane B.M. (2008). Cone penetration test (CPT) methods for end-bearing assessment of open- and closed-ended driven piles in siliceous sand. Canadian Geotechnical Journal, 45(8), 1130–1141.

Yang Z.X., Jardine R.J., Zhu B.T., Foray P. and Tsuha C.H. C. (2010). Sand grain crushing and interface shearing during displacement pile installation in sand. *Geotechnique*, 60(6), 469–482.

Zhou J., Xie Y., Zuo Z.S., Luo M.Y. and Tang X.J. (1982), Prediction of limit load of driven pile by CPT, 2nd Eur. Symp. On Penetration testing, Amsterdam, 957–963.

Cone Penetration Testing 2022 – Gottardi & Tonni (eds)
© 2022 the Author(s), ISBN 978-1-032-35624-2

Material Point Method simulations of cone penetration and CPT interpretation

R. Salgado
Purdue University, West Lafayette, USA

V. Bisht
Itasca Consulting Group, Minneapolis, USA

M. Prezzi
Purdue University, West Lafayette, USA

ABSTRACT: It is now possible to realistically simulate cone penetration in either sand or clay using the Material Point Method (MPM). To do that, the following requirements must be met: (1) a realistic constitutive model must be used; (2) stress integration of the constitutive model must be accurate; (3) the explicit scheme for the solution of the governing equation must be accurate, robust and efficient; and (4) shear strain localization must be correctly captured. The paper provides an overview of how to correctly perform such simulations and then discusses the implications for the future of CPT interpretation of the availability of accurate, reliable cone penetration simulation methods.

1 INTRODUCTION

1.1 *The cone penetration test*

One of the attractions of the cone penetration test ("CPT") as originally conceived is its simplicity. A penetrometer with a simple, axisymmetric shape is pushed into the ground, and the resistance to penetration at its tip is measured. Given the simplicity of the test, one would expect that a theoretical basis for its interpretation would have been in place for some time. Unfortunately, what is simple in concept is not necessarily so once implemented (Salgado 2012). The complexity appears when we try to develop methods of analysis for realistic interpretation of CPT results. For example, when we attempt to estimate fundamental state variables or direct functions of state variables – such as relative density ("D_R") for sand or undrained shear strength ("s_u") for clay – we find out that there have not been rigorous theories to guide us.

The two main sources of difficulties in cone penetration simulation are: (1) the modeling of the soil response itself and (2) the method of solving the global equation of motion. To address (1), researchers have developed constitutive models that are now accurate and reliable, at least in the absence of cycling. To address (2), researchers have explored a variety of approaches, which are detailed next.

1.2 *Methods of cone penetration simulation*

To remedy the gap in the theoretical basis for CPT interpretation, efforts started early. Initial attempts relied on modeling the soil as a rigid-plastic material with either a Tresca or Mohr-Coulomb yield surface and applying the notion of a limiting equilibrium to an axially loaded penetrometer and a surrounding soil mass. The Durgunoglu & Mitchell (1976a) model was the most successful of these early models, but fell short when people tried to use it. A major shortcoming of the model was its reliance on perfect plasticity and an interpretation based on estimating the friction angle of the soil. As discussed elsewhere (e.g., Salgado 2012), the mobilized friction angle varies across the soil domain in any boundary-value problem involving real soil, so any method based on such an over-simplification of soil response is fatally flawed.

The next main thrust in developing a framework for CPT simulations was cavity expansion theory (e.g., Salgado et al. 1997; Salgado and Prezzi 2007; Salgado and Randolph 2001). Cavity expansion theory was an approximation to the problem based on the requirement that the cone must expand a cylindrical cavity in the soil in order to advance. With the exception of Salgado and Randolph (2001), these efforts concentrated on the use of simple elastoplastic soil models with Tresca or Mohr-Coulomb yield criteria, although with some modifications attemping to capture the effects of soil nonlinearities. It produced useful relationships (e.g., Salgado and Prezzi 2007) that can be used in CPT interpretation, but it could not truly simulate a process that is much more involved than a simple cavity expansion process.

The strain path method (Teh and Houlsby 2009) was proposed for analysis of cone penetration in clay, and it was a reasonable approximation to the penetration process, but its impact was again limited by a simple elasto-plastic model with a Tresca yield criterion.

Early application of the Finite Element Method ("FEM") also met with difficulties. In geomechanics applications, FEM has traditionally followed a Lagrangian approach. This means that the nodes of a mesh are tracked throughout the analysis, and it is through node displacements that solutions are obtained. Deformation follows from relative node displacements, and stress follows from deformation. After considerable deformation has occurred, mesh elements may be so distorted that accurate solutions are no longer possible. In fact, a solution may not even be possible, with the analysis crashing instead.

The Arbitrary Lagrangian-Eulerian ("ALE") FEM approach (Belytschko and Kennedy 1978) and the material point method ("MPM") (Schreyer et al. 1994) were proposed to avoid this limitation. In ALE FEM, mesh distortion is limited by remeshing or node repositioning after some threshold level of deformation has happened. When nodes are repositioned or a new mesh formed, we need to map the state varia-bles from quadrature points on the old mesh to quadrature points on the new mesh. The remapping can lead to stress states that lie outside the yield surface for plasticity models with single yield surface or outside the bounding surface for bounding surface mod-els. Remapping can be challenging with bounding surface models or complex models with several variables to map. In contrast to ALE FEM, MPM does not require remapping of variables because the state variables are carried by the same material points throughout the computations. The material point method is discussed next.

2 THE MATERIAL POINT METHOD

2.1 *Fundamentals of the method*

The material point method can be best understood as a variant of the finite element method in which Gauss points are replaced by so called material points. A material point ("MP") may be a point (as in classical MPM), but it may also be a small area (for two-dimensional analysis) or small volume (for three-dimensional analysis). In contrast with Gauss points, material points may move within and even across elements. This key difference between FEM and MPM has a number of computational consequences.

The most important and most useful implication of the use of material points instead of Gauss points as used in ALE FEM is that the approach is a pure Lagrangian approach, with material point displacements being the basis for the solution. Whereas grid nodes do move during a computation step, they can be repositioned to their original locations before the next computation time step. It may also be advatangeous to consider parts of the grid to either translate in some direction or to compress one-dimensionally. This means that, whether the grid remains static, moves in some direction or is unidirectionally compressed in MPM, it does not get distorted as in FEM, escaping therefore the difficulties created by mesh distortion.

The second most important implication of the use of material points instead of Gauss points in an analysis is that the absolute rigor of FEM with respect to integration of the constitutive model is, to some extent, lost. This happens mainly because Gauss integration is no longer possible. This is not fatal, but a number of remedial measures must be taken to limit any error from this partial loss of rigor.

The next sub-section reviews the MPM formulation used for the analyses whose results are presented in this paper and the key steps in an MPM computation time step.

2.2 *The MPM formulation*

The MPM formulation used in the simulations discussed later is based on the uniform generalized interpolation material point method ("uGIMP" or, in this paper, also "GIMP") variant of MPM discussed in detail by Bisht and Salgado (2018), Bisht et al. (2021b) and Salgado and Bisht (2021).

Figure 1 shows how space and matter can be discretized for a cone penetration problem. Space can be discretized in the form of a computational grid, and the soil and cone can be discretized using material points. The grid can be a rectangular grid with elements sufficiently small around the cone penetrometer. Material point discretization is achieved derivatively: the initial number of material points per element is fixed across the grid (e.g., four material points per element), and finer discretization is therefore achieved where the grid is itself made finer.

How fine the grid must be depends on whether shear strain localization is expected. The smallest element size in any zone in which shear bands are expected to form should have a width equal to the width of the shear band. We discuss this in greater detail later.

There are two other aspects of discretization that should be briefly discussed. First, the grid can be set up so that part of it moves with the cone penetrometer during penetration and another part, at the bottom of the grid, compresses vertically. Second, when material points flow into a grid element for which they are too large, they are split in a way that conserves energy, momentum and mass. What these two measures accomplish is that, when the cone moves down, together with the fine grid around it, material points located in the coarser grid elements below it will subdivide as they enter the fine grid around the cone, maintaining the required level of material discretization and a sufficient number of material points per element for adequate quadrature.

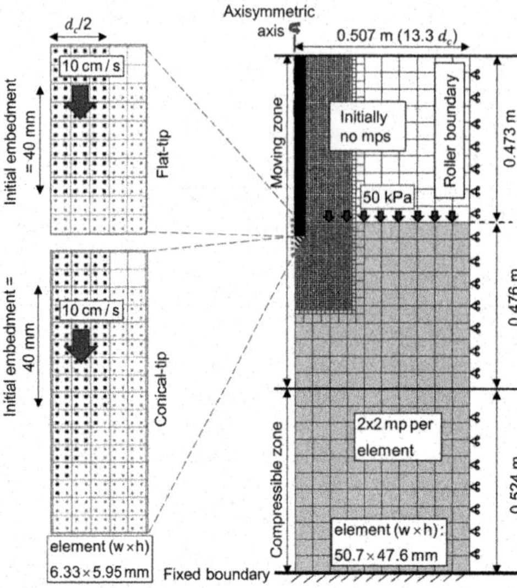

Figure 1. Spatial and material discretization for simulation of penetration of a flat-tip and a conical-tip penetrometer in sand (after Bisht 2021).

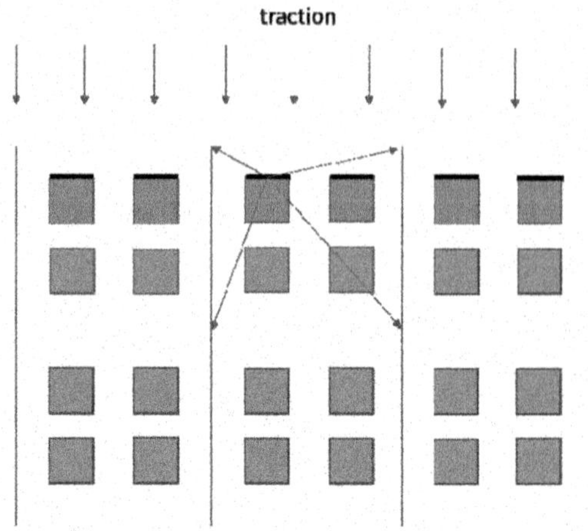

Figure 2. Application of tractions to a body in uGIMP MPM: the tractions are first applied on the edges of the material points (rep-resented by thick lines) and then, using finite element-element shape functions, mapped to the nodes of the element containing the material point.

Displacement or velocity boundary conditions are straightforward in MPM: the correct displacement or velocity is directly applied on the outermost grid nodes corresponding to that boundary. Application of traction boundary conditions is more involved, requiring distribution of tractions to the material point boundaries aligned with the boundary of the body for every material point so positioned, and then redistribution of these tractions to the nodes of any element containing any part of the material point (Bisht and Salgado 2018). See Figure 2.

To model a semi-infinite soil medium using a finite grid, silent boundaries are required. Silent boundaries let energy dissipate just as it would by radiation in a semi-infinite medium. Kellezi (2000) proposed a "cone boundary" that assumes that a load applied on a free surface of a half-space would propagate within a volume expanding with increasing depth, which can be approximated as a cone. At the boundary, there is both a viscosity term to simulate the energy dissipation through radiation and a stiffness term. Bisht and Salgado (2018) adapted it for MPM, and it is their formulation that is used here.

MPM formulations for solving the global equation of motion have typically been explicit. The governing equation for a general analysis applicable to simple models and any drained simulation is:

$$\dot{p}_i^{(I)} = f_i^{(I)EXT} - f_i^{(I)INT} \tag{1}$$

The rate of the momentum $p_i^{(I)}$ *pi(I)*, external force $f_i^{(I)EXT}$ and internal force $f_i^{(I)INT}$ at node *I*, all appearing in (1), are given by (Bisht et al. 2021a):

$$\dot{p}_i^{(I)} = \sum_P S^{(IP)} \dot{p}_i^{(P)} \tag{2}$$

$$f_i^{(I)EXT} = \int_\Gamma N^{(I)} t_i d\Gamma + \sum_P S^{(IP)} m^{(p)} b_i \tag{3}$$

and

$$f_i^{(I)INT} = \sum_P S_j^{(IP)} \sigma_{ij}^{(P)} V^{(P)} \tag{4}$$

In Equations (2) through (4), σ_{ij} is the Cauchy stress, *bi* is the body force per unit mass, $m^{(P)}$ is the mass of the material point, Γ is the boundary of the domain Ω of the problem and $t_i = \sigma_{ij} n_j$ is the traction on an arbitrary point on Γ, with n_j being the unit vector normal to Γ at that point. Also appearing in Equations (2) through (4) are the finite element shape function $N^{(I)}$ at node *I*, the material point shape function $S^{(IP)}$, which maps the value of a variable at the centroid of a material point *P* to node *I*, and its gradient $S_j^{(IP)}$. These last two are defined as:

$$S^{(IP)} = \frac{1}{V^{(P)}} \int_{\Omega^{(P)}} N^{(I)} \chi^{(P)} d\Omega \tag{5}$$

and

$$S_j^{(IP)} = \frac{1}{V^{(P)}} \int_{\Omega^{(P)}} N_j^{(I)} \chi^{(P)} d\Omega \tag{6}$$

where $\chi^{(P)}$ is the particle characteristic function, usually taken to be uniformly equal to 1 across the MP and zero outside it:

$$\chi^{(P)} = \begin{Bmatrix} 1 & r \in \Omega^{(P)} \\ 0 & r \notin \Omega^{(P)} \end{Bmatrix} \tag{7}$$

One reason for using an explicit solution scheme in MPM are the difficulties in inverting ill-conditioned stiffness matrices that result when there are only a few material points in some element or elements (Wang et al. 2016).

The Update Stress Last (USL) explicit time integration scheme, in which the stresses in the material point are computed after momentum in the grid is updated (Wallstedt and Guilkey 2008), is used here.

Explicit time integration is conditionally stable, which means that time increments must be less than some limit for the computations to converge. For nonlinear problems, the time step Δt is usually taken as:

$$\Delta t = \kappa \min\left(ele/\sqrt{E_c/\rho}\right) \tag{8}$$

where κ is the Courant number (Courant et al. 1967), ele_{min} is the minimum element size, E_c is a measure of material stiffness (the Young's modulus or the constrained modulus are common choices, depending on the problem), and ρ is the material density. For dynamic problems, $\kappa \approx 0.4$ yields suitable results for the USL approach, but higher values of κ may be used for quasi-static problems (Wallstedt and Guilkey 2008). For quasi-static problems, such as cone penetration, the time step Δt can be increased by mass scaling to reduce the overall computation time. In mass scaling, the material density ρ is artificially increased by a mass scaling parameter η:

$$\rho_{scaled} = \eta\rho \tag{9}$$

This allows the use of a larger time step Δt_{ms}:

$$\Delta t_{ms} = \sqrt{\eta}\Delta t \tag{10}$$

2.3 *Volumetric locking*

In numerical solutions to problems in mechanics involving incompressible or nearly incompressible materials, a phenomenon known as volumetric locking may occur. Volumetric locking is the insufficiency of variables provided by a grid discretization to allow solution of all equations. This results from the additional equation involving strain tensor components resulting from the imposition of the incompressibility constraint on an integration point. Incompressibility or near incompressibility appears in Tresca solids in the plastic range and at critical state for any soil model. It often appears as strong spatial oscillations in the values of fundamental variables, such as stress, which appears as a checkerboard pattern when results are plotted as contour plots.

In the finite element method, volumetric locking can be avoided by using higher-order elements, which lead to a sufficient number of variables to overcome any excess equations from the incompressibility constraint. This is not an option in MPM because shape functions for higher-order elements can take negative values for some regions in an element, which in turn can lead to negative mass at the corresponding nodes from a material point located in such regions. The attending complications are discussed in detail in Bisht et al. (2021a).

An effective way to deal with volumetric locking in MPM is the use of the so-called non-linear $\overline{\mathbf{B}}$ method (Hughes 1980). The method is based on the fact that volumetric locking can be prevented by reducing the number of constraints imposed by incompressibility in each element. For example, for Q4 elements, volumetric locking can be prevented if a reduced, single-point quadrature rule is used (see Sloan and Randolph (1982)). The use of single Gauss point in reduced integration for Q4 elements is sufficient to prevent volumetric locking, but this strategy is not applicable in MPM.

For MPM, Bisht et al. (2021a) used a large-deformation formulation of the $\overline{\mathbf{B}}$ method with the deformation gradient $\mathbf{F}$ split into a volumetric and a deviatoric component, with the volumetric component calculated only at the element center. The computational algorithm that results requires that the gradient $S_{,j}^{(IP)}$ of the material point shape functions (material point mapping functions) must be calculated also at the element center. A modified deformation gradient $\overline{F}^{(P)}$ at the material point P, obtained from combining the deviatoric component of the deformation gradient with the volumetric component of it at the element center, is then used in stress integrations.

2.4 *A computation time step in an MPM simulation*

During a time step in an MPM simulation, we follow the steps outlined below.

1. Start with initialization of all variables at material points and grid nodes and set the time t to 0.
2. Determine an appropriate time increment Δt using Eq. (10).
3. For each material point P:
   (a) Find every node I around the material point P to which nonzero values of material point variables must be transferred, i.e., nodes for which $\int_{\Omega^{(P)}} N^{(I)}\chi^{(P)} d\Omega \neq 0$
   (b) Compute mapping functions $S^{(IP)}$ using Eq. (5) and their gradients $S_{,j}^{(IP)}$ using Eq. (6). If the node is a hanging node, transfer mapping

functions and their gradients to surrounding nodes as explained by Bisht et al. (2021a).
   (c) For simulations in incompressible soil, compute mapping function gradients $S_{,j}^{(IP)}(C)$ at element center C.

4. For each node I:
   (a) Compute nodal mass $m^{(I)} = \sum_P m^{(P)} S^{(IP)}$ and momentum $p_i^{(I)} = \sum_P m^{(P)} v_i^{(P)} S^{(IP)}$
   (b) Compute nodal external $\overline{f}_i^{(I)EXT}$ (Eq. (3)) and internal forces $\overline{f}_i^{(I)INT}$ (Eq. (4)).
   (c) Compute the rate of change of momentum $\underline{p}_i^{(I)}$ using Eq. (1) and update nodal momentum using $p_i^{(I)} \leftarrow p_i^{(I)} + \dot{p}_i^{(I)} \Delta t$
   (d) If the node is on a velocity boundary, set $p_i^{(I)} = mv_i^*$, where v_i^* is the prescribed velocity.
   (e) Compute contact forces $f_i^{(I)CT}$ using the contact algorithm proposed by Bardenhagen et al. (2000, 2001) and update the nodal momentum using $p_i^{(I)} \leftarrow p_i^{(I)} + f_i^{(I)CT} \Delta t$

5. For each material point P:
   (a) Update material point velocity $v_i^{(P)} \leftarrow v_i^{(P)} + \Delta t \sum_I S^{(IP)} \dot{p}_i^{(I)} / m^{(I)}$ and position $x_i^{(P)} \leftarrow x_i^{(P)} + \Delta t \sum_I S^{(IP)} p_i^{(I)} / m^{(I)}$
   (b) Compute the velocity gradient $v_{i,j}^{(P)}$ and deformation gradient $F_{ij}^{(P)}$
   (c) For simulations in incompressible soil, compute the velocity gradient $v_{i,j}^{(P)}(C)$ at the element center and compute the modified deformation gradient $\overline{F}_{ij}^{(P)}$
   (d) Using a suitable objective rate formulation, update the stresses and internal variables using the algorithms proposed in Nazem et al. (2009).
   (e) Update the material point volume $V^{(P)} \leftarrow \left(\det F_{ij}^{(P)}\right) V^{(P)}$

6. Adjust (translate/compress/reset) the back-ground grid.
7. Update time $t \leftarrow t + \Delta t$. If $t > t_{end}$, go to step 2; otherwise exit.

An objective time rate is required in large-deformation, large-rotation analyses (Lubliner 2008). There is also a specific objective rate formulation that must be used to obtain the correct solution of a bound-ary-value problem (Lim, J., Salgado, R. and Prezzi, M., paper in preparation).

3 VALIDATION OF MPM SIMULATIONS OF RELATED GEOMECHANICS BOUNDARY VALUE PROBLEMS

3.1 *Limit bearing capacity*

If an MPM formulation is capable of modeling cone penetration, then it must be able to also model a bearing capacity problem. The Prandtl punch problem, in which a flat plane-strain punch (a strip footing) is forced into a Tresca half-space from the surface, is a good test of the quality of the formulation because the solution to it is known exactly. But limit analysis solutions also exist bracketing the bearing capacity of footings, strip or square, embedded to various depths D in Tresca soil (Salgado et al. 2004). The values of limit unit bearing capacity q_{bL} are given in Table 1 for strip and Table 2 for square footings. We can see very good agreement between MPM and the benchmark results, including the value for a strip footing on the surface of the half-space, which is known to be $2 + \pi$ times the yield stress c of the Tresca material and is well approximated by the MPM solution.

The MPM analyses were performed for very stiff Tresca soils. This allowed development of the limit bearing capacity for small footing deflection, before any significant change to the geometry of the problem could occur. This is important because this is a feature of the limit analysis solutions used to benchmark the MPM values of q_{bL}.

In only three of the ten cases considered in the two tables was the MPM value outside, and only margin-ally outside, the range defined by the lower and upper bounds from limit analysis.

Table 1. Value of limit unit bearing capacity q_{bL} of a strip footing in Tresca soil from MPM compared to values obtained using formation gradient.

D/*B*	L	U	MPM
0.0	5.132	5.203	5.213
0.1	5.448	5.548	5.471
0.2	5.696	5.806	5.761
0.4	6.029	6.133	5.977
1.0	6.562	6.657	6.535

B = footing width.

Table 2. Value of limit unit bearing capacity qbL of a square footing in Tresca soil from MPM compared to values obtained using limit analysis.

D/*B*	L	U	MPM
0.0	5.856	6.227	6.149
0.1	6.491	7.140	7.063
0.2	6.897	7.523	7.426
0.4	7.303	8.104	8.002
1.0	8.771	9.429	9.296

B = footing width.

3.2 *Footing loaded to large settlements*

Rather than simply producing the correct values of limit bearing capacity, the MPM formulation must also reproduce the load response of a footing to large settlements. This problem has been studied using several methods, including sequential limit analysis (da Silva et al. 2011), ALE FEM (Kardani et al.

2015; Nazem et al. 2006, 2009), a form of ALE FEM known as "remeshing and interpolation technique with small strain" (RITSS) (Wang et al. 2013) and MPM (Iaconeta et al. 2019; Sołowski and Sloan 2015; Woo and Salgado 2018).

In the validation simulations, a 1-m wide footing was modeled as a rigid material, and the soil was modeled as a weightless Tresca material with $E = 100$ kPa, $v = 0.49$, $\rho = 1.0 \text{g/cm}^3$ and $s_u = 1$ kPa (the same parameter values used by Wang et al. 2013). The problem domain was discretized using an irregular background grid with a minimum element size equal to $B/20$ near the base of the footing and a maximum element size equal to $B/80$ away from the footing. Initially, each element contained 4 material points in a 2×2 configuration. The size of the domain was $10B$ in both the horizontal and vertical directions when measured from the center of the base of the footing. The Courant number κ was taken as 0.3. To simulate quasi-static conditions, the vertical penetration velocity increased linearly from 0 to 0.02 m/s over 1 s and then remained constant at 0.02 m/s after 1 s.

Comparison of the MPM formulation of Bisht et al. (2021a), noted in the legend as "B-bar GIMP," with a variety of other methods is shown in Figure 3. The favorable comparison with the high-quality ALE FEM simulation (labeled "RITSS" in the figure) is an important validation result.

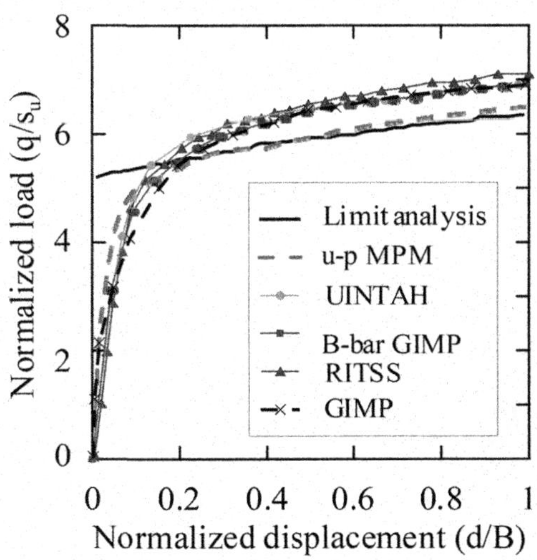

Figure 3. MPM GIMP simulations of a footing pushed to large penetration in an elasto-plastic Tresca soil compared with the solution of various other methods of analysis (after Bisht et al. 2021a).

4 CONE PENETRATION IN SAND

4.1 *Constitutive model*

Simulation of cone penetration in sand using MPM is less complex than in clay largely because drained penetration can be assumed a priori, there being therefore no need to perform coupled flow-deformation analyses. Coupled analyses are required for clay, as we will discuss later.

To properly simulate penetration in sand, a realistic sand constitutive model is required. The model must be able to accurately reproduce soil element response in triaxial compression, triaxial extension and simple shear loading. A model satisfying this requirement is that of Loukidis and Salgado (2009). The analyses reported here were performed using this model.

Model parameters and the test data required for their determination are listed in Table 3. Also given in the table are the values of model parameters for Ot-tawa sand. It is for Ottawa sand that MPM simulations were performed. Although the number of parameters (22) may seem daunting, only 5 parameters (h_1, h_2, e_{lim}, μ and k_h) are determined using a trial and error approach. The remaining 17 can be directly determined by fitting the model equations to the data. Most parameters can be determined using routine tests such as drained and undrained TXC, while others can be reasonably assumed based on the corresponding values for similar sands.

Table 3. Constitutive model input parameters for Ottawa sand (after Loukidis and Salgado 2009).

	Symbol	Value	Test(s)
Small-strain("elastic") parameters	v	0.15	compression w/unloading
	C_g	611	BE or RC
	n_g	0.437	BE or RC
	γ_1	0.00065	RC or TX
	α_1	0.47	TXCU
Critical state	Γ_c	0.780	TXC
	λ	0.081	TXC
	ξ	0.196	TXC
	M_{cc}	1.21	TXC
Bounding surface	k_b	1.9	TXC
Dilatancy	D_0	1.31	TXC
	k_d	2.2	TXC
Plastic modulus	h_1	2.2	TXC
	h_2	0.240	TXC
	e_{lim}	0.81	TXC
	μ	1.2	TXCU
Stress-induced anisotropy	c_1	0.71	TXE
	c_2	0.78	SS
	n_s	0.35	SS
Inherent anisotropy	a	0.31	TXE
	k_h	0.39	TXE
Yield surface radius	m	0.05	-

TXC: triaxial compression; TXE: triaxial extension; TXCU: undrained triaxial compression; SS: simple shear; RC: resonant column; and IC: isotropic compression.

4.2 *Cone penetration simulation scheme*

A fixed time step size – determined using Eq. (10) – was used for all cone penetration simulations. For us to use Eq. (10), certain variables and parameters must be evaluated. As will be discussed in detail in the next sub-section, the minimum element size ele_{min} used for the simulations in this study was 5.95 mm, and the relative density D_R for all tests was ~90% (implying a dry density $\rho$of 1.72 ton/m_3). Using a reasonably high mean stress p' of 10 MPa and a void ratio e of 0.524, corresponding to a relative density D_R of 90%, the small-strain shear modulus G_{max} is equal to 800 MPa using the Hardin and Richart (1963) correlation. Assuming $\kappa = 0.7$ and $\eta = 25$, a scaled time step $\Delta t_{ms} = 2.0 \times 10^{-5}$s was computed. Simulations performed using smaller κ and smaller η values produced nearly identical results. A fixed time step $\Delta t_{ms} = 2.0 \times 10^{-5}$s was therefore considered acceptable and was used in the simulations.

Penetration was simulated by imposing on the cone penetrometer successive displacement increments. The same rate of penetration is ideally used in computations as in reality, but to reduce computation time, a greater penetration velocity was used in the simulations. This is possible so long as penetration remains quasi-static, so that there is no rate dependence in the value of q_c. Table 4 details the parameters for the MPM analyses reported in this paper.

Table 4. General numerical scheme used for the MPM simulations of cone penetration in sand (after Bisht 2021).

Feature	Implementation
Type of MPM	uGIMP
Time integration	Explicit: $\Delta t = 2 \times 10^{-5}$s Mass scaling factor: 25
Mesh	Structured irregular grid divided into moving and compressible regions; Q4 element with initially four material points; element size approximately equal to the width of the shear band
Penetration velocity	V_{cone} = 0–10 cm/s for $t \leq$ 1s; 10 cm/s for $t >$ 1s
Initial stress field	Initial vertical stress known; K_0 = 0.45 assumed
Contact model	Coulomb friction model with μ_f = 0.3
Stress integration algorithm	Explicit with automatic sub-stepping and error control
Objective stress rate and strain measure	Modified kinematic logarithmic rate; Hencky strain
Volumetric locking	Locking not observed
Constitutive model	Loukidis and Salgado (2009) sand model; Parameters calibrated to Ottawa sand used

ρ: density; Δt = time step size; Q4: linear quadrilateral elements; K_0: coefficient of lateral earth pressure at-rest; mps.: material points; V_{cone}: penetration velocity of cone; and μ_r coefficient of friction.

The grid used to perform the simulations resembles that shown in Figure 1. It has a moving grid region that moves down with the cone penetrometer and a compressible region at the bottom (Woo and Salgado 2015). The nodes at the bottom of the compressible grid do not move, but those above them do. To capture shear bands that develop near the cone base and cone shaft, the smallest element size was chosen to be roughly equal to the shear band width observed in sands (e.g., Alshibli and Hasan 2008; Alshibli and Sture 1999; Desrues and Viggiani 2004; Salgado et al. 2017; Tehrani et al. 2017).

4.3 *Validation using calibration chamber tests*

The calibration chamber with a viewing window to enable digital image correlation (DIC) analysis of penetration tests was used to perform two cone penetration tests in dense samples subjected to two different levels of vertical effective stress.

The chamber, shown in Figure 4, is a half-cylindrical calibration chamber with a diameter Dc of 1680 mm and a height of 1200 mm. The tests were performed under BC3 calibration chamber boundary conditions (Ghionna and Jamiolkowski 1991; Salgado et al. 1998) by using an inflatable air-rubber bladder at the top of the sample and rigid lateral walls. Three cameras with a resolution of 5 megapixels (SV5M10 complementary metal-oxide semiconductor CMOS cameras, EPIX, Buffalo Grove, Illinois) were used to take images during testing (GalvisCastro et al. 2019; Salgado et al. 2016; TovarValencia et al. 2018).

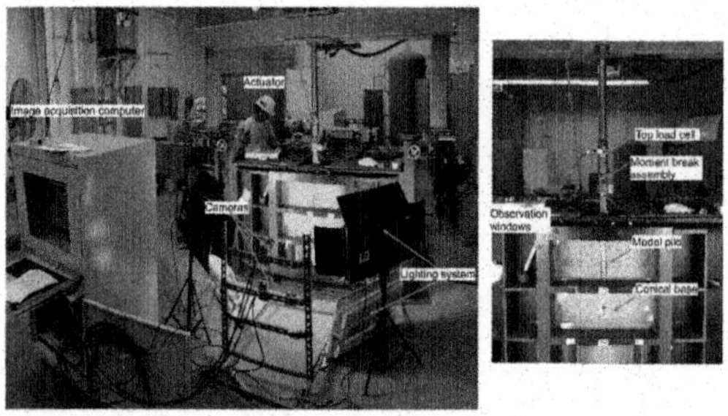

Figure 4. DIC calibration chamber at COFFEE, Purdue University: front view, showing viewing windows.

Figure 5 shows a comparison of measured q_c with the q_c values computed using MPM for two values of vertical effective stress at the level of the cone. The MPM simulations capture the gradual rise of q_c from the time the cone penetrometer enters the soil, which is also observed in the real experiment, and also the plateau that forms in the q_c versus depth plot once boundary effects are no longer a factor. The experiments have not been repeated for confirmation, but the comparisons of the plots from the

MPM simulation with those obtained from the calibration chamber test are encouraging.

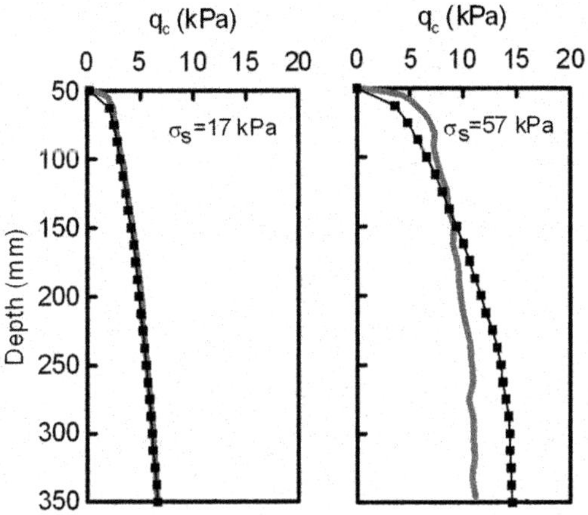

Figure 5. Comparison of MPM simulations of cone penetration tests in a calibration chamber with the results of those tests (modified after Bisht 2021).

What a DIC calibration chamber enables us to do that a regular calibration chamber does not is the measurement of displacement fields. Figure 6 shows a comparison of the horizontal and vertical displacement increments measured using a DIC analysis of images collected during a small downward move of the cone penetrometer to those computed using MPM. The comparison shows that MPM can effectively capture displacement fields around the advancing cone penetrometers in addition to also approximating closely the cone resistance values.

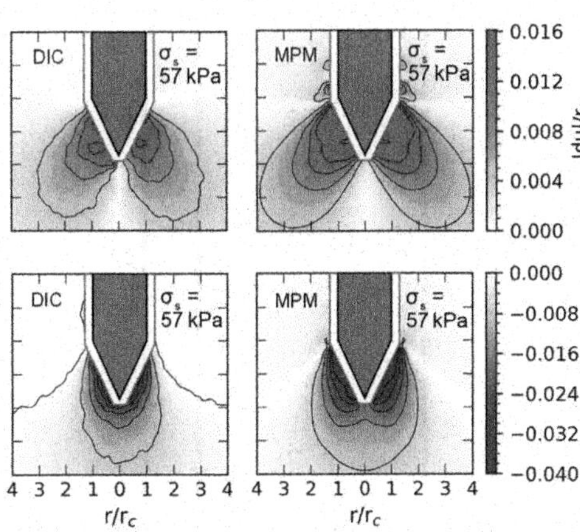

Figure 6. Comparison of vertical and horizontal displacement increments – d*u* and d*v* – obtained from DIC analysis of the chamber test and MPM simulations of the same test (modified after Bisht 2021).

5 CONE PENETRATION IN CLAY

5.1 *Formulation*

Simulation of cone penetration in clay requires a coupled, hydro-mechanical, effective stress analysis. Such an analysis allows the calculation of effective stress and pore pressure everywhere in the soil domain as the penetration progresses. It is accomplished in the analysis discussed here using a single material point to represent both the soil matrix and water and an explicit scheme to solve the governing equations, which are in terms of the velocity of the soil matrix and the velocity of water as the primary variables (Bisht et al. 2021b). Incompressibility constraints in the soil matrix are resolved using the non-linear B-bar method, and pore pressures are computed at element centers.

A coupled formulation can be used to handle penetration under partial drainage conditions. Undrained conditions need not be assumed. If we do assume undrained conditions, a simplification of the governing equations is possible. The governing equations are derived from consideration of momentum of the water, momentum of the soil matrix, and interaction of the soil particles and water though a drag resistance between the two due to a relative velocity.

5.2 *Validation*

The MPM formulation has been validated using, among other benchmarks, the one-dimensional consolidation theory of Terzaghi (1943). We show that here for illustration of the quality of the analysis framework, but point readers to Bisht et al. (2021b) for a more extended discussion of the accuracy and robustness of the numerical schemes.

Figure 7(a) shows a soil wall with height equal to 1.0 m and width equal to 0.075 m. An MPM grid was superposed on the wall. The grid is a structured regular grid with linear quadrilateral (Q4) elements with dimensions 0.025 m × 0.025 m. Each element contained four material points initially. The Young's modulus E of the soil is 10,000 kPa, its Poisson's ratio is 0.3, soil particle density is 2.67 ton/m^3, the initial porosity is 0.4, and the hydraulic conductivity is 10^{-3} m/s. The density of water is 2.67 ton/m^3 and its bulk modulus is 2.2 GPa. An initial pore pressure of 10 kPa was assigned to the soil. A fixed boundary condition was applied at the bottom, and roller boundaries restricting horizontal movement were applied to the sides. Impermeable boundary conditions were applied at the bottom boundary and the two side boundaries. The consolidation test was performed by applying a 10 kPa traction at the top surface and allowing the water to drain from the top. A fixed time step size equal to 1×10^{-5}s was used in the MPM simulation of the compression of the soil wall.

Figure 7(b) shows that the coupled MPM formulation closely matches the results obtained using the traditional Terzaghi solution.

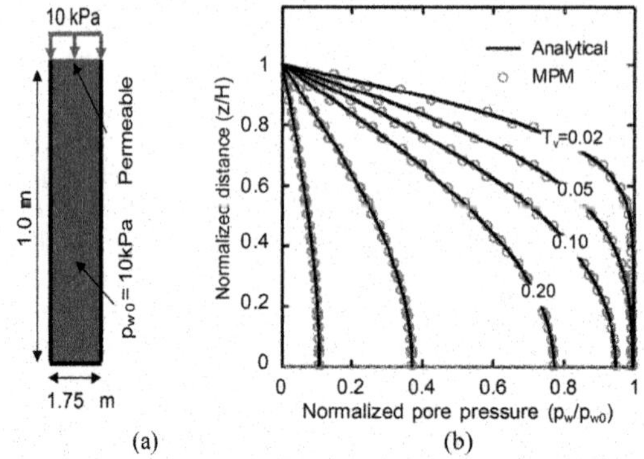

Figure 7. One-dimensional consolidation test: (a) slide of soil undergoing 1D consolidation; and (b) a comparison of the results obtained using the coupled MPM formulation and Terzaghi's solution after (Bisht et al. 2021b).

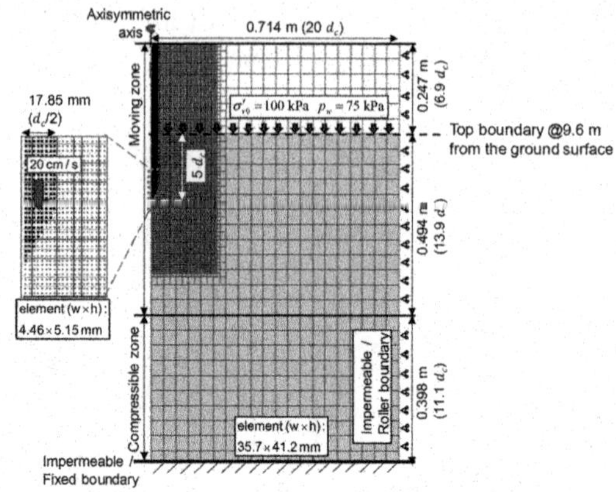

Figure 8. Grid and material point discretization for simulation of cone penetration at a depth of 9.6 m for the cone penetration tests of Landon (2007) (after Bisht et al. 2021b).

5.3 *Simulation of a real CPT in clayey soil*

We use the two CPTs performed by Landon (2007), which were located approximately 10 m apart, to test the MPM formulation for clay. The cone resistances measured in these two tests at a depth of 9.6 m were approximately 580 kPa and 730 kPa. Soil samples were collected from several depths. Based on testing done on these samples, the soil state at a depth of 9.6 m was estimated to be defined by an initial vertical effective stress σ'_{v0}= 100 kPa, initial vertical total stress σ_{v0}= 175 kPa, initial horizontal effective stress σ'_{h0}= 65 kPa, and overconsolidation ratio OCR = 2.2. The groundwater level was at a depth of 1.7 m. The hydraulic conductivity varied from 10^{-10} to 10^{-9} m/s.

The constitutive model used in the simulations is that of Chakraborty et al. (2013). The model contains a total of 22 parameters, most of which, as in the case of the sand model discussed earlier, can be determined through simple inspection and curve fitting of equations to data obtained from routine laboratory tests. The analyses were performed using the calibrated values of Boston Blue Clay ("BBC") that are given in Chakraborty et al. (2013), except that strain-rate independence is assumed because the problem is quasistatic.

Figure 8 shows both space and material point discretization of the simulations. The default four material points per elements were used. The same general approach followed for the grid in the analysis of cone penetration in sand was repeated for clay. A structured irregular grid with both a moving and a compressible mesh zone was used to discretize the do-main. A standard cone with diameter d_c = 35.7 mm and cone tip height h_c = 30.9 mm (apex angle = 60°) was assumed. Axisymmetric conditions were assumed, and thus only half the cone was discretized.

The goal of the analysis was not to produce the entire history of penetration from the ground surface, but penetration just into the clay layer. Accordingly, the initial position of the cone for the analysis was $5d_c$ below the upper boundary of the clay layer, which is at a depth of 9.6 m from the ground surface.

The soil was discretized using a mesh of dimensions $20d_c$ in width by $25d_c$ in height. The mesh was built with linear quadrilateral elements, each initially containing 2 × 2 material points.

Figure 9 shows the results of the MPM simulation compared to the two values measured in each CPT sounding at a depth of 9.6m. The comparison is very encouraging.

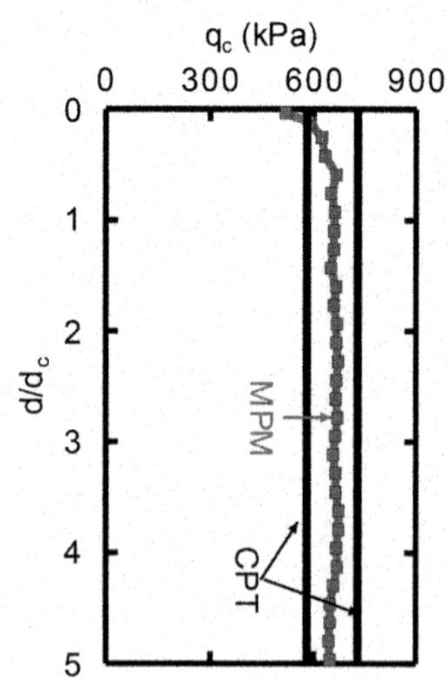

Figure 9. Comparison of cone resistance q_c, during a penetration path $5d_c$ long, with the values measured in two CPT soundings (after Bisht et al. 2021b).

6 IMPLICATIONS FOR FUTURE OF CPT INTERPRETATION

6.1 *Importance of MPM in future development of CPT interpretation approaches*

This paper has shown that MPM, combined with realistic constitutive models, is capable of accurate predictions of cone resistance in both sandy and clayey soils. What does that mean for CPT interpretation in the future? It depends on how far in the future, and it depends on what type of interpretation we are pursuing.

6.2 *Limitation of MPM in the present and in the near term*

At present, MPM with finite-size material points, and even, to a lesser degree, with material points that are actual points, is computationally expensive if accurate, realistic simulations are desired. It is important to understand that much faster simulations can be done, and can be found in the literature, but not with the level of discretization required for accuracy. At the time of this writing, simulation of an entire CPT starting from the ground surface is cost-prohibitive in an engineering design setting. However, cloud computing is expected to ease the computational burden significantly in the next several years.

This means that routine application of MPM to simulate entire soundings and the richness of interpretation that would result from that lies in the more distant future. However, it is entirely possible to compute cone resistance locally. For example, if the interest is in a specific layer with a certain soil existing with a certain state, then that is a simulation that is practically possible to perform. The computations done earlier for a clay layer illustrate this.

The other use of MPM is to perform simulations ahead of time for various combinations of soil intrinsic and state variables, and then use the results of these simulations to derive correlations of cone resistance with these variables or to feed them to AI tools. There are two types of relationships that can be developed: direct and indirect relationships.

Indirect relationships constitute the basic tool for pure CPT interpretation: from q_c and other measurements made during a CPT, obtain the values of the soil state variables. Direct relationships allow you to obtain estimates of quantities of direct interest in design, such as unit shaft and base resistance of a pile or liquefaction resistance from cone resistance and any other CPT measurements.

6.3 *Indirect interpretation of CPT results*

In indirect CPT interpretation methods, we are interested in evaluating soil state from CPT measurements. The thinking is somewhat akin to that of solving a system of equations (Salgado 1993). If there are enough equations, the unknown variables can be determined. The "equations" in this case are whatever relationships are available between the knowns (q_c and any other measurement) and the unknowns. The unknowns are, in sands, variables such as the relative density or void ratio, the effective stress state and soil fabric. In other words, the unknowns would be state variables. The relationships would be equations developed based on MPM simulations between cone resistance and the state variables. Parameters in these equations would be the soil intrinsic variables, those related to factors like mineralogy, particle size distribution, and the like, as well as critical state parameters. Intrinsic variables would be determined from routine laboratory tests.

Perhaps the greatest impact of the availability of reliable MPM simulation capabilities will be on interpretation in clay. Relationships already exist relating q_c to relative density and lateral effective stress, for example, in sands (Salgado and Prezzi 2007). However, interpretation of CPT results in clay has remained linked to a Tresca-linked view of clay, with relationships between q_c and *s*u instead of between q_c and void ratio, effective stress and fabric. In these traditional relationships, a cone factor N_k is defined to relate q_c to s_u, but how should s_u be defined in such a relationship? Should it be that obtained under triaxial compression, triaxial extension or simple shear loading? Additionally, somewhat awkward relationships must be resorted to relating N_k to a rigidity index defined in terms of a modulus and a shear strength that are again not defined. An interpretation method leading to the values of state variables would be preferable.

6.4 *Direct interpretation of CPT results*

Direct interpretation occurs in a design application context, instead of a soil mechanics or soil characterization context. To develop a direct CPT method, it is necessary that the soil response for the specific design problem can be correlated with q_c and/or other CPT measurement. If the design problem can be itself simulated numerically, we have an entirely science-based correlation that can be used in design. Otherwise, the correlation will be semi-empirical, as liquefaction potential assessment methods must still be. Table 5 contains three of the most important design problems to which the CPT is routinely applied. Foundation loading can already be effectively simulated using the finite element method to settlement levels of interest in design (see, e.g., Han et al. 2017; Loukidis and Salgado 2011; Salgado et al. 2017). MPM can be used for problems in which significant mesh distortion will render FEM analyses not feasible. Liquefaction cannot as yet be simulated. The main obstacle is the unavailability of a constitutive model that can properly simulate the soil response under repeated cycling, so methods will for now and in the near future remain semi-empirical.

Table 5. Direct interpretation of CPT results.

Design problem	Target variables	Basis of relationships
Shallow foundation design	Limit unit bearing capacity q_{bL}; settlement w in service	Computational simulations using MPM or FEM
Deep foundation design	Ultimate unit base resistance $q_{b,ult}$; limit unit shaft resistance q_{sL}	Computational simulations using MPM or FEM
Liquefaction potential assessment	Cyclic resistance ratio CRR	Semi-empirical because liquefaction cannot yet be accurately simulated

7 CONCLUSIONS

The main purpose of this paper was to review our ability to simulate the cone penetration process. For realistic simulation of cone penetration in both sand and clay, it is necessary to have (1) realistic, robust constitutive models; (2) realistic, robust constitutive model stress integration schemes; and (3) accurate, efficient, robust numerical schemes for integration of the governing equation for the cone penetration process. The paper focused on prong 3 of the simulation framework: the use of the u-GIMP MPM formulation to simulate the penetration process.

MPM replaces Gauss points in Lagrangian FEM formulations by material points. The material points are not affixed to the finite element constituting the mesh or grid into which a body is discretized but rather can move through the grid, even crossing elements. The material points in fact are the discretization of the body, whereas the grid is the discretization of space.

There are certain consequences from this fundamental difference between the two methods. Two of the most consequential are: (1) integration is not as accurate as in FEM; (2) some techniques used in FEM to avoid pathologies, such as volumetric locking, are not effective in MPM; (3) because the grid can always be reset between computation time increments, mesh distortion does not impact calculations, and MPM simulations can progress to very large displacement levels.

The paper illustrated some of the validation exercises that the MPM computational framework discussed here has been subjected to and successfully passed. These validations have shown that accurate cone penetration simulations are possible. The impact of this capability is yet to come. In the future, the ability to accurately, realistically simulate cone penetration and other processes involved in the direct or indirect interpretation of CPT results will lead to high-quality interpretation methods. Other factors in interpretation, such as spatial variability of soil state variables and soil type, will also be better taken into account with the more advanced command of the mechanics of cone penetration.

REFERENCES

Alshibli, K. A. & Hasan, A. 2008. Spatial variation of void ratio and shear band thickness in sand using X-ray computed tomography. *Géotechnique* 58(4): 249–257.

Alshibli, K. A. & Sture, S. 1999. Sand Shear Band Thickness Measurements by Digital Imaging Techniques. *Journal of Computing in Civil Engineering* 13(2): 103–109.

Bardenhagen, S. G., Brackbill, J. U., & Sulsky, D. 2000. The material-point method for granular materials. *Computer Methods in Applied Mechanics and Engineering* 187(3– 4): 529–541.

Bardenhagen, S. G., Guilkey, J. E., Roessig, K. M., Brackbill, J. U., Witzel, W. M., & Foster, J. C. 2001. An improved contact algorithm for the material point method and application to stress propagation in granular material. *CMES - Computer Modeling in Engineering and Sciences* 2(4): 509–522.

Belytschko, T. B. & Kennedy, J. M. 1978. Computer models for subassembly simulation. *Nuclear Engineering and Design* 49(1–2): 17–38.

Bisht, V. 2021. Cone penetration analysis using the Material Point Method. Purdue University.

Bisht, V. & Salgado, R. 2018. Local transmitting boundaries for the generalized interpolation material point method. *International Journal for Numerical Methods in Engineering* 114(11): 1228–1244.

Bisht, V., Salgado, R., & Prezzi, M. 2021.a. Simulating penetration problems in incompressible materials using the material point method. *Computers and Geotechnics* Elsevier Ltd 133(December 2019): 103593.

Bisht, V., Salgado, R., & Prezzi, M. 2021.b. Material Point Method for Cone Penetration in Clays. *Journal of Geotechnical and Geoenvironmental Engineering* 147(12): 1–16.

Bisht, V., Salgado, R., & Prezzi, M. 2021.c. Simulating penetration problems in incompressible materials using the material point method. *Computers and Geotechnics* 133: 103593.

Chakraborty, T., Salgado, R., & Loukidis, D. 2013. A twosurface plasticity model for clay. *Computers and Geotechnics* 49(765): 170–190.

Courant, R., Friedrichs, K., & Lewy, H. 1967. On the Partial Difference Equations of Mathematical Physics. *IBM Journal of Research and Development* 11(2): 215–234.

Desrues, J. & Viggiani, G. 2004. Strain localization in sand: an overview of the experimental results obtained in Grenoble using stereophotogrammetry. *International Journal for Numerical and Analytical Methods in Geomechanics* 28(4): 279–321.

Galvis-Castro, A. C., Tovar-Valencia, R. D., Salgado, R., & Prezzi, M. 2019. Effect of loading direction on the shaft resistance of jacked piles in dense sand. *Geotechnique* 69 (1).

Ghionna, V. N. & Jamiolkowski, M. 1991. A critical appraisal of calibration chamber testing of sands. *Proc., 1st Int. Symp. on Calibration Chamber Testing* Elsevier, Amsterdam, The Netherlands 13–39.

Han, F., Salgado, R., Prezzi, M., & Lim, J. 2017. Shaft and base resistance of non-displacement piles in sand. *Computers and Geotechnics* 83: 184–197.

Hardin, B. O. & Richart, J. F. E. 1963. Elastic Wave Velocities in Granular Soils. *Journal of the Soil Mechanics and Foundations Division* 89(1): 33–65.

Hughes, T. J. R. 1980. Generalization of selective integration procedures to anisotropic and nonlinear media. *International Journal for Numerical Methods in Engineering* 15(9): 1413–1418.

Iaconeta, I., Larese, A., Rossi, R., & Oñate, E. 2019. A stabilized mixed implicit Material Point Method for non-linear incompressible solid mechanics. *Computational Mechanics* Springer Berlin Heidelberg 63(6): 1243–1260.

Kardani, M., Nazem, M., Carter, J. P., & Abbo, A. J. 2015. Efficiency of high-order elements in large-deformation problems of geomechanics. *International Journal of Geomechanics* 15(6): 04014101.

Kellezi, L. 2000. Local transmitting boundaries for transient elastic analysis. *Soil Dynamics and Earthquake Engineering* 19(7): 533–547.

Landon, M. M. 2007. Development of a non-destructive sample quality assessment method for soft clays.

Loukidis, D. & Salgado, R. 2009. Modeling sand response using two-surface plasticity. *Computers and Geotechnics* Elsevier Ltd 36(1–2): 166–186.

Loukidis, D. & Salgado, R. 2011. Effect of relative density and stress level on the bearing capacity of footings on sand. *Géotechnique* 61(2): 107–119.

Lubliner, J. 2008. *Plasticity Theory*. Dover Publications, INC.

Nazem, M., Carter, J. P., Sheng, D., & Sloan, S. W. 2009. Alternative stress-integration schemes for largedeformation problems of solid mechanics. *Finite Elements in Analysis and Design* 45(12): 934–943.

Nazem, M., Sheng, D., & Carter, J. P. 2006. Stress integration and mesh refinement for large deformation in geomechanics. *International Journal for Numerical Methods in Engineering* 65(7): 1002–1027.

Of, N. & Yu, L. A. Y. B. H. S. 2000. of S Teady C One P Enetration. *Manager* (July): 594–605.

Salgado, R. 1993. Analysis of Cone Penetration Resistance in Sand. University of California, Berkeley.

Salgado, R. 2012. The mechanics of cone penetration: Contributions from experimental and theoretical studies. *Geotechnical and Geophysical Site Characterization 4* CRC Press, Porto de Galinhas, Brazil 131–153.

Salgado, R. & Bisht, V. 2021. Advances in the solution of geotechnical boundary-value problems. *Computers and Geotechnics* Elsevier Ltd 138(July): 104183.

Salgado, R., Han, F., & Prezzi, M. 2017. Axial resistance of nondisplacement piles and pile groups in sand. *Rivista Italiana di Geotecnica* 51(4): 35–46.

Salgado, R., Lyamin, A. V., Sloan, S. W., & Yu, H. S. 2004. Twoand Three-dimensional Bearing Capacity of Foundations in Clay. *Géotechnique* Ice Virtual Library 54(5): 297–306.

Salgado, R., Mitchell, J. K., & Jamiolkowski, M. 1997. CAVITY EXPANSION AND PENETRATION RESISTANCE IN SAND By R. Salgado,' Associate Member, ASCE,. *Journal of Geotechnical and Geoenvironmental Engineering* 123(April): 344–354.

Salgado, R., Mitchell, J. K., & Jamiolkowski, M. 1998. Calibration chamber size effects on penetration resistance in sand. *Journal of Geotechnical and Geoenvironmental Engineering* 124(9): 878–888.

Salgado, R. & Prezzi, M. 2007. Computation of cavity expansion pressure and penetration resistance in sands. *International Journal of Geomechanics* 7(4): 251–265.

Salgado, R. & Randolph, M. F. 2001. Int. J. Geomech. 2001.1:175-192. *International Journal of Geomechanics* 1(2): 175–192.

Salgado, R., Tovar, R. D., Tehrani, F. S., Castro, A. G., Han, F., & Prezzi, M. 2016. Effect of surface roughness on the shaft resistance of non-displacement piles embedded in sand. *Géotechnique* 66(5): 386–400.

Schreyer, H. L., Sulsky, D., & Chen, Z. 1994. A particle method for history-dependent materials. *Computer Methods in Applied Mechanics and Engineering* 118: 179–196.

da Silva, M. V., Krabbenhoft, K., Lyamin, A. V., & Sloan, S. W. 2011. Rigid-plastic large-deformation analysis of geotechnical penetration problems. *13th International Conference of the International Association for Computer Methods and Advances in Geomechanics (IACMAG)* 42–47.

Sloan, S. W. & Randolph, M. F. 1982. *Numerical prediction of collapse loads using finite element methods. International Journal for Numerical and Analytical Methods in Geomechanics*.

Sołowski, W. T. & Sloan, S. W. 2015. Evaluation of material point method for use in geotechnics. *International Journal for Numerical and Analytical Methods in Geomechanics* 39(7): 685–701.

Teh, C. I. & Houlsby, G. T. 2009. An analytical study of the cone penetration test in clay. *Géotechnique* 41(1): 17–34.

Tehrani, F. S., Han, F., Salgado, R., & Prezzi, M. 2017. Laboratory study of the effect of pile surface roughness on the response of soil and non-displacement piles. *Geotechnical Frontiers 2017* American Society of Civil Engineers, Reston, VA 256–264.

Terzaghi, K. 1943. *Theoretical Soil Mechanics*. John Wiley & Sons, Inc., Hoboken, NJ, USA.

Tovar-Valencia, R., Galvis-Castro, A., Salgado, R., & Prezzi, M. 2018. Effect of surface roughness on the shaft resistance of displacement model piles in sand. *Journal of Geotechnical and Geoenvironmental Engineering* 144(3).

Wallstedt, P. C. & Guilkey, J. E. 2008. An evaluation of explicit time integration schemes for use with the generalized interpolation material point method. *Journal of Computational Physics* 227(22): 9628–9642.

Wang, B., Vardon, P. J., Hicks, M. A., & Chen, Z. 2016. Development of an implicit material point method for geotechnical applications. *Computers and Geotechnics* Elsevier Ltd 71: 159–167.

Wang, D., Randolph, M. F., & White, D. J. 2013. A dynamic large deformation finite element method based on mesh regeneration. *Computers and Geotechnics* Elsevier Ltd 54: 192–201.

Woo, S. I. & Salgado, R. 2015. Bounding surface modeling of sand with consideration of fabric and its evolution during monotonic shearing. *International Journal of Solids and Structures* Elsevier Ltd 63: 277–288.

Woo, S. I. & Salgado, R. 2018. Simulation of penetration of a foundation element in Tresca soil using the generalized interpolation material point method (GIMP). *Computers and Geotechnics* Elsevier Ltd 94: 106–117.

Cone Penetration Testing 2022 – Gottardi & Tonni (eds)
© 2022 the Author(s), ISBN 978-1-032-35624-2

Practical use of shear wave velocity measurements from SCPTU in clays

M. Long
University College Dublin (UCD), Ireland

ABSTRACT: The main objective of this paper is to promotes use of shear wave velocity (V_s) measurements from SCPTU testing as a compliment to standard CPTU data. Recent developments in the technique are described. Uncertainties in the methods have been well researched and several methods have been proposed to deal with these uncertainties. Nonetheless V_s measurements seem reliable in isotropic soil conditions. V_s profiles for a range of soft to firm clays world-wide are presented. Clear links between these profiles are identified based on fundamental soil properties. Stiff overconsolidated clays do not follow the same patterns due to pure stress anisotropy and their fissured nature. Correlations between V_s and CPTU data and with a variety of soil properties can be very powerful. However these correlations should ideally be local and only applied to other soils and areas with great caution. An exception might be for correlation between V_s and preconsolidation stress pc'. Use of SCPTU data to help assess sample disturbance and for the prediction of settlement of footings is described. Some future outlooks including dealing with anisotropy are discussed.

1 INTRODUCTION

Shear wave velocity (V_s) is a fundamental measurement in all solids for example steel, concrete, wood, soil and rock (Mayne, 2000). Because of this broad range of application, V_s values are an attractive means of characterising a range of natural geomaterials. Water is not able to tolerate shear waves. Therefore measurements in soil are not influenced by the presence of groundwater (unlike P-waves).

Over the last decades, V_s measurements have gained popularity in geotechnical engineering practice. Advances in cost effective and efficient methods of determination of V_s focused attention on this parameter, which originally was mainly used for seismic hazard assessment or dynamic analyses. However, its use has be extended to general site characterisation studies, ground movement analyses for tunnels and excavation problems, determination of strength and compressibility parameters by empirical correlation, prediction of the behaviour of deep and shallow foundations, assessment of sample disturbance effects and in the quality control of ground improvement schemes among other applications.

V_s can be measured by a wide range of intrusive and non-intrusive methods, see L'Heureux and Long (2017) for example. In this paper the focus will be on the use of the seismic piezocone test (SCPTU) for V_s measurement.

The paper will deal with some basics on shear waves including shear wave anisotropy. The measurement technique will be briefly described together with some recent developments in the method. Uncertainties in the measurement technique will be considered. The meaning of numerical V_s values will be discussed and typical V_s profiles for various clays will be presented and compared.

Use of V_s profiles for site characterisation will be described. Correlations between V_s and other SCPTU measurements and with a range of soil properties will be considered. An emphasis on estimating preconsolidation stress from V_s. Finally use of V_s for sample disturbance assessment and for prediction of settlement of shallow foundations will be outlined.

The paper will focus on clay soils only (with some comparison made with peat). It will not deal with earthquake problems such as liquefaction.

It is stressed that the objective of the paper is to promote increasing use of SCPTU V_s measurements as a compliment, not as a replacement, to other CPTU measurements. The author supports the suggestion of Mayne et al. (2009), in their Key Note lecture to the 17th ICSMGE in Alexandria, who proposed that in site characterisation studies the standard approach should be in the use of the SCPTU (or seismic dilatometer, SDMT) as up to five separate measurements are captured at each depth tested.

2 SOME BASIC CONSIDERATIONS

2.1 *Shear waves*

Engineer's interest in V_s determination has largely been driven by the need to obtain values of the small strain shear modulus (G_{max}), as this is an important parameter for a range of geotechnical design applications. This usually involves strains of 10^{-4} to 10^{-3} % and less.

According to elastic theory G_{max} may be calculated from the shear wave velocity using the following equation:

$$G_{\max} = \rho V_s^2 \tag{1}$$

where: G_{max} is the shear modulus (in Pa), V_s is in m/s, and ρ is the total mass density (in kg/m^3).

It is important to remember that for practical use, the G_{max} value needs to be reduced to give the strain level relevant to a particular problem. This point will be addressed later.

2.2 *Anisotropy of shear wave velocity*

Anisotropy of shear wave velocity/stiffness may be significant in many clays. This is particularly the case for overconsolidated materials. Butcher and Powell (1995) and others have suggested that to distinguish between shear wave velocities with different propagation and polarisation directions, subscripts can be used to donate these. For example $V_{s\text{-}he}$ denotes a vertically propagating, horizontally polarised shear wave velocity such as would be measured in a downhole or SCPTU test. Similarly $V_{s\text{-}hv}$ or $V_{s\text{-}hh}$ would be measured in cross-hole testing.

An example for the well-researched Italian clay site at Augusta (saline area) is shown on Figure 1. The materials below 15 m comprise a medium stiff, overconsolidated (overconsolidation ratio, OCR = 2 to 6) Pleistocene marine clay with low to medium plasticity (Lo Presti et al., 1998). $V_{s\text{-}vh}$ values from seismic dilatometer (SDMT) and downhole testing are very similar but they are some 100 m/s less than the cross-hole $V_{s\text{-}hv}$ values.

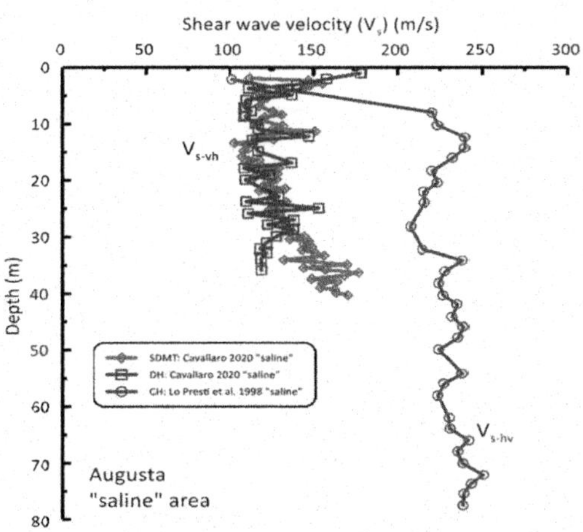

Figure 1. Anisotropy of shear wave velocity for August (saline area) research site in Italy. Data replotted from Lo Presti et al. (1998) and Cavallaro (2020).

For simplicity in this paper the SCPTU measurement will be denoted as V_s unless otherwise stated.

3 SCPTU MEASUREMENT TECHNIQUE

3.1 *Traditional approach*

The SCPTU was first introduced in 1984 at the University of British Columbia, see Rice (1984), Campanella et al. (1986) and Robertson et al. (1986). Some schematics to illustrate the test set-up are shown on Figure 2.

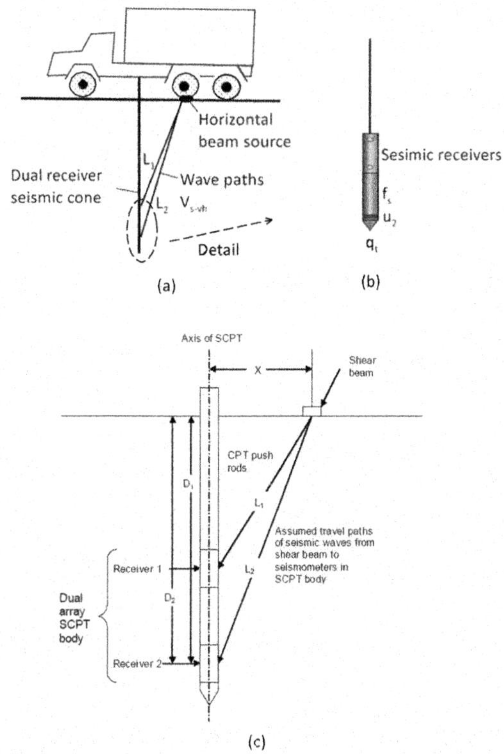

Figure 2. Set-up for CPTU test (a) typical test arrangement (Powell, 2005), (b) details of cone (Mayne, 2000) and (c) schematic for V_s determination (Butcher et al., 2005).

Ideally the test should be carried out according to ASTM (2019) or ISO (2014). Some detailed guidance for the execution of the test is given by Butcher et al. (2005). A standard cone penetrometer is equipped with one to three horizontally aligned seismic sensors. Originally one sensor was used but most modern equipment has two sets of sensors comprising three component geophones. Butcher and Powell (1996) discuss some problems associated with a single sensor SCPTU. Some commercial companies successfully use equipment with one accelerometer instead of a geophone.

Recordings are made during a pause in the cone penetration typically every 0.5 m. The seismic signals are generated by striking a horizontal beam which is coupled to the ground, usually by the weight of the testing vehicle. Hammers of mass about 10 kg are typically used. The beam should be aligned parallel to the axis of the receivers. Assuming straight ray paths V_s is determined by (Figure 2):

$$V_s = \frac{L_2 - L_1}{t_2 - t_1} \tag{2}$$

where: t_2 and t_1 are the shear wave travel time from source to receiver

Several methods exist for picking the shear wave arrival time such as manually picking the first arrival time, the first major cross-over method or by use of cross correlation. In a dual array seismic cone, the wave traces from each of the two sensors can be used to get the "true interval" travel time between the sensors. Many authors (Lunne et al., 1997b) have stressed the advantage of the "true interval" set-up stating that it gives more accurate results than the "pseudo interval" method where only one sensor is used. A particular advantage is the same source signal, rather than two successive blows, is used in the calculations.

Similar techniques are used in offshore SCPTU testing, see for example Peuchen et al. (2002), Looijien et al. (2017) or Lunne et al. (2013) with the signal being produced by different techniques such as using a seismic cap (Campanella et al., 1986), hydraulic underwater shear wave box (de Lange et al., 1990) or underwater hammers (Paoletti et al., 2013) or the twin hammers of the PROD system (Nguyen et al., 2015).

3.2 *SDMT*

The seismic dilatometer (SDMT) is the combination of the standard flat dilatometer (DMT) with a similar seismic module for measuring V_s as employed in the SCPTU (Marchetti et al., 2008, Marchetti, 2015). The SDMT employs two geophones spaced 0.5 m apart and therefore allows a more accurate "true- interval" two-receiver test configuration.

3.3 *Recent developments in techniques*

Typically modern dual sensors are spaced at between 0.5 m and 1.0 m. The system used by the Norwegian Public Roads Administration allows the sensors to be set at either 0.5 m or 1.0 m (Valsson et al., 2020).

Typically V_s measurements are made at 0.5 to 1 m intervals, the latter often corresponding to rod breaks. Although the process is more time consuming and expensive, McGillivray and Mayne (2004), (McGillivray and Mayne, 2008), Mayne (2005b) and others advocate the use of frequent interval SCPTU (or SDMT) with measurements taken every 0.2 m. An example of such a set of readings compared to conventional single sensor measurements from the famous Treporti test embankment, Venice are shown on Figure 3 (McGillivray and Mayne, 2004). The site is underlain by a complex mixture of silts, sands and clays (Simonini et al., 2007). These data clearly demonstrates that enhanced depth resolution can be achieved with this technique.

Mayne and McGillivray (2008), McGillivray and Mayne (2008), Ku and Mayne (2012), Ku et al. (2013a) and Ku et al. (2013b) introduce the devel-opment of continuous interval seismic piezocone testing (CiSCPTU). Here V_s is measured at continuously during cone penetration using a specially developed autoseis source. The system employs an electro-mechanical auto hammer to deliver impact type seismic waves that are continuously recorded while the device is penetrating. As well as producing a detailed V_s profile, the significant advantage of this system is that there is no halting in penetration during the SCPTU procedure and therefore no effect on the resulting q_t, f_s, and u_2 readings.

Examples from this system are also given by Agaiby et al. (2016) and Mayne and Woeller (2015).

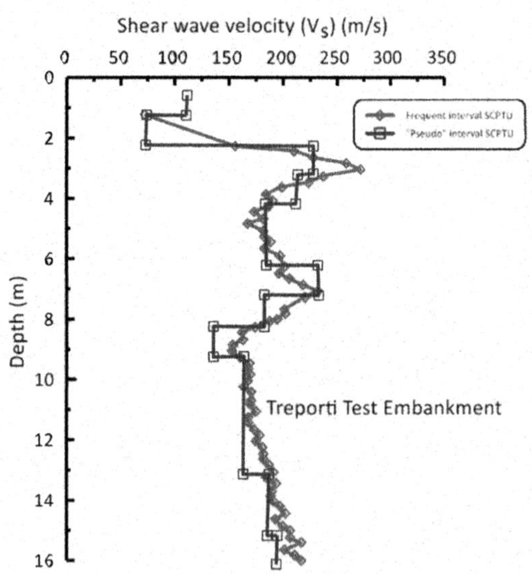

Figure 3. Frequent interval and "pseudo" interval SCPTU from Treporti test embankment (McGillivray and Mayne, 2008).

3.4 *Uncertainties in measurements*

In engineering practice it is often commonplace to provide a single V_s profile versus depth for a specific location to the design engineer with no description of the uncertainty involved (Stolte and Cox, 2020). There are two types of uncertainty associated with SCPTU measurements, aleatory variability and epistemic uncertainty (Hallal and Cox, 2019).

Aleatory variability is connected to spatial variability of the ground across the site area and can be dealt with by performing multiple tests across a site with perhaps the "gaps" in the data infilled by another geophysical technique such as electrical resis-tivity tomography (ERT).

An epistemic uncertainty refers to the deficiencies by lack of knowledge or uncertainty. Some epistemic uncertainties associated with dual-receiver SCPTU testing are as follows (Peuchen et al., 2020, Stolte and Cox, 2020, Styler and Weemees, 2016, Parasie et al., 2022):

- choice of signal source,
- uncertainties related to timing,
- lateral source offset distance,
- choice of testing interval,
- vertical depth control,
- assumption of isotropic ground conditions,
- assumption of straight-line ray path of the signal,
- identification of signal arrival time,
- external noise (expressed as signal to noise ratio.SNR),
- near field effects especially at shallow depths (< 5m) e.g. from surface waves or acoustic reflections,
- conversion of signal arrival time to shear wave velocity,
- wave interference due to presence of rods/ penetrometer.

In offshore SCPT testing some additional uncertainties could be considered (Peuchen and Gomez Meyer, 2021) such as:

- sea wave interference effects,
- influence of the physical seabed frame used to support the measuring devices (particularly over the top 5 m of a profile),
- ambient noise.

With respect to the lateral distance between the source and the receiver it would seem that if the distance is too great the wave form may be closer to that of $V_{s\text{-}hh}$ rather than the conventional $V_{s\text{-}vh}$ (Personal communication P.W. Mayne). ASTM (2019) suggests the lateral source offset should be in the range 1 m to 6 m. Hallal and Cox (2019) suggest that this distance is limited to a maximum of 1.5 m.

Hallal and Cox (2019) suggest that ideally the testing interval should be less than 1 m and that issues with data reduction may occur of there is a stiff layer close to the ground surface.

The length of the wave path between the source and receiver is often assumed to be a straight line (ASTM, 2019). Alternatively the refracted wave path method has been proposed (Joh and Mok, 1998, Kim et al., 2004). This technique is based on refraction caused by stiffness differences between layers. Bang et al. (2014) propose an improvement on this system which they term the mean refracted wave path method.

Much work has been done on assessing different ways of obtaining shear wave travel times and converting these data into shear wave velocity profiles, see for example Baziw (1993), Baziw (2002) Pidlisecky and Haines (2011), Stolte and Cox (2020) or Valsson et al. (2020).

In engineering practice, for example in seismic response analyses, upper and lower bound estimates of V_s versus depth are often used to account for epistemic uncertainty (Stolte and Cox, 2020). However this approach has recently been criticised especially where uncertainty in V_s is assumed to be depth independent (Teague and Cox, 2016). Several researchers, e.g. Garofalo et al. (2016) have shown that V_s uncertainty is more pronounced nearer to the surface than with depth.

Stolte and Cox (2020) provide a suggested means of quantifying the epistemic uncertainty by involv-ing various wave travel time and velocity analysis methods. These authors also state that, although the conventional single V_s profile may be sufficient for design engineers, any site report should clearly state the assumptions that were made and should report on how the wave travel time was obtained and the report should detail the actual results so the engineer can check the resulting V_s values.

Parasie et al. (2022) suggest a technique for estimating an "uncertainty budget" for V_s/G_{max} based on an inventory of items similar to those listed above. They give an example of two cases for SCPTU testing in overconsolidated clay and dense sand and determine that the uncertainty budget in both cases is reasonable.

To end this section on a positive note it has been found that, despite the uncertainties discussed above, there have been many published accounts of V_s measurements where different systems were used by different operators but returned very similar output. This is particularly the case for sites which are un-derlain by relatively homogenous, isotropic materials. More care needs to be taken in interpreting data form non-homogenous or layered or highly anisotropic soils. The example shown on Figure 4 for the well characterised Fucino site in Italy. The site is underlain by a thick deposit of soft cemented clay (Soccodato, 2003).

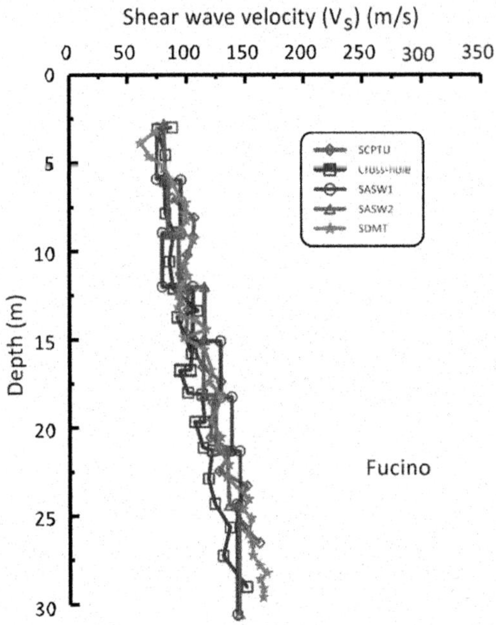

Figure 4. V_s profiles for Fucino site, Italy. Data from Burghignoli et al. (1991), Foti et al. (2006) and Soccodato (2003).

V_s measurements by SCPTU, SDMT, cross-hole and SASW (spectral analysis of surface waves) all give similar results, certainly for the purposes of an engineering type analysis.

Table 1. V_s values for clay soil classification.

Material	NIBS/EN	Poulos (2021)
	V_s (m/s)	V_s (m/s)
Very soft clay	< 100	85 - 105
Soft clay	{100 to	106 - 135
Medium stiff clay	180}	136 - 185
Stiff clay	180 - 360	186 - 275
Very stiff clay	360 - 800	276 - 365

4 SHEAR WAVE VELOCITY VALUES

Before looking at V_s profiles from various countries in detail it is worth considering the possible meaning/classification of numerical V_s values. For seismic assessments the average V_s over the top 30 m of a site (V_s,30) is often considered.

Both the US (NIBS, 2003) and European (EN, 2004) guidelines quote the same range of V_s values for site classification as summarised on Table 1.

Recently for general engineering classification Poulos (2021) suggested the range of values for site classification also given on Table 1.

The range of values and their allotted meaning is very similar in both cases. For the purposes of this paper an upper bound V_s value for stiff to very stiff clays of 350 m/s will be considered, and all plots will be drawn where possible with this as the maximum value on the x-axis. Similarly the value V_s = 100 m/s will be considered

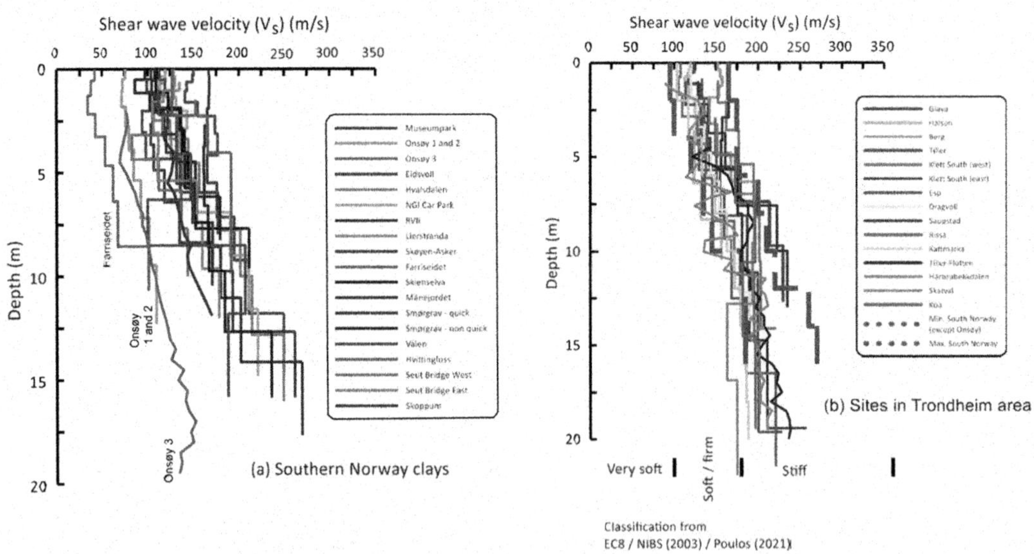

Figure 5. V_s values for Norwegian clays (a) south Norway clays and (b) clays from Trondheim area. Data from L'Heureux and Long (2017) with additional data from Paniagua et al. (2017), L'Heureux et al. (2019) and Dahl and Haugen (2022).

as the upper bound for very soft clays for illustration purposes.

These values need to be treated with caution and can be misleading as will be described later.

5 TYPICAL V_S PROFILES FOR CLAYS

5.1 *Norwegian clays*

A summary of all the available V_s data for Norwegian clays is given on Figure 5, with sites from South-East Norway shown on Figure 5a and those from Mid-Norway/Trondheim area on Figure 5b. The data is mostly taken from L'Heureux and Long (2017) with some additional data as indicted.

Some of the data was obtained using the MASW method (multichannel analysis surface waves) tech-nique. As has been shown in the above paper and elsewhere, the results do not seem to be affected by the technique used or the directions of propagation and polarisation of the waves. This is likely to be due to the largely isotropic nature of these materials.

All the Southern Norway sites show a very similar trend between V_s and depth and differ only in the value of V_s close to the surface. Teachavorasinskun and Lukkunaprasit (2004) found a similar pattern for soft Bangkok clay and they expressed the relationship in the form:

$$V_{sz} = V_{sg} + mz \tag{3}$$

where:

$V_{sz} = V_s$ at any depth z (m/s)

$V_{sg} = V_s$ close to the ground surface (m/s)

m = slope of the line of V_s versus depth (units m/s.m)

Some exceptions are the very soft, high water content and organic, clays at Onsøy and especially Farriseidet which show much lower values of V_s. In general the values increase from about 125 m/s near the surface to 225 m/s at 20 m depth classifying the materials as soft becoming medium stiff to stiff with depth (Table 1) consistent with other measurements on the soils. For example anisotropically consolidated undrained triaxial compression (CAUC) tests on the clay from the NGTS (Norwegian GeoTest Site) at Tiller-Flotten show undrained shear strength (su) values increasing from about 25 kPa near the surface to about 115 kPa at 20 m depth (L'Heureux et al., 2019).

Data for the Trondheim and Mid-Norway sites show similar values to those from Southern Norway.

The data fits in a very tight band within the limits for the Southern Norway clays. The fact that the soils have similar values is not surprising given the similar depositional and stress history and mineralogy of the sediments.

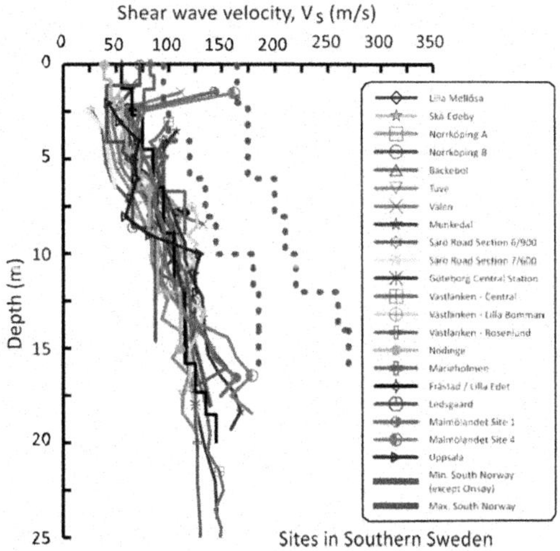

Figure 6. V_s values for Swedish clays. Data from Long et al. (2017) with additional data from Kania et al. (2000), (Dannewitz et al., 2005), Comina et al. (2017) and Granskär (2018).

5.2 *Southern Swedish clays*

A summary of all the available data from southern Sweden is shown on Figure 6. These data are taken from Long et al. (2017) augmented with additional data as shown.

The data are in general very similar and fall into a relatively narrow band with V_s values typically increasing from 50 m/s at the surface to about 125 m/s at 20 m depth. A similar pattern to the Norwegian data can be seen with the values gradually increasing with depth. However the measured values are significantly less than those from Norway. Also the slope of the Swedish profiles with depth are not as steep as those of the Norwegian sites. These lower values are compatible with the higher water content, greater organic content and lower unit weight and lower silt content of the Swedish clays compared to the Norwegian sediments as will be explored later.

The range of V_s values encountered suggest that the materials are very soft becoming soft to firm (Table 1), again consistent with the results of other testing.

5.3 *Clays from Finland*

Only limited data is available for V_s profiles from sites in Finland. SCPTU data has been published for the soft clay sites at Perniö, Lempäälä and Masku from southern Finland These are shown on Figure 7. Two profiles from each site are shown and in general they are very similar.

The profiles shown on Figure 7 are compared to the range of values for Swedish clays from

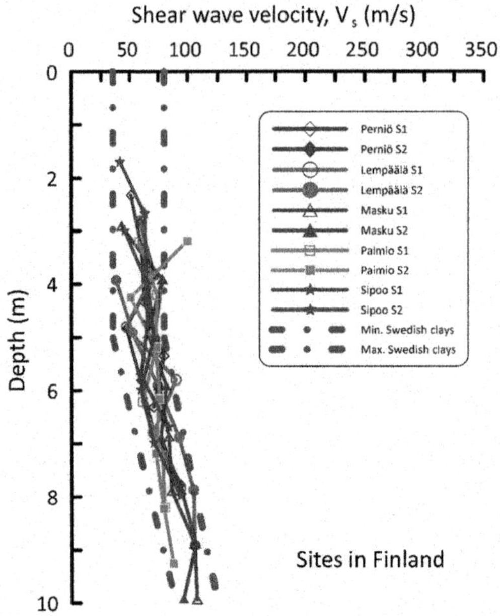

Figure 7. V_s values for Finnish clays (Di Buò et al., 2016, Di Buò et al., 2018, Di Buò et al., 2019).

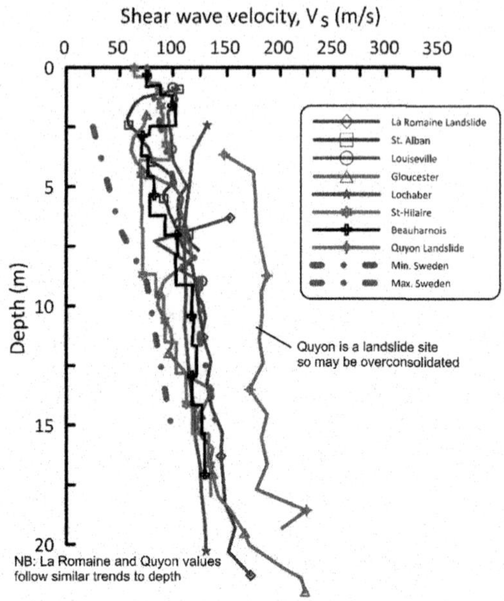

Figure 8. V_s profiles for Eastern Canadian sensitive clays Data from Bouchard et al. (2017), Lefebvre et al. (1994), Leroueil et al. (2003), Mayne et al. (2019), Fabien-Ouellet etal. (2014), Elbeggo et al. (2021) and Agaiby (2018).

Figure 6. Overall the range of values measured for the sites in Finland are very like those from Sweden and fall within the range of values measured for the Swedish clays. One possible reason leading to consistent V_s versus depth profiles is to be found in the similarities between the basic clay properties (water content, bulk density and plasticity) (Long and D'Ignazio, 2020).

5.4 *Eastern Canadian clays*

A compilation of some Eastern Canadian data is shown on Figure 8. Note the geology of the Eastern Canadian area is complex and a more detailed discussion of SCPTU data in this area has recently been published by (Salsabili et al., 2022). The profiles chosen here correspond to well-known geotechnical research sites.

Many of the profiles fall within the bounds of the southern Sweden sites. An exception is the profile from the Quyon Landslide site and perhaps the City of Ottawa data.

5.5 *Peats*

Trafford and Long (2020) developed a small lightweight probe for measuring the shear wave velocity in peat. The motivation was to allow testing on site where it would not be easy to gain access with traditional SCPTU equipment.

The device is very similar to the "pseudo interval" SCPTU described above. V_s measurements, together with parallel water content measurements, for four peat sites in four different countries are show on Figure 9. Water content values are often in the range 1000% to 2000%. Not surprisingly the V_s values are very low, typically 15 m/s. There seems to be a clear trend of lower V_s corresponding to the highest water content.

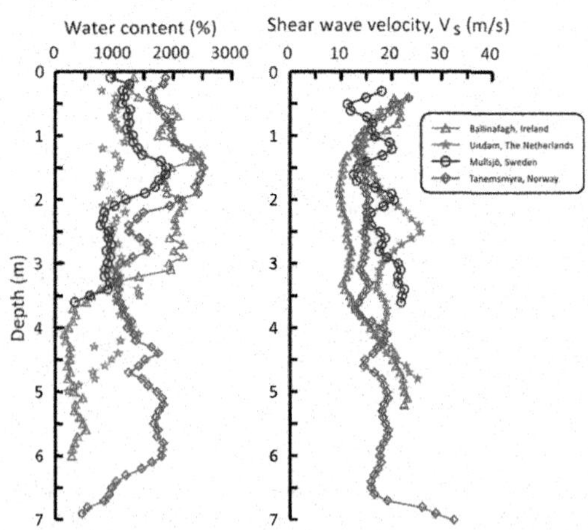

Figure 9. Water content and V_s profiles for four peat sies from four countries (x-axis ranged changed from earlier figures).

5.6 *London clay*

Perhaps most well characterised stiff clay in the world is London Clay (Hight et al., 2003). Some profiles for V_s-vh from SCPTU and other testing in London Clay are shown on Figure 10. Due to the stiffness of the clay and the presence of stiff "clay- stone" bands the tests often need to be terminated at relatively shallow depths. The SCPTU data for the Bradwell B nuclear power station, measured by In Situ Site Investigations Ltd., and reported by LessiCheimariou et al. (2019) are particularly interesting as the tests were able to penetrate to the base of the London clay due to use of a technique of continuing the tests in grouted up boreholes once the original tests were terminated.

On Figure 10 the V_s profiles for soft to medium stiff southern Norwegian clays and stiff to very stiff London clay. Intuitively one would expect the London clay V_s values to be much higher than those of the Norwegian soils. However, as can be seen, the range of values recorded are more or less identical.

The reason for the low values in the London clay is likely to be linked to the fissured nature of these soils (Hight et al., 2003) and due to pure stress ani-sotropy with the horizontal propagating waves pass-ing along layers of higher stiffness (Butcher and Powell, 1996, Piriyakul and Haegeman, 2008).

However it is clear that correlations derived for example between V_s and su at one location or for one specific soil type need to be applied only with great caution to another soil type.

5.7 *Other UK stiff overconsolidated clays*

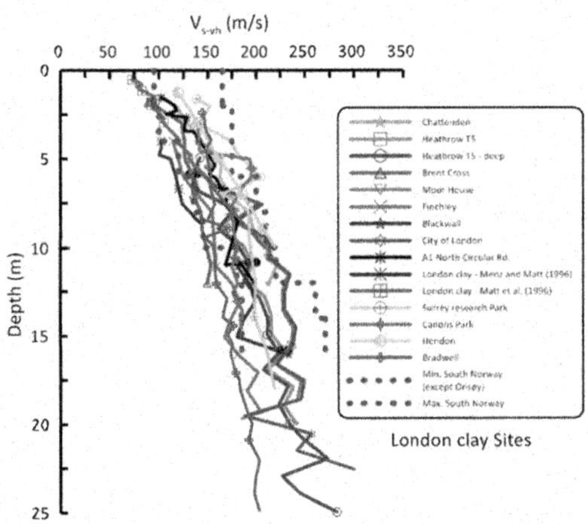

Figure 10. Comparison of V_s-vh values for London Clay and range for South Norway clays (L'Heureux and Long, 2017). London clay data from Powell and Butcher (1991), Butcher and Powell (1995), Gordon et al. (1996), Matthews et al. (1996), Menzies and Matthews (1996), Hight et al. (2003), Hight et al. (2007), Clayton (2011) and Lessi-Cheimariou et al. (2019).

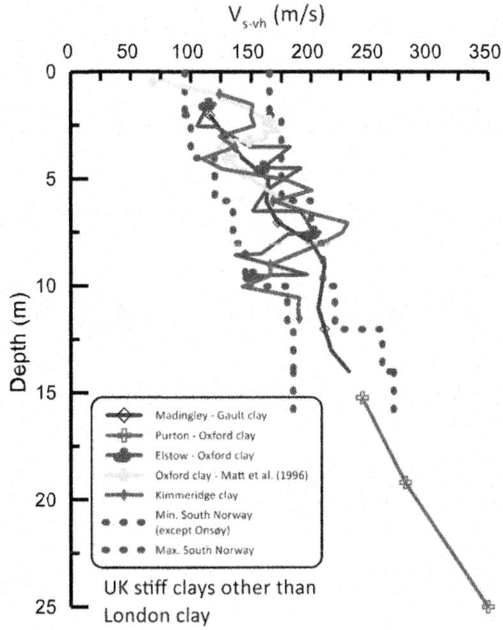

Figure 11. V_{s-vh} values for other UK stiff overconsolidated clays (other than London clay). Data from Butcher and Powell (1995), Hosseini-Kamal et al. (2014), Bates and Philips (2000) and Matthews et al. (1996).

Data for UK stiff clays other than London clay are shown on Figure 11. A picture similar to that of London clay emerges with surprisingly low V_{s-vh} values for these stiff to hard heavily overconsolidated materials. Again the values are similar to those of the South Norway soft to firm clays. Many of these materials have very high V_{s-hh} values and it is likely that the low V_{s-vh} values are due to stress anisotropy and possibly also due to fissuring of the materials.

5.8 *Comparison/link between various clay types*

Given the consistence of the V_s profiles for different areas it is worth considering the basic engineering properties of the clays in an attempt to identify the links between the V_s values.

According to Leroueil and Hight (2003) and Hardin (1978) the empirical equation describing the influence of the controlling factors on G_{max} (and V_s) can then be written as follows:

$$G_{max} = SF(e)(\sigma'_v \sigma'_h)^n p_a^{(1-2n)} \tag{4}$$

where: S is a dimensionless parameter characterising the considered soil; F(e) is a void ratio function; σ_V' and σ_h' are the vertical and horizontal effective stresses respectively; n is a parameter indicating the influence of stress; and p_a is atmospheric pressure.

Table 2. Summary of material properties for selected study sites. w = water content, Ip = plasticity index, Org. = organic content OCR = overconsolidation ratio, S_t = fall cone sensitivity. Main references Gundersen et al. (2019), L'Heureux et al. (2019), Wood (2016), Di Buò et al. (2019) and Trak et al. (1980).

Site	w (%)	Clay (%)	Ip (%)	Org. (%)	OCR	S_t
Onsøy 3	40-80	50-70	25-50	2.5-4	1.1-2.0	5-8
Tiller-Flotten	30-50	45-70	8-20	Very low	1.5-2.0	up to 350
Göteborg CS	60-90	70-90	27-40	2-5	1.5-2.0	12-30
Perniö	80-110	50-90	30-40	1-2	1.5-2.0	40-60
St. Alban	60-90	45-81	5-30	0.9	2.2	14-22
Peat	1000-2000	Very low	n/a	90-100	Very low	n/a

A summary of the key properties of the clays from well characterized test sites in Norway, Sweden, Canada and Finland are given on Table 2, together with some typical properties of peat. Two sites from Norway were chosen namely the NGTSsites at Onsøy and Tiller-Flotten.

The Tiller-Flotten site is characteristic of many Norwegian clays and is significantly different from the other sites under study. These Norwegian sites have relatively lower water content and plasticity and higher density (1.7 – 1.9 Mg/m^3 compared to 1.6 – 1.7 Mg/m^3) than the Canadian, Finnish and Swedish sites.

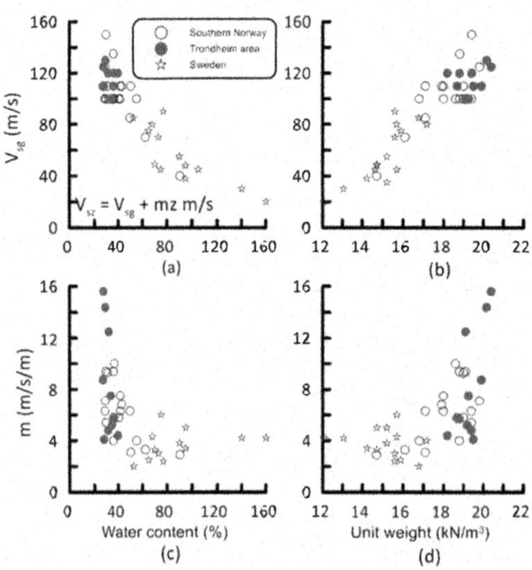

Figure 12. Coefficients V_{sg} and m (Equation 3) as a function of water content and unit weight for sites in Norway and Sweden.

Also Tiller-Flotten has very low organic content compared to the other sites. The Onsøy site parameters are much closer to those of the Swedish, Finnish and Canadian sites. All sites under consideration have similar clay content and stress history.

In order to explore this further, the coefficients V_sg and m in Equation 3 are plotted against average water content (w) and unit weight (γ) for each site (over the interval where V_s data is available) on Figure 12 (L'Heureux and Long, 2017, Long et al., 2017). It can be seen that both parameters decrease with increasing w and increase with increasing (γ) as would be expected. The trend between the parameters is reasonably good and these relationships could therefore be used for first order estimates of V_s or for controlling site measurements.

6 USE OF V_S IN SOIL BEHAVIOUR TYPE CHARTS

Robertson (1990) introduced the well-known soil behaviour charts based on normalised cone resistance (Qt), normalised sleeve friction (Fr) and the pore pressure parameter (Bq)

$$Q_t = \frac{q_t - \sigma_{v0}}{\sigma_{v0}'} \tag{5}$$

$$F_r = \frac{f_s}{q_t - \sigma_{v0}} \tag{6}$$

$$B_q = \frac{u_2 - u_0}{q_t - \sigma_{v0}} \tag{7}$$

where:

q_t = corrected cone end resistance
f_s = sleeve friction
u_2 = pore pressure measured behind cone
u_0 = ambient pore pressure
σ_{v0}/σ_{v0}' = total/effective initial vertical stress

Later Robertson et al. (1995) proposed a complementary SCPTU soil behaviour chart based on Q_t and normalised small strain shear modulus (G_{max}/q_t). This chart was intended mostly for identifying "unusual" soils such as highly compressible sands, cemented and aged soils and clays with either high or low void ratio.

Robertson (2015) suggested that $I_G = G_{max}/q_t$ can be considered to be a small strain rigidly index as it relates small strain stiffness to a measure of soil strength. He proposed a new chart relating Q_{tn} to I_G which could be used to identify soils with microstructure such as cementation/bonding or aging. Note Q_{tn} is an update to Q_t using a variable stress exponent (Robertson, 2009).

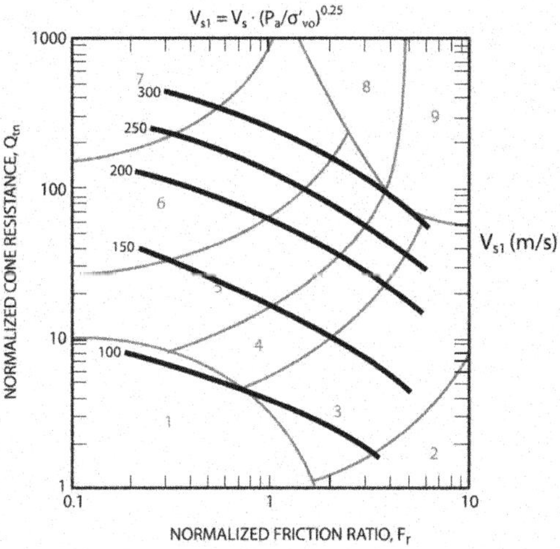

Figure 13. Contours of normalised shear wave velocity superimposed on the normalized Q_{tn} - F_r soil behavior type chart (Robertson, 2009).

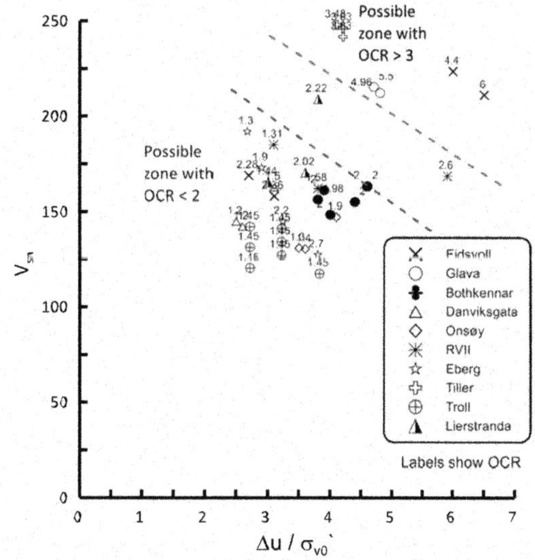

Figure 14. V_{sn} versus $\Delta u/\sigma_{v0}'$ for some Norwegian and UK soft clays. Adapted from Long and Donohue (2010).

Based on 20 years' experience of SCPTU tests in California combined with results from the literature Robertson (2009) also developed a series of contours of normalised shear wave velocity (V_{sn}) and superimposed these on a base of the Q_{tn} – F_r chart as shown on Figure 13. The soils involved were mostly of Holocene age unbonded silica-based soils which tended to plot in the central part of the chart. Some older Pleistocene age soils plot toward the upper right-hand part.

$$V_{sn} = \frac{V_s}{\left(\frac{\sigma_{vo}'}{p_a}\right)^n} \tag{8}$$

Robertson (2009) and others have chosen the exponent n in Equation 8 = 0.25 and have used the expression V_{s1} for V_{sn}. Ku and Mayne (2013) and Moon and Ku (2016a) and others have pointed out that n = 0.25 is based on limited laboratory data on clean quartz and silica sands and suggested n is chosen on site or area specific conditions.

Several authors have suggested that the CPTU u2 measurement is perhaps the most reliable of the normal three sets of data collected (Lunne et al., 2018). Therefore, as V_s is also relatively easy to measure reliably in isotropic material, it seems logical to attempt to develop a characterisation chart based on these two data sets.

On Figure 14 V_{sn} data are plotted against $\Delta u/\sigma_{v0}'(=u_2-u_0/\sigma_{v0}')$ for some Norwegian and UK soft soils.

Here n was chosen to be = 0.5. A tentative division can be made between the lightly overconsolidated material (OCR < 2) and the moderately overconsolidated soils (OCR >3).

Although there appears to be some promise in the use of the charts shown on Figure 13 and 14 and other similar charts, it is felt that some more work, and some more data, is needed to bring these into civil engineering practice.

It is also stressed that these charts are designed to complement the more well-known CPTU soil behaviour charts and not to replace them.

7 CORRELATION BETWEEN V_S AND OTHER SCPTU MEASURMENTS

Various researchers have studied relationships between CPTU parameters and V_s in clayey soils. These studies have explored relationships between in situ V_s and various parameters such as CPTU tip resistance (q_c), corrected tip resistance (q_t), cone net resistance (q_{net}), sleeve friction (f_s), pore pressure parameter (B_q), effective stress (σ_v'), water content (w) and void ratio (e).

Many empirical relationships between V_s and these CPTU parameters have been published. For example L'Heureux and Long (2017) summarise 21 such relationships for clayey soils. Since the publication of the 2017 paper the author is aware of at least ten further newly developed relationships. These include those by Holmsgaard et al. (2016) for silty soils in Denmark, Shahri and Naderi (2016) for clay soils in south-west Sweden, Zhang and Tong (2017), Zou et al. (2017), Tong et al. (2018) and Duan et al. (2019a) all for Jiangsu clays in China, McGann et al. (2015) for silty soils in Christchurch.

New Zealand and Salsabili et al. (2022) for post-glacial sediments in Southern Québec. In addition Ahmed (2017) and Karray and Hussien (2017) derived generalised relationship for a variety of soils.

Robertson (2015) suggests that even though qt and V_s are controlled by the same factors there is no unique correlation between q_t and V_s. However it should be possible to obtain a reasonable relationship if the database used is limited to soils of similar mineralogy, stress history, age and cementation.

An example from the Skatval site in Mid-Norway is shown on Figure 15 (Paniagua et al., 2017). Data from this site was not included in the original Norwegian clay database of L'Heureux and Long (2017). However the correlation developed by these authors, as follows on Equation 9, fits very well with the measured data for this new site.

$$V_s = 8.35(q_{net})^{0.22}(\sigma_{v0}')^{0.357} \quad (9)$$

In contrast the relationship developed for Southern Swedish clays by Shahri and Naderi (2016), given in Equation 10, significantly underestimates the measured V_s values

$$V_s = 3.839(q_t)^{0.5151}(1 + B_q)^{0.174} \quad (10)$$

The message here is that very good correlations can be developed between CPTU parameters and V_s but these should be used locally and they should only be used with great care for soil types other than those for which they were intended.

Figure 16. (a) Sherbrooke block sampler and (b) mini-block sampler. Image from (Emdal et al., 2016).

8 CORRELATION BETWEEN V_S AND SOIL PROPERTIES

8.1 *General*

Much work has been done in the past in correlating V_s with various soil properties derived from laboratory testing. An important consideration is that, for the correlation to be reliable, the quality of the soil samples must be very high. To that end in this paper soil samples extracted by high quality samplers such as the Sherbrooke block (Lefebvre and Poulin, 1979), the mini-block sampler (Emdal et al., 2016) (Figure 16) or the Laval sampler (La Rochelle et al., 1981) only are used.

It is also important that the laboratory tests and the derivation of the resulting parameters is carried out in a consistent manner.

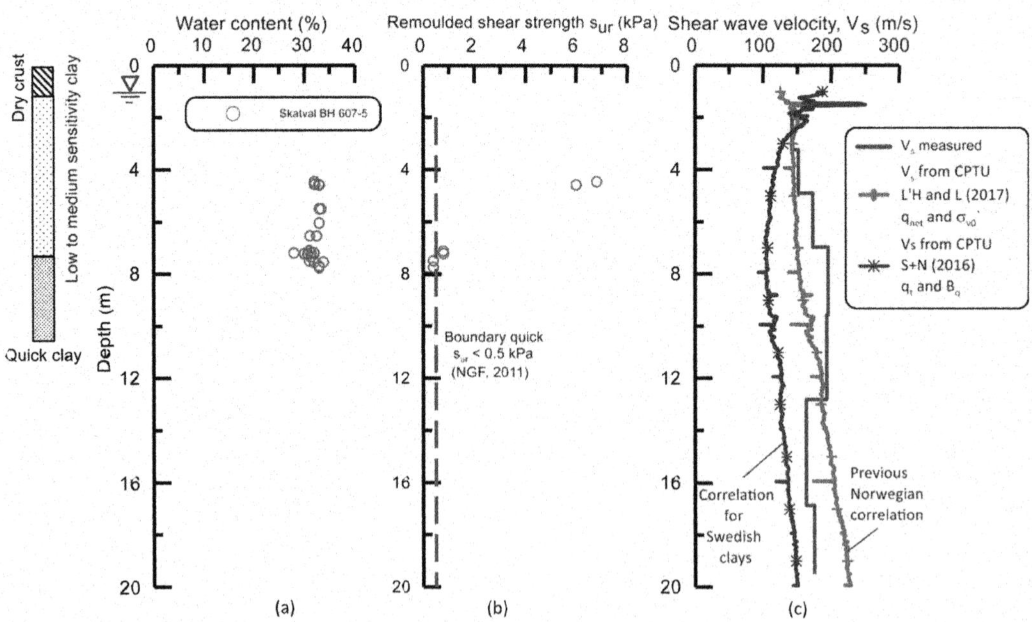

Figure 15. Skatval site Norway – predicted and measured V_s.

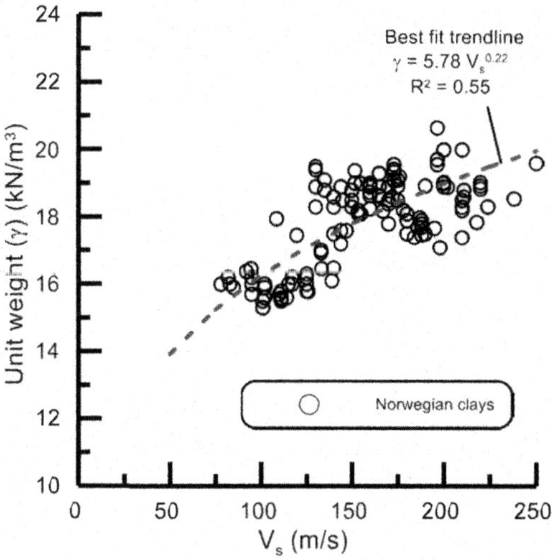

Figure 17. Unit weight versus V_s for Norwegian clays. Data from L'Heureux and Long (2017).

8.2 *Unit weight from V_s*

An important relationship between geophysical parameters and soil index parameters is that between V_s and unit weight (γ). Various authors such as Mayne (2007) and Moon and Ku (2016a) have developed relationships between V_s and γ using a world-wide database of soil tests. They have shown how the resulting relationship will depend amongst other factors on soil type, in situ stress and plasticity.

Different equations are suggested for different soil types and the fit between the measured and predicted data can be improved by normalisation of V_s or by introduction of additional parameters such as the plasticity index (Ip) into the relationships. Although these relationships are very useful it is perhaps more worthwhile to develop local correlations. An example of such a correlation for Norwegian clays is shown in Figure 17.

8.3 *Undrained shear strength from V_s*

Similar to V_s-CPTU relationships much work has been done on deriving empirical relationships between V_s and undrained shear strength s_u. L'Heureux and Long (2017) summarise 16 such relationships. Since the publication of the 2017 paper the author is aware of at least four other publications on the topic by Powell et al. (2016), Moon and Ku (2016b), Duan et al. (2019a) and Duan et al. (2019b).

Undrained shear strength has no unique value but depends on the age of the sample, test type, rate of loading and other factors. Therefore comparisons can only be made on results of tests carried out on exactly the same manner.

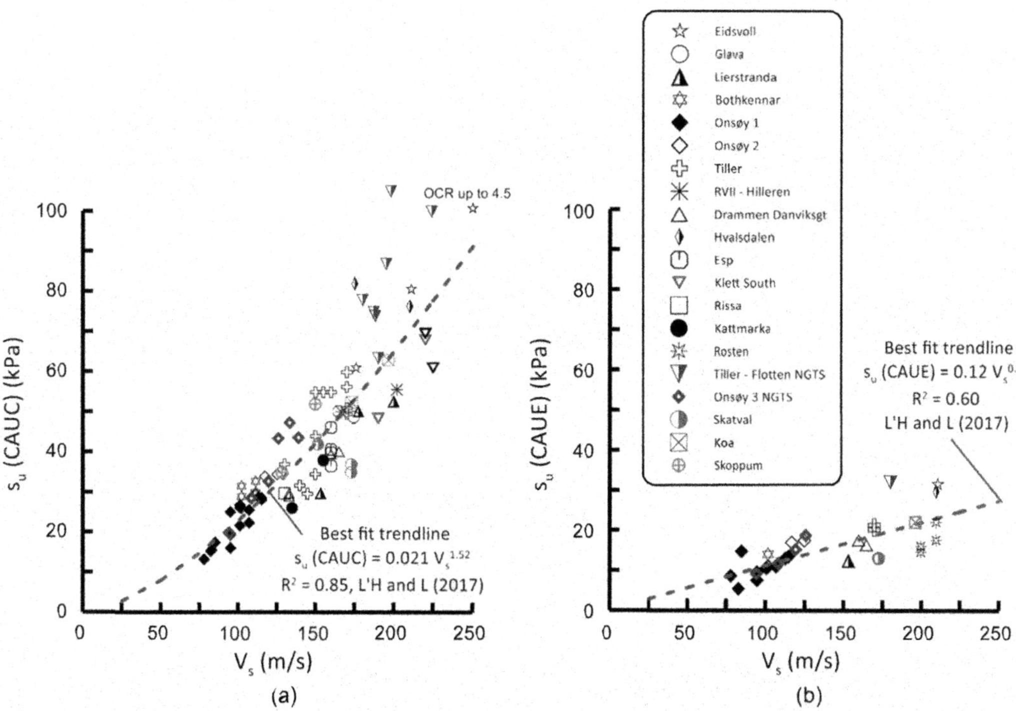

Figure 18. Norwegian clay relationship between su and V_s for (a) CAUC tests and (b) CAUE tests. Original 2017 data from L'Heureux and Long (2017) is in black and white font. Recent data is in colour font. Recent data is from Gundersen et al. (2019), L'Heureux et al. (2019), Paniagua et al. (2017) and Dahl and Haugen (2022).

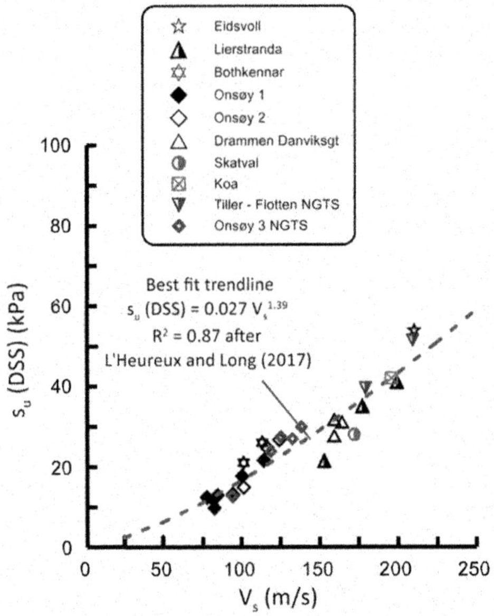

Figure 19. Norwegian clay relationship between s_u and V_s for DSS tests. Original 2017 data from L'Heureux and Long (2017) is in black and white font. Recent data is in colour font. Recent data is from Gundersen et al. (2019), Long et al. (2019) and Paniagua et al. (2017).

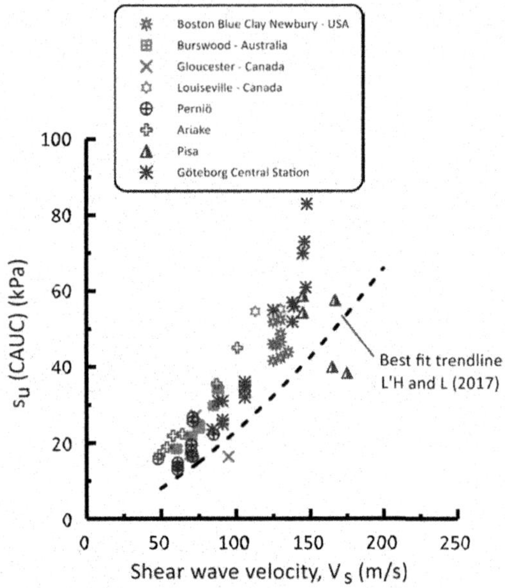

Figure 20. Relationship between su and V_s for CAUC tests on world-wide clays for testing on high quality samples. Data is from Landon (2007), Tanaka et al. (1998), Tanaka et al. (2001), Di Buò et al. (2019), Lo Presti et al. (2003) and Wood (2016).

In Norway it is common practice to carry out triaxial testing after the sample has been first consolidated anisotropically to the best estimate of its in situ stress. Shearing can subsequently be by compression (CAUC tests) or by extension (CAUE tests).

The su values obtained from CAUC and CAUE triaxial tests on high quality samples of Norwegian clay are plotted against in situ shear wave velocity in Figure 18a and 18b respectively. A similar set of data for direct simple shear tests (DSS) are shown on Figure 19.

The"best fit" empirical relationship derived in the 2017 study by L'Heureux and Long (2017) is also shown on Figure 18 and 19. The data used in the 2017 study, which was used to derive the empirical relationships, is shown in black and white font. New data available since the 2017 study are shown in colour. It can be seen that in all three cases the new data fit the old relationships very well and it can be concluded that the 2017 relationships still apply.

A similar set of CAUC data but in this case for various world-wide clays is shown on Figure 20. All the samples were of high quality as described above.

In this case the 2017 relationship for Norwegian clays does not work and in fact would underestimate the measured su.

The clear message here is that these V_s-s_u correlations can be very powerful and useful but they should only be used for the test type and materials for which they were originally derived and they should only be applied to other materials/other test types with great caution.

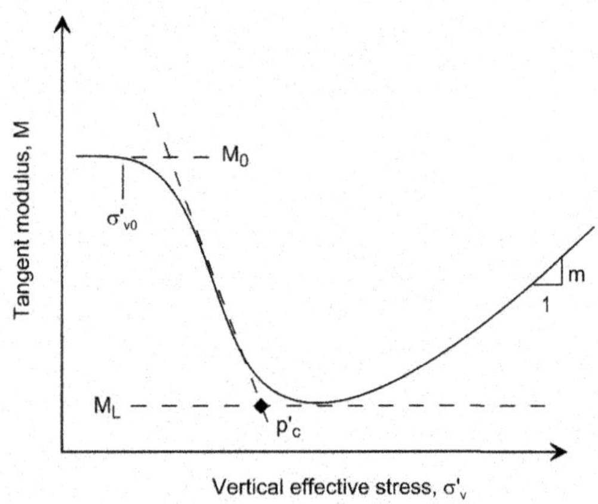

Figure 21. Classical Janbu plot of 1D compression stiffness versus effective stress.

8.4 *Preconsolidation pressure from V_s*

In this and the following sections V_s measurements are compared to the classical 1D compression parameters often used in Scandinavian practice (Janbu, 1963, Janbu, 1969). The classical Janbu plot of 1D compression stiffness against stress is shown

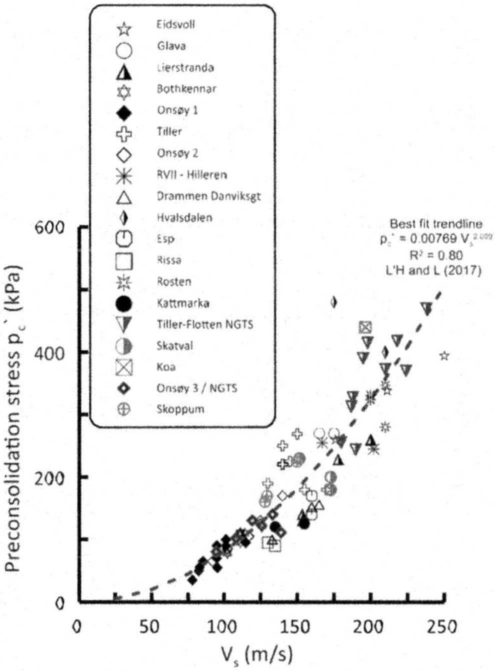

Figure 22. Norwegian clay relationship between p'$_c$ and V$_S$. Original 2017 data from L'Heureux and Long (2017) is in black and white font. Recent data is in colour font. Recent data is from Gundersen et al. (2019), L'Heureux et al. (2019), Paniagua et al. (2017) and Dahl and Haugen (2022).

on Figure 21. Janbu (1963) used the resistance concept to interpret one dimensional consolidation in an oedometer test. He defined the tangent modulus (or the constrained modulus), M, as the ratio of the change in stress ($\Delta\sigma'$) to the change in strain ($\Delta\varepsilon$) for a particular load in increment, i.e.:

$$M = \frac{\Delta\sigma'}{\Delta\varepsilon} \tag{11}$$

As for su above high quality data preconsolidation stress (p'$_c$) for Norwegian clays published by L'Heureux and Long (2017) has been augmented by more recent data (Figure 22). Again, as found for s_u, the recent data fit the 2017 correlations very well and support the idea that local V$_s$ correlations based on high quality data can be very powerful.

The relationship derived for Norwegian clays is tested against data for high quality samples of Swedish clay on Figure 23. It can be seen that there is a reasonable relationship between the two parameters for Swedish clay but the data falls above the line derived for Norwegian clay.

At the 2017 International Workshop on Landslides in Sensitive Clays (IWLSC) in Trondheim,

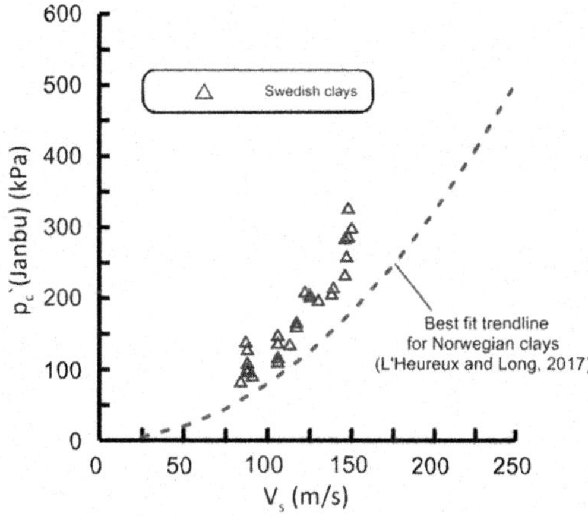

Figure 23. Swedish clay relationship between p'$_c$ and V$_S$. Data from Wood (2016).

Norway, Professor Guy Lefebvre of the University of Sherbrooke suggested to the author that plots similar to that on Figure 22 and Figure 23 could be made more universal by normalising the V$_s$ value by effective stress. This seems a logical suggestion given the clear differences in the basic properties of the two soil types as demonstrated on Table 2.

Data from various countries, where high quality block sample data were available, are shown on Figure 24a. The V$_s$ data were normalised using Equation 8 with a stress exponent n chosen to = 0.5. Var-ious other exponents, including n = 025, were also trailed but no difference in the output was found. Full details of this process are given in a companion paper to this conference by Long and L'Heureux (2022).

The normalisation by effective stress appears to be unsuccessful in unifying the data.

To take the stress history of the materials into account the measured V$_s$ data have been normalised by the preconsolidation stress (p'$_c$) on Figure 24b. As can be seen on Figure 24b this form of normalization As for su above high quality data preconsolidation stress (p'$_c$) for Norwegian clays published by L'Heureux and Long (2017) has been augmented by more recent data (Figure 22). Again, as found for su, the recent data fit the 2017 correlations very well and support the idea that local V$_s$ correlations based on high quality data can be very powerful. was very successful in harmonising the fifteen sets of data. All fifteen profiles are very similar and show an average V$_{snp}$ value of about 96 m/s. Taking this average V$_{snp}$ value the following equation can be obtained to relate V$_s$ and p'$_c$.

$$p_c' = 0.0108V_s^2 \tag{12}$$

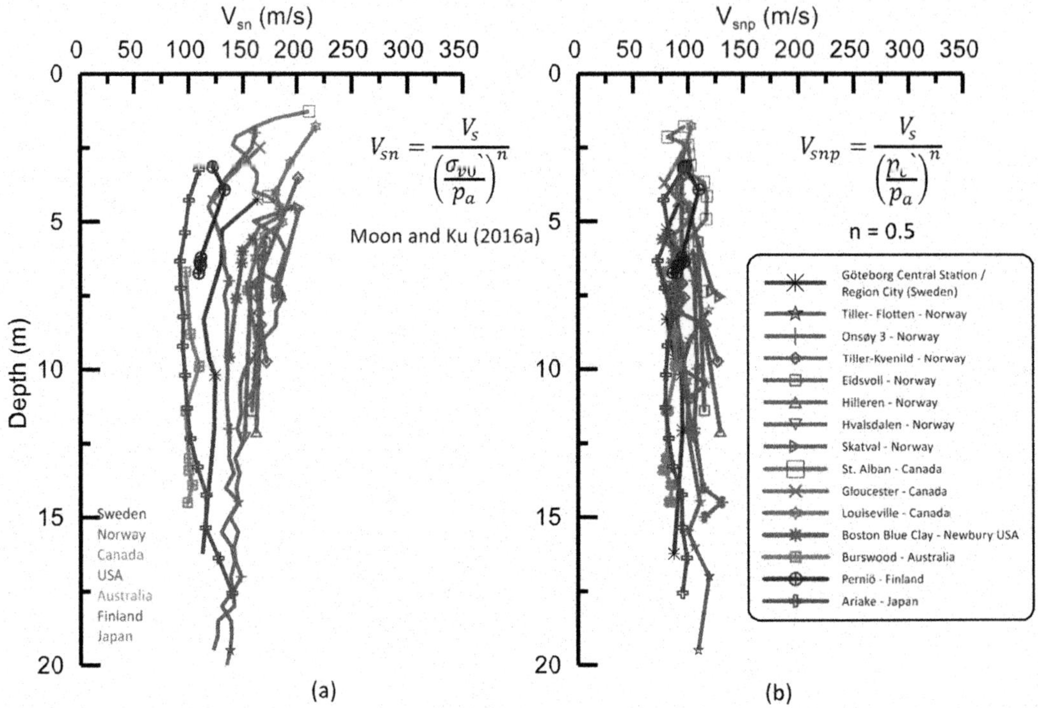

Figure 24. Normalised V_s profiles (a) by vertical effective stress and (b) preconsolidation stress. Data from Wood (2016), Gundersen et al. (2019), L'Heureux et al. (2019), (L'Heureux and Long, 2017), (Paniagua et al., 2017), Landon (2007), Tanaka et al. (1998), Tanaka et al. (2001) and Di Buò et al. (2019).

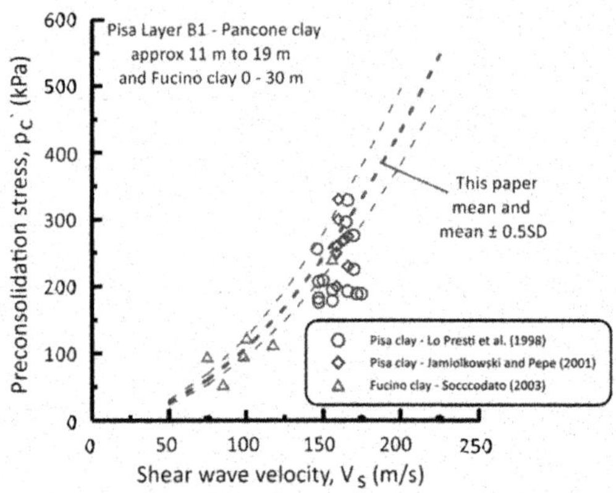

Figure 25. Italian clay relationships between p'_c and V_s trialed against formulae derived in this paper. Data from Lo Presti et al. (1998), Jamiolkowski and Pepe (2001) and Soccodato (2003).

This form of power equation supports and justifies some previous similar empirical equations that have been developed. These include the general relationship developed by Mayne et al. (1998) as shown on Equation 13, that derived for Norwegian marine clays by L'Heureux and Long (2017) (Equation 14) and by Duan et al. (2019a) for Jiangsu clays in China (Equation 15).

$$p_c' = 0.106V_s^{1.47} \tag{13}$$

$$p_c' = 0.00769V_s^{2.009} \tag{14}$$

$$p_c' = 0.1097V_s^{1.3575} \tag{15}$$

Equation 12 is trialed against data for two Italian clays on Figure 25. The mean relationship together with those representing ±0.5 times the standard deviation are shown. Italian data was not included in the original derivation in Figure 24. The Fucino clay data fits the proposed relationship very well. The fit is reasonable for the complex Pancone clay found beneath Pisa Tower.

8.5 *Other one-dimensional consolidation parameters from V_s*

Relationships have also been proposed between V_s and other 1D consolidation parameters, for example M0 and ML (Figure 21). Such data for Perniö clay from Finland and Göteborg Central Station clay from Sweden are compared to relationships for Norwegian clays (L'Heureux and Long, 2017) on Figure 26. In both cases there is a clear relationship between the parameters but the previously published Norwegian clay relationships would underestimate

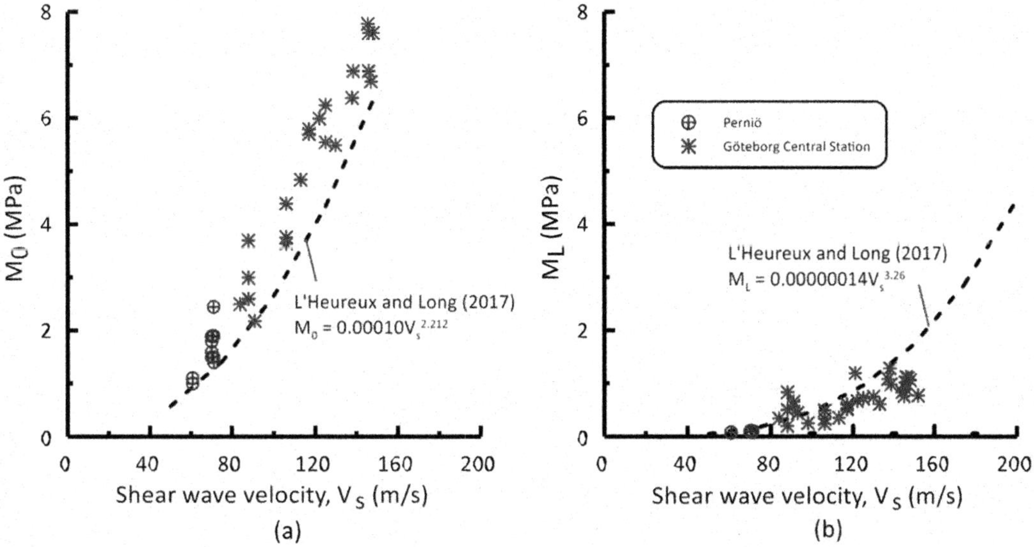

Figure 26. Relationship between (a) M_0 and (b) M_L versus V_s for Swedish (Wood, 2016) and Finnish (Di Buò et al., 2019) clays.

and overestimate respectively the values for Finnish and Swedish clays.

9 USE OF V_S FOR SAMPLE DISTURBANCE ASSESSMENT

Various authors have studied the use of shear wave velocity measurements to assess sample disturbance effects. The overall idea is to compare in situ V_s, as measured by SCPTU for example, to V_s measured on soil samples. Some examples of this work, specifically on clay soils, include that of Landon et al. (2007), Ferreira et al. (2011), Ramsey (2020), Gu et al. (2021) and Long et al. (2021). Other authors have used a combination of V_s and suction measurements to assess sample disturbance, e.g. Donohue and Long (2010) and Horng et al. (2010).

Landon et al. (2007) developed a series of criteria for assessing sample disturbance based on the ratio of V_s measured on the sample (in unconfined conditions) to that of the in situ V_s (e.g. from seismic piezocone, SCPTU). However these criteria were based on tests on low sensitivity, relative high water content and high plasticity clays from Onsøy, Norway, Newbury, USA and Burswood, Australia. The specific guidelines and criteria for assessing sample disturbance from this work is summarised on Table 3. They are compared to the well known method of Lunne et al. (1997a) which involves measuring the change in void ratio relative to the initial void ratio ($\Delta e/e_0$) when consolidating a sample back to its in situ stress in the lab.

It is important to note that Landon's criteria were primarily intended to provide a quick technique for assessing sample dis turbance and to act as a rapid screening tool for assessing sample disturbance prior to advanced laboratory testing. This is of particular concern in offshore site investigations given the cost of these investigations and the need to maximise data quality. They were not intended to replace the more rigorous techniques.

Table 3. Sample quality designation (SQD) from Landon et al. (2007) and Lunne et al., (1997a).

SQD		Lunne et al. with OCR 1-2 $\Delta e/e_0$	Lunne et al. with OCR 2-4 $\Delta e/e_0$	Sample quality (SQ)
SQ1	0.6	< 0.04	< 0.03	Very good/ excellent
SQ2	0.6	0.04-0.07	0.03-0.05	Good to fair
SQ3	0.35–0.6	0.07-0.14	0.05-0.10	Poor
SQ4	< 0.35	> 0.14	> 0.10	Very poor

The sample disturbance assessment criteria of Landon et al. (2007) are tested against recently derived data from three Norwegian sites on Figure 27. The data are plotted in the form of $V_{s,0}/V_{s,insitu}$ against $\Delta e/e_0$obtained from CAUC or CAUE tests or CRS tests on Figures 27a and 27b. The sample shear wave velocity was measured using bender el-ments at zero confining stress ($V_{s,0}$).

For both plots it can be seen that although the data under study here is classified as either SQ1 or

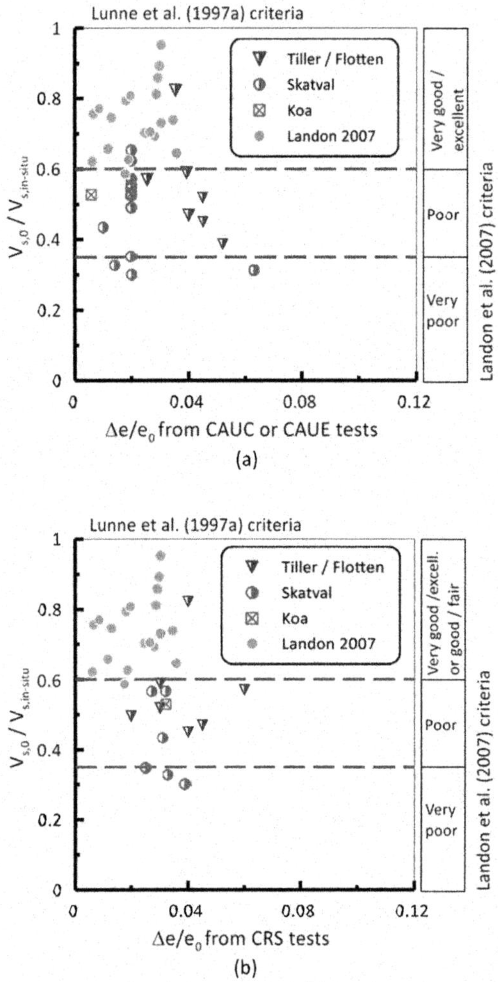

Figure 27. Comparison of shear wave velocity ratio and normalised void ratio for (a) CAUC and CAUE tests and (b) CRS tests. Data from Landon (2007), L'Heureux et al. (2019), Paniagua et al. (2017) and Long et al. (2021).

SQ2 by the criteria of Lunne et al. (1997a) consistent with the observations of the stress – strain and stress path plots, it falls well below that of the data of Landon (2007) and would be classified as "poor" according to the latter criteria.

The clays at Tiller-Flotten, Skatval and Koa differ significantly from those studies by Landon (Table 2). These clays have lower water content, lower plasticity and higher sensitivity compared to Landon's clays.

Valsson et al. (2020) and Dahl and Haugen (2022) carried out similar studies on low sensitivity but high plasticity Onsøy clay and low plasticity sensitive clay from Skoppum. Their results are not directly comparable with those presented here as the bender elementtests were carried out after the samles had been consolidated to the best estimate of the-in situ stress. Nonetheless their conclusion were similar to those found here in that the low plasticity clay from Skoppum showed a significant discrepancy between the field and laboratory tests whereas the results for Onsøy were similar for both techniques.

It is concluded that the criteria Landon et al. (2007) are too of strict for the materials under study here. It is suggested that a lower $V_{s,0}/V_{s,insitu}$ limit of 0.35 (compared to 0.6) might be more appropriate for low plasticity sensitive clays considered here. However insufficient data exits here to give a definitive recommendation.

10 USE OF V_S FOR SETTLEMENT PREDICTION

Several researchers have used V_s measurements for the purpose of estimating settlement of foundations. The technique has mostly been used for foundations in sandy soils, e.g. Lehane et al. (2008) but has also been used in soft clay Mayne (2000) and Omar et al. (2011) and soft clayey silt (Mayne, 2005a). Small strain shear modulus (G_{max}) can be calculated directly from V_s (Equation 1). For the ground beneath foundations the strains will be larger and therefore the G_{max} value needs to be adjusted downwards to be compatible with the actual strains encountered.

Perhaps the most common technique for this purpose is the use of the hyperbolic stiffness degradation formula suggested by Fahey and Carter (1993) as follows

$$\frac{G}{G_{max}} = 1 - f\left(\frac{\tau}{\tau_{max}}\right)^g \tag{16}$$

where τ and τ_{max} are the applied shear stress and shear strength respectively and f and g are fitting parameters. The ratio τ/τ_{max} is analagous to the inverse of factor of safety. Values of f = 1 and g = 0.3 appear to give reasonable estimates for unstructured and uncemented geomaterials and provide a general fit to experimental data (Mayne et al., 2009).

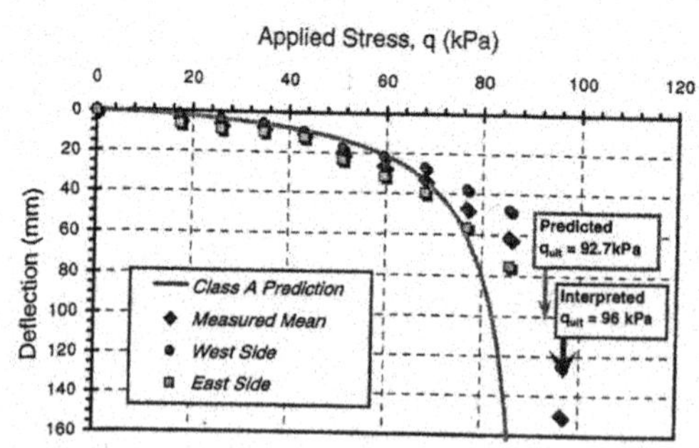

Figure 28. Prediction of settlement of footing on soft silt using V_s (Mayne, 2005a).

Some details for the Class "A" footing response prediction on a trial structure founded on soft silt (known locally as "sleech") in Belfast, Northern Ireland are shown in Figure 28. Mayne (2005a) used the measured V_s values (Lehane, 2003) and the stiffness degradation formula describe in Equation 16 above to obtain a Class "A" estimate of the footing settlment. It can be seen that the predicted and measured data are in good agreement, especially to an applied stress of about 80 kPa, at which point the onset of bearing capacity failure occurs.

11 FUTURE OUTLOOKS AND CHALLENGES TO SOLVE

Despite significant developments in the equipment, data acquisition systems and the cumulative experience gained over the past 35 years since the introduction of the SCPTU some significant challenges remain to be solved. Perhaps of these the most significant are (Personal communication from Nico Parasie, Fugro):

- measurement uncertainty,
- spatial variability/aleatory uncertainty,
- anisotropy.

Issues related to measurement uncertainty have been well researched and are dealt with in Section 3.2 above.

11.1 *Spatial variability*

Spatial variability issues, in sites with widely varying ground conditions, can only be dealt with by carrying out a relatively large number of tests or by introducing another technique to "fill the gaps" between the SCPTU testing.

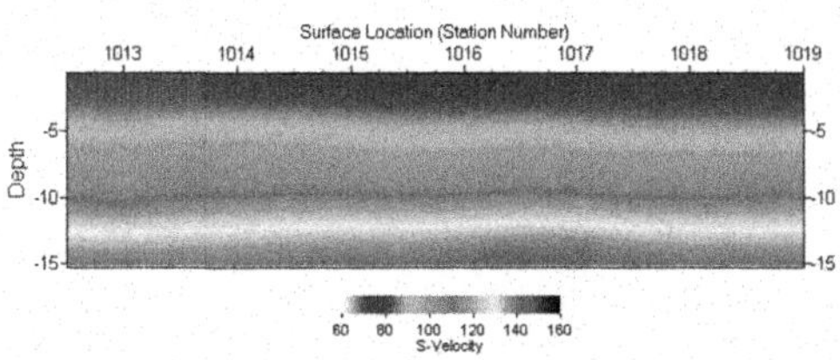

Figure 29. 2D stiffness profile for Onsøy 2 site in Norway. Image courtesy of Dr. Shane Donohue, UCD. Note horizonatal and vertical scales the same.

An example of such an approach would be to combine SCPTU with a surface wave technique such as MASW. A 2D stiffness profile for the highly uniform Onsøy 2 site in Norway is shown on Figure 29. This profile was produced by "continuous" MASW where a streamer of geophones mounted on wooden blocks were pulled along and repeated hammer shots were imparted to produce regular seismic signals. The resulting series of 1D profiles were then combined to produce the image.

For offshore investigations a recent development has been to produce images similar to that on Figure 29 using fibre optic/distributed acoustic sensing (DAS) techniques, see for example (Trafford et al., 2021).

Another option is to combine SCPTU with ERT as mentioned above.

11.2 *Anisotropy*

A significant input in the design of engineering systems such as retaining walls and piles subject to lateral loading is the horizontal ground stiffness. In particular for large monopiles, used for example in offshore wind turbine construction, the behaviour of the system is heavily dependent on the horizontal ground stiffness at small strain. A technique for obtaining such input parameters from SCPTU has been described by McAdam et al. (2020) based on work in the PISA project. Wang et al. (2022) have proposed a simple rotational spring model for the design of laterally loaded rigid piles in sand with V_s as a key input parameter.

Strictly speaking the required input parameter in these cases is $V_{s\text{-}hh}$. Unfortunately SCPTU yields $V_{s\text{-}vh}$. In unform relatively homogenous softer soils $V_{s\text{-}hh}$ and $V_{s\text{-}vh}$ have the same values. However for more overconsolidated soils this is not always the case as has been illustrated above for example for UK stiff clays and some Italian clays.

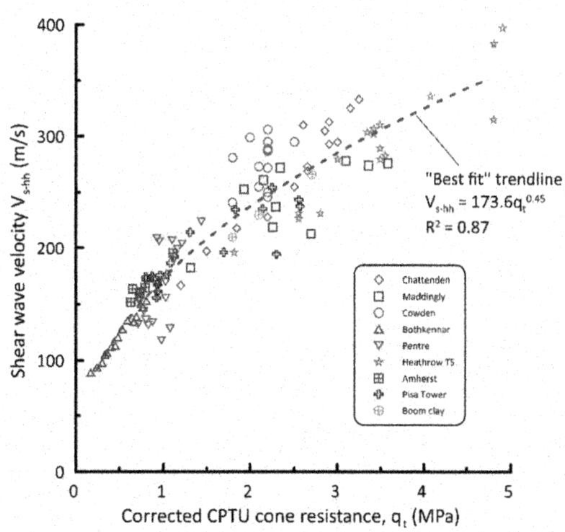

Figure 30. $V_{s\text{-}hh}$ versus q_t. Data from Powell and Butcher (2004) and Powell et al. (2016) augmented with additional data from Lo Presti et al. (2003), Jamiolkowski and Pepe (2001), DeGroot and Lutenegger (2003) and Ku and Mayne (2014).

Powell and Butcher (2004) and Powell et al. (2016) have studied the relationship between corrected once resistance (q_t) and the various in situ derived shear moduli G_{vh}, G_{hv} and G_{hh} using a database of high quality tests. These authors found that qt is most strongly influenced by horizontal stiffness and stress in the ground. They proposed that G_{hh} or $V_{s\text{-}hh}$ could therefore be obtained from correlation with q_t.

The data presented by these authors, augmented by some additional results, are plotted using shear wave velocity rather than shear modulus on Figure 30. It can be seen that the relationship between the two parameters is very good and the resulting "best fit" power function has a good R^2 value of 0.87. As previously suggested by Powell et al. (2016), a suggestion for future work is therefore that $V_{s\text{-}vh}$ is obtained from SCPTU and $V_{s\text{-}hh}$ is obtained from the same test by correlation with q_t.

12 CONCLUSIONS

The main objective of this paper was to summarise the practical and useful application of V_s measurements from SCPTU testing. The intention is to promote the use of SCPTU testing and to highlight the application of V_s data as a compliment to standard CPTU data in future studies and practical projects. It is not intended that V_s measurements replace standard CPTU data. Some conclusions from this work are as follows:

- V_s is a fundamental measure in all solids. Therefore it is an attractive way of characterising a wide range of geomaterials.
- Use of V_s has gained popularity in recent times due to developments in geophysical testing systems and analytical methods.
- Application of the SCPTU testing technique is well covered in guidelines and codes of practice.
- Some descriptions of the equipment used have been given together with some recent developments in the technique including the frequent interval testing and continuous interval testing.
- Uncertainties in the methods have been well researched and understood and several methods have been proposed in the literature suggesting ways of dealing with these uncertainties.
- Despite these uncertainties V_s measurements seem repeatable in isotropic soil conditions using a variety of techniques and methods.
- V_s profiles for sites in southern Norway, mid Norway, Sweden, Finland, Canada and peat have been presented. These data often fall in tight bands.
- There appears to be clear links between these profiles depending on fundamental soil properties such as water content, density, plasticity and also stress history. This has resulted in tentative soil behaviour type charts involving V_s being proposed. This approach seems promising but further work is required on the development and application of these charts.
- An important exception to this is data for stiff overconsolidated clays, such as the UK stiff clays / London clay. For these soils the $V_{s\text{-}vh}$ values are unexpected low. This is due to the natural stress anisotropy and also partially due to the fissured nature of the materials.
- V_s-CPTU correlations can be very powerful. New data for the same soil type in the same area fit the older correlations very well. However these correlations may not work for other areas. Therefore correlations should ideally be locally developed and only applied to other soils and areas with great caution.
- A similar finding was made for soil properties such as s_u and M_0 where once again strong correlations can be developed but these correlations may not apply to other areas
- An exception might be for correlation between V_s and preconsolidation stress p'_c where it seems that high quality data from several areas may fit the same trend.
- SCPTU data can also be used to assess sample disturbance effects by comparing in situ V_s values to measurements made on samples. However it would seem that previously published assessment criteria derived for relatively high plasticity clays may not apply to low plasticity clays.
- V_s has been successfully used to directly predict settlement of footings on clays.
- Some future outlooks include further work on understanding test uncertainties and anisotropy. It is suggested that for future studies $V_{s\text{-}vh}$ could be obtained from SCPTU and $V_{s\text{-}hh}$ by correlation with q_t.

ACKNOWLEDGMENTS

The work presented here largely comes from long term collaboration between the author and colleagues in the Norwegian Geotechnical Institute (NGI), and the Norwegian University for Science and Technology (NTNU). In particular the author is grateful to Jean-Sebastien L'Heureux, Priscilla Paniagua and Tom Lunne.

John Powell of BRE/Geolabs has been very helpful in providing ideas and data in particular with respect stiff overconsolidated clays.

From the academic world Professors Paul Mayne of Georgia Tech and Barry Lehane of the University of Western Australia (UWA) were both very helpful in providing ideas and guidance and in sharing data.

Several colleagues in the geotechnical industry including Nick Ramsey, Joek Peuchen and Nico Parasie of Fugro, Darren Ward and Louisa Dhimitri of In Situ Site Investigation Ltd. and Joe Hobbs of Lankelma Ltd. provided useful data and guidance.

REFERENCES

Agaiby, S. S. (2018) Advancements in the interpretation of seismic piezocone tests in clays and other geomaterials. PhD Thesis, School of Civil and Enironmental Engineering, Georgia Institute of Technology.

Agaiby, S. S., Cargill, E., Ku, T. & Mayne, P. W. (2016) Continuous-interval seismic piezocone testing in Piedmont residuum. IN Lehane, B. M., Acosta-martĺnez, H. E. & Kelly, R. B. (Eds.) *5th International Conference on Geotechnical and Geophysical Site Characterisation (ISC'5)* Gold Coast, Australia, Australian Geomechanics Society.

Ahmed, S. A. (2017) Correlating the shear wave velocity with the cone penetration test. *2nd World Congress on Civil, Structural, and Environmental Engineering (CSEE'17)*. Barcelona, Spain.

ASTM (2019) Standard test methods for downhole seismic testing, ASTM D7400-19. West Conshohocken, PA, USA, ASTM International.

Bang, E.-S., Cho, S.-J. & Kim, D.-S. (2014) Mean refracted ray path method for reliable downhole seismic data interpretations. *Soil Dynamics and Earthquake Engineering- Elsevier Press*, 65.

Bates, C. R. & Philips, D. R. (2000) Multi-component seismic surveying for near surface investigations: examples from central Wyoming and southern England. *Applied Geophysics*, 22, 257–272.

Baziw, E. J. (1993) Digital filtering techniques for interpreting seismic cone data. *Journal of Geotechnical Engineering ASCE*, 119, 98–108.

Baziw, E. J. (2002) Derivation of seismic cone interval velocities utilizing forward modeling and the downhill simplex method. *Canadian Geotechnical Journal*, 39, 1181–1192.

Bouchard, S., Ali, H., Leboeuf, D., Leroueil, S. & Cascante, G. (2017) Dynamic properties of sensitive clay deposits. IN Thakur, V., L'heureux, J.-S. & Locat, A. (Eds.) *2nd International Workshop on Landslides in Sensitive Clays (IWLSC), Chapter 15 in Landslides in Sensitive Clays, Advances in Natural and Technological Hazards Research 46*. Trondheim, Norway, Springer International Publishing AG.

Burghignoli, A., Cavalera, L., Chieppa, V., Jamiolkowski, M., Mancuso, C., Marchetti, S., Pane, V., Paoliani, P., Silvestri, F., Vinale, F. & Vittori, E. (1991) Geotechnical Characterization of Fucino Clay. *10th European Conferece for Soil Mechanics and Foundation Engineering (ECSMFE)*. Fierenze, Italy, Balkema, Rotterdam.

Butcher, A. P., Campanella, R. G., Kaynia, A. M. & Massarsch, K. R. (2005) Seismic cone downhole procedure to measure shear wave velocity - a guideline. ISSMGE TC10: Geophysical Testing in Geotechnical Engineering. ISSMGE Bulletin April 2015 issue.

Butcher, A. P. & Powell, J. J. M. (1995) The effects of geological history on the dynamic stiffness in soils. *XIth European Conference on Soil Mechanics and Foundation Engineering (ECSMFE)*. Copenhagen, Danish Geotechnical Society (DGF).

Butcher, A. P. & Powell, J. J. M. (1996) Practical considerations for field geophysical techniques used to assess ground stiffness *Proceedings of the Conferences on Advances in Site Investigation Practice* London, U.K., Thomas Telford London.

Campanella, R. G., Robertson, P. K. & Gillespie, D. (1986) Seismic cone penetration tests. *Use of In-Situ Tests in Geotechnical Engineering, ASCE GSP 6*, 116–130.

Cavallaro, A. (2020) The use of Cone Penetration Tests (CPT) for the study of the dynamic characteristics of the soils. *2020 IMEKO TC-4 International Conference on Metrology for Archaeology and Cultural Heritage*. Trento, Italy.

Clayton, C. R. I. (2011) Stiffness at small strain: research and practice. *Géotechnique*, 61, 5–37.

Comina, C., Krawczyk, C. M., Polom, U. & Socco, L. V. (2017) Integration of SH seismic reflection and Love-wave dispersion data for shear wave velocity determination over quick clays. *Geophysical Journal International*, 210, 1922–1931.

Dahl, M. & Haugen, E. (2022) Evaluation of selected procedures for estimating shear wave velocity with SCPTu and bender element. *20th International Conference on Soil Mechanics and Geotechnical Engineering (ICSMGE)*. Sydney, Australia, International Society for Soil Mechanics and Geotechnical Engineering (ISSMGE).

Dannewitz, N., Eriksson, H., Mattsson, H., Larsson, R. & Holm, G. (2005) Utvecklingsprojekt 11357 - sesimisk kontrollmetod för kc-pelare - sluttrapport 2005-05-31.

De lange, G., Rawlings, C. G. & Willet, N. (1990) Comparison of shear moduli from offshore seismic cone tests and resonant column and piezoceramic bender element laboratory tests. *Society of Underwater Technology (SUT)*, 16, 13–20.

Degroot, D. J. & Lutenegger, A. J. (2003) Geology and engineering properties of Connecticut valley varved clay. IN Tan, T. S., Phoon, K. K., Hight, D. W. & Leroueil, S. (Eds.) *In Characterization and Engineering Properties of Natural Soils: Proceedings of the International Workshop*. Singapore A.A. Balkema, Rotterdam.

DI buò, B., D'ignazio, M., Selānpāā, J., Haikola, M., Lānsivaara, T. & Di sante, M. (2019) Investigation and geotechnical characterisation of Perniö clay, Finland. *AIMS Geosciences*, 5, 591–616.

DI buò, B., D'ignazio, M., selānpāā, J. & Lānsivaara, T. (2016) Preliminary results from a study aiming to improve ground investigation data. *17th Nordic Geotechnical Meeting, NGM 2016, Reykjavik*. Reykjavik, Iceland.

DI buò, B., Selānpāā, J., Lānsivaara, T. & D'ignazio, M. (2018) Evaluation of existing CPTu-based correlations fo rthe deformation properties of Finnish soft clays. IN Hicks, M. A., Pisanò, F. & Peuchen, J. (Eds.) *CPT18*. Delft, The Netherlands.

Donohue, S. & Long, M. (2010) Assessment of sample quality in soft clay using shear wave velocity and suction measurements. *Géotechnique*, 60, 883–889.

Duan, W., Cai, G., Liu, S. & Puppala, A. J. (2019a) Correlations between shear wave velocity and geotechnical parameters for Jiangsu clays of China. *Pure and Applied Geophysics*, 176, 669–684.

Duan, W., Cai, G., Liu, S., Puppala, A. J. & Chen, R. (2019b) In-situ evaluation of undrained shear strength from seismic piezocone penetration tests for soft marine clay in Jiangsu, China. *Transportation Geotechnics*, 20, 100253.

Elbeggo, D., Ethier, Y., Dubé, J. S. & Karray, M. (2021) Critical Insights in laboratory shear wave

velocity correlations of clays. *Canadian Geotechnical Journal*, Accepted manuscript.
Emdal, A., Gylland, A., Amundsen, H. A., Kåsin, K. & Long, M. (2016) The mini-block sampler. *Canadian Geotechnical Journal*, 53, 1235–1245.
EN (2004) EN ISO 1998-1:2004: Eurocode 8: Design of structures for earthquake resistance - Part 1: General rules, seismic actions and rules for buildings European Committee for Standardisation (CEN), Brussels, Belgium.
Fabien-ouellet, G., Fortier, R. & Giroux, B. (2014) Joint acquisition and processing of seismic reflections and surface waves in a sensitive clay deposit in the Outaouais region (Québec), Canada. IN L'heureux, J.-S., Locat, A., Leroueil, S., Demers, D. & Locat, J. (Eds.) *Advances in Natural and Technological Hazard Research 36, From Geoscience to Risk Management, 1st International Workshop on Landslides in Sensitve Clays (IWLSC)*. Québec, Springer.
Fahey, M. & Carter, J. P. (1993) A finite element study of the pressuremeter in sand using a nonlinear elastic plastic model. *Canadian Geotechnical Journal*, 30, 348–362.
Ferreira, C., Vianna Da Fonseca, A. & Nash, D. F.T. (2011) Shear wave velocities for sample quality assessment on a residual soil. *Soils and Foundations - Japanese Geotechnical Society*, 51, 683–692.
Foti, S., Lancellotta, R., Marchetti, D., Monaco, P. & TOTANI, G. (2006) Interpretation of SDMT tests in a transversely isotropic medium. IN Failmezger, R. A. & Anderson, J. B. (Eds.) *Second International Conference on the Flat Dilatometer*. Washington, D.C., USA.
Garofalo, F., Foti, S., Hollender, F., Bard, P. Y., Cornou, C., Cox, B. R., Dechamp, A., Ohrnberger, M., Verron, V., Sicilia, D., Teague, D. & Vergniault, C. (2016) Interpacific project: Comparison of invasive and non-invasive methods for seismic site characterization. Part II: Intra-comparison between surface wave and borehole methods. *Soil Dynamics and Earthquake Engineering - Elsevier Press*, 82, 241–254.
Gordon, M. A., Clayton, C. R. I., Thomas, T. C. & Matthews, M. C. (1996) The selection and interpretation of seismic geophysical methods for site investigation. *Proceedings of the Conferences on Advances in Site Investigation Practice* London, U.K., Thomas Telford London.
Granskär, J. (2018) Evaluation of SCPT-surveys as method for accessing dynamic modulus, MSc Thesis Luleå University of Technology, Department of Civil, Environmental and Natural Resources Engineering.
Gu, X., Jiang, W., Qian, J., Huang, M. & Shen, P. (2021) Sampling disturbance evaluation based on the shear wave velocity measured in laboratory and field tests. *6th International Conference on Geotechnical and Geophysical Site Characterisation (ISC'6)*. Budapest, Hungarian Geotechnical Society.
Gundersen, A. S., Hansen, R. C., Lunne, T., L'heureux, J.-S. & Strandvik, S. O. (2019) Characterization and engineering properties of the NGTS Onsøy soft clay site. *AIMS Geosciences*, 5, 665–703.
Hallal, M. M. & Cox, B. R. (2019) Theoretical evaluation of the interval method commonly used for downhole seismic testing. *ASCE Geo-Congress 2019: 8th International Conference on Case Histories in Geotechnical Engineering*. Philadelphia, Penn., USA, American Society of Civil Engineering, Reston, Va.
Hardin, B. O. (1978) The nature of stress – strain behaviour for soils. *In Proceedings ASCE Speciality Conference on Earthquake Engineering and Soil Dynamics, Pasadena, California.*
Hight, D. W., Gasparre, A., Nishimura, S., Minh, N. A., Jardine, R. J. & Coop, M. R. (2007) Characteristics of the London clay from the Terminal 5 site at Heathrow Airport. *Géotechnique*, 57, 3–18.
Hight, D. W., Mcmillan, F., Powell, J. J. M., Jardine, R. J. & Allenou, C. P. (2003) Some characteristics of London Clay. IN Tan, T. S., Phoon,K. K., Hight, D. W. & Leroueil, S. (Eds.) *Proceedings International Workshop on Charaterisation and Engineering Properties of Natural Soils*. Singapore, Balkema, Rotterdam.
Holmsgaard, R., Ibsen, L. B. & Nielsen, B. N. (2016) Interpretation of seismic cone penetration testing in silty soil. *Electronic Journal of Geotechnical Engineering, EJGE*, 21, 4759–4479.
Horng, V., Tanaka, H. & Obara, T. (2010) Effects of sampling tube geometry on soft clayey sample quality evaluated by nondestructive methods. *Soils and Foundations Japanese Geotechnical Society*, 50, 93–107.
Hosseini-Kamal, R., Coop, M. R., Jardine, R. J. & Brosse, A. (2014) The post-yield behaviour of four Eocene-to-Jurassic UK stiff clays. *Géotechnique*, 64, 620–634.
ISO (2014) ISO 19901-8: International Organization for Standardization, Petroleum and Natural Gas Industries -Specific Requirements for Offshore Structures – Part 8: Marine Soil Investiga-tions. . Geneva, ISO/CEN.
Jamiolkowski, M. & Pepe, M. C. (2001) Vertical yield stress of Pisa clay from piezocone tests. *Journal Geotechnical and Geoenvironmental Engineering, ASCE*, 127, 893–897.
Janbu, N. (1963) Soil compressibility as determined by oedometer and triaxial tests. *Proceedings of the 3rd European Conference on Soil Mechanics and Foundation Engineering*. Wiesbaden, Germany, Deutsche Gesellschaft für Erd-und Graundbau e.V.
Janbu, N. (1969) The resistance concept applied to deformations of soils. *Proceedings of the 7th International Soil Mechanics and Foundation Engineering Conference*. Mexico City, A.A. Balkema, Rotterdam.
Joh, S. H. & Mok, Y. J. (1998) Development of an inversion analysis technique for downhole testing and continuous seismic CPT. *Journal of the Korean Geotechnical Society*, 14, 95–108.
Kania, A., Madshus, C. & Zackrisson, P. (2000) Ground vibration from high-speed trains: prediction and counter measure. *Journal Geotechnical and Geoenvironmental Engineering, ASCE*, 126, 531–537.
Karray, M. & Hussien, M. N. (2017) Shear wave velocity as a function of cone penetration resistance and grain size for Holocene-age cemented soils: a new perspective. *Acta Geotechnica*, 12, 1129–1158.
Kim, D. S., Bang, E. S. & Kim, W. C. (2004) Evaluation of various downhole data reduction methods to obtain reliable V_s profile. *ASTM Geotechnical Testing Journal*, 27, 585–597.
Ku, T. & Mayne, P. W. (2012) Frequent-interval SDMT and continuous SCPTu for detailed shear wave velocity profiling in soils. *Geotechnical Engineering Journal of the SEAGS & AGSSEA*, 43, 34–40.

Ku, T. & Mayne, P. W. (2013) Evaluating the in situ lateral stress coefficient (K0) of soils via paired shear wave velocity modes. *ASCE Journal of Geotechnical and Geoenvironmental Engineering*, 139, 775–787.

Ku, T. & Mayne, P. W. (2014) Stress history profiling in soils using OCD–G0 anisotropy relationship. *Proceeding Institute of Civil Engineers, Geotechnical Engineering*, 167, 476–490.

Ku, T., Mayne, P. W. & Cargill, E. (2013a) Continuous-interval shear wave velocity profiling by auto-source and seismic piezocone tests. *Canadian Geotechnical Journal*, 50, 382–390.

Ku, T., Weemees, I., Cargill, E., Mayne, P. W. & Woeller, D. (2013b) Post-processing continuous shear wave signals taken during cone penetrometer testing. *ASTM Geotechnical Testing Journal*, 36, 1–11.

L'heureux, J.-S., Lindgård, A. & Emdal, A. (2019) The Tiller-Flotten research site: Geotechnical characterisation of a sensitive clay deposit. *AIMS Geosciences*, 5, 831–867.

L'heureux, J.-S. & Long, M. (2017) Relationship between shear wave velocity and geotechnical parameters for Norwegian clays. *Journal of Geotechnical and Geoenvironmental Engineering ASCE*, 04017013–1 – 04017013–20.

La Rochelle, P., Sarrailh, J., Tavenas, F., Roy, M. & Leroueil, S. (1981) Causes of sampling disturbance and design of a new sampler for sensitive soil. *Canadian Geotechnical Journal*, 18, 52–66.

Landon, M. E. (2007) Development of a non destructive sample quality assessment method for soft clays. *PhD Thesis, Department of Civil and Environmental Engineering, University of Massachusetts, Amherst.* PhD Thesis, Department of Civil and Environmental Engineering, University of Massachusetts, Amherst.

Landon, M. E., De Groot, D. J. & Sheahan, T. C. (2007) Nondestructive sample quality assessment of a soft clay using shear wave velocity. *ASCE Journal of Geotechnical and Geoenvironmental Engineering*, 133, 424–432.

Lefebvre, G., Beloueuf, D., Rahhal, M. E., Lacroix, A., Warde, J. & Stokoe, K. H. (1994) Laboratory and field determinations of small-strain shear modulus for a structured Champlain clay. *Canadian Geotechnical Journal*, 31, 61–70.

Lefebvre, G. & Poulin, C. (1979) A new method of sampling in sensitive clay. *Canadian Geotechnical Journal*, 16, 226–233.

Lehane, B. M. (2003) Vertically loaded shallow foundarions on soft clayey silt. *Institution of Civil Engineers Journal of Geotechnical Engineering*, 156, 17–26.

Lehane, B. M., J.P., D. & Schneider, J. A. (2008) Settlement prediction for footings on sand. IN BURNS, S. E., Mayne, P. W. & Santamarina, J. C. (Eds.) *Deformation Characteristics of Geomaterials*. Atlanta, Georgia, IOS Press, Amsterdam.

Leroueil, S., Hamouche, K., Tavenas, F., Boudali, M., Locat, J., VIrely, D., Roy, M., LA Rochelle, P. & Leblond, P. (2003) Geotechnical characterisation and properties of a sensitive clay from Québec. IN Tan, T. S., Phoon, K. K., Hight, D. W. & Leroueil, S. (Eds.) *Proceedings International Workshop on Charaterisation and Engineering Properties of Natural Soils*. Singapore, Balkema, Rotterdam.

Leroueil, S. & Hight, D. W. (2003) Behavior and properties of natural soils and soft rocks. IN Tan, T. S., Phoon, K. K., Hight, D. W. & Leroueil, S. (Eds. *Proceedings International Workshop on Charaterisation and Engineering Properties of Natural Soils*. Singapore, Balkema, Rotterdam.)

Lessi-Cheimariou, A., Tromans, I., Hadlow, N., Floyd, M. & Pateman, J. (2019) A novel technique for deep sesimic cone tests in challenging ground condiitons. *2019 SECED Conference - Earthquake risk and engineering towards a resilient world*. Grenwich, London, Society for Earthquake and Civil Engineering Dynamics.

Lo Presti, D. C. F., Jamiolkowski, M. & Pepe, M. (2003) Geotechnical characterization of the subsoil of Pisa Tower IN Tan, T. S., Phoon, K. K., Hight, D. W. & Leroueil, S. (Eds.) *In Characterization and Engineering Properties of Natural Soils: Proceedings of the International Workshop*. Singapore A.A. Balkema, Rotterdam.

Lo Presti, D. C. F., Pallara, O., Maugeri, M. & Cavallaro, A. (1998) Shear modulus and damping of a stiff marine clay from in situ and laboratory tests. IN Robertson, P. K. & Mayne, P. W. (Eds.) *Proceedings 1st. International Conference on Geotechnical Site Characterisation*. Atlanta, Georgia, Balkema, Rotterdam.

Long, M. & D'ignazio, M. (2020) Shear wave velocity as a tool for characterising undrained shear strength of Nordic clays. *18th Nordic Geotechnical Meeting*. Helsinki, Finland (Conferece held virtually January 2021), IOP Conf. Series: Earth and Environmental Science 710 (2021) 012008.

Long, M. & Donohue, S. (2010) Characterisation of Norwegian marine clays with combined shear wave velocity and CPTU data. *Canadian Geotechnical Journal*, 47, 709–718.

Long, M. & L'heureux, J.-S. (2022) Shear wave velocity - SCPTU correlations for sensitive marine clays. *5th International Symposium on Cone Penetration Testing (CPT'22)* Bologna, Italy.

Long, M., L'heureux, J.-S., Fiskvik Bache, B. K., Lund, A. K., Hove, S. E., Sødal, K. G., Amundsen, H. A., Nordal, S. & Montafia, A. (2019) Site characterisation and some examples from large scale testing at the Klett quick clay research site. *AIMS Geosciences*, 5, 344–389.

Long, M., Paniagua, P. & L'heureux, J.-S. (2021) Assessing sample disturbance in low plasticity sensitive clays using shear wave velocity. *6th International Conference on Geotechnical and Geophysical Site Characterisation (ISC'6)*. Budapest, Hungarian Geotechnical Society.

Long, M., Wood, T. & L'heureux, J.-S. (2017) Relationship between shear wave velocity and geotechnical parameters for Norwegian and Swedish sensitive clays. IN Thakur, V., L'heureux, J.-S. & Locat, A. (Eds.) *2nd International Workshop on Landslides in Sensitive Clays (IWLSC), Chapter 6 in Landslides in Sensitive Clays, Advances in Natural and Technological Hazards Research 46*. Trondheim, Norway, Springer International Publishing AG.

Looijien, P., Cooper, A. G. & Doan, D. H. (2017) Seismic cone testing using seafloor drill technology. *79th EAGE Conference and Exibition*. Paris, France, European Association of Geoscientists and Engineers (EAGE).

Lunne, T., Berre, T. & Strandvik, S. (1997a) Sample disturbance in soft low plasticity Norwegian clay. IN Almeida, A. (Ed.) *Symposium on Recent Developments in Soil and Pavement Mechanics*. Rio de Janeiro, Balkema, Rotterdam.

Lunne, T., Kelleher, P. J., Banimahd, M., Degroot, D. J., Nguyen, H. Q. & Senders, M. M. (2013) Offshore site characterisatin of small strain modulus using a seabed drilling system. *Offshore Technology Conference, Houston, Texas*.

Lunne, T., Robertson, P. K. & Powell, J. J. M. (1997b) *Cone Penetration Testing in Geotechnical Practice*, Blackie Academic and Professional, London.

Lunne, T., Strandvik, S., Kåsin, K., L'heureux, J.- S., Haugen, E., Uruci, E., Veldhuijzen, A., Carlson, M. & Kassner, M. (2018) Effect of cone penetrometer type on CPTU results at a soft clay test site in Norway. IN Hicks, M. A., Pisanó, F. & Peuchen, J. (Eds.) *CPT18*. Delft, The Netherlands.

Marchetti, D. (2015) Applications and Recent Developments of the Flat Dilatometer (DMT) and Seismic Dilatometer (SDMT). *XVIth European Conference on Soil Mechanics and Geotechnical Engineering (ECSMGE)* Edinburgh, Scotland, Thomas Telford Ltd.

Marchetti, S., Monaco, P., Totani, G & Marchetti, D. (2008) In situ tests by seismic dilatometer (SDMT). *From Research to Practice in Geotechnical Engineering, A Geotechnical Special Publication Honoring John H. Schmertmann*. ASCE GeoInstitute.

Matthews, M. C., Hope, V. S. & Clayton, C. R. I. (1996) The use of surface waves in the determination of ground stiffness profiles. *Institution of Civil Engineers Journal of Geotechnical Engineering*, 119, 84–95.

Mayne, P. W. (2000) Enhanced geotechnical site characterization by seismic piezocone penetration tests. *Invited lecture, Fourth International Geotechnical Conferece*, . Caoiro University, Egypt.

Mayne, P. W. (2005a) Class "A" footing response prediction from sesimic cone tests. IN Di Benedetto, H., Doanh, T., Geoffroy, H. & Sauzéat, C. (Eds.) *3rd International Symposium on Deformation Characteristics of Geomaterials*. Lyon, France, Swets and Zeitlinger, Lisse.

Mayne, P. W. (2005b) Keynote: Integrated ground behavior: in-situ and laboratory tests. IN Di Benedetto, H., Doanh, T., Geoffroy, H. & Sauzéat, C. (Eds.) *International Symposium on Deformation Characteristics of Geomaterials*. Lyon, Taylor and Francis.

Mayne, P. W. (2007) Cone Penetration Testing: A Synthesis of Highway Practice, National Cooperative Highway Research Program Synthesis 368. Washington, D.C., Transport Research Board, National Cooperative Highway Research Program.

Mayne, P. W., Cargill, E. & Miller, B. (2019) Geotechnical characteristics of sensitive Leda clay at Canada test site in Gloucester, Ontario. *AIMS Geosciences*, 5, 390–411.

Mayne, P. W., Coop, M. R., Springman, S. M., Huang, A. B. & Zornberg, J. G. (2009) Geomaterial behavior and testing - Keynote lecture. IN Hamza, M. (Ed.) *17th International Conference on Soil Mechanics and Geotechnical Engineering (ICSMGE)*. Alexandria, Egypt, Millpress/IOS Press Rotterdam.

Mayne, P. W. & Mcgillivray, A. V. (2008) Improved shear wave measurements using autoseis sources. *Deformational Characteristics of Geomaterials*. Amsterdam, The Netherlands, Millpress/IOS Press.

Mayne, P. W., Robertson, P. K. & Lunne, T. (1998) Clay stress history evaluated from seismic piezocone tests. IN Robertson, P. K. & Mayne, P. W. (Eds.) *Proceedings 1st. International Conference on Geotechnical Site Characterisation*. Atlanta, Georgia, Balkema, Rotterdam.

Mayne, P. W. & Woeller, D. J. (2015) Advances in seismic piezocone testing. *XVIth European Conference on Soil Mechanics and Geotechnical Engineering (ECSMGE)* Edinburgh, Scotland, Thomas Telford Ltd.

Mcadam, R. A., Byrne, B. W., Houlsby, G. T., Beuckelaers, W. J., Burd, H. J., Gavin, K. G., Igoe, D. J. P., Jardine, R. J., Martin, C. M., Muir Wood, A., Potts, D. M., Gretlund, J. S., Tarborda, D. M. G. & Zdravkovic, L. (2020) Monotonic laterally loaded pile testing in a dense marine sand at Dunkirk. *Géotechnique*, 70, 986–998.

Mcgann, C. R., Bradley, B. A., Taylor, M. L., Wotherspoon, L. M. & Cubrinovski, M. (2015) Applicability of existing empirical shear wave velocity correlations to seismic cone penetration test data in Christchurch New Zealand. *Soil Dynamics and Earthquake Engineering - Elsevier Press*, 75, 76–86.

Mcgillivray, A. V. & Mayne, P. W. (2004) Seismic piezocone and seismic flat dilatometer tests at Treporti. IN Viana Da Fonseca, A. & MAYNE, P. W. (Eds.) *2nd International Conference on Geotechnical and Geophysical Site Characterisation, ISC'2*. Porto, Millpress.

Mcgillivray, A. V. & Mayne, P. W. (2008) An automated seismic source for continuous-push shear wave velocity profiling with SCPT and frequent-interval SDMT. IN Huang, A. B. & Mayne, P. W. (Eds.) *3rd International Conference on Geotechnical and Geophysical Site Characterization - ISC'3*. Taipei, Taylor and Francis, London.

Menzies, B. & Matthews, M. C. (1996) The continuous surface wave system: a modern technique for site investigation. *Special Lecture at the Indian Geotechnical Conference*. Madras, Geotechnical Data Systems (GDS), Egham, Surrey, UK.

Moon, S.-W. & Ku, T. (2016a) Development of global correlation models between in situ stress-normalized shear wave velocity and soil unit weight for plastic soils. *Canadian Geotechnical Journal*, 53, 1600–1611.

Moon, S.-W. & Ku, T. (2016b) Empirical estimation of soil unit weight and undrained shear strength from of shear wave velocity measurements. IN Lehane, B. M., Acosta-Martínez, H. E. & Kelly, R. B. (Eds.) *5th International Conference on Geotechnical and Geophysical Site Characterisation (ISC'5)* Gold Coast, Australia, Australian Geomechanics Society.

Nguyen,H. Q., Degroot, D. J., Meidani, M., Vanneste, M. & Lunne, T. (2015) Measurement and interpretation of downhole sesimic probe data for estimating shear wave velocity in deep-water environments. IN MAYER, V. (Ed.) *International Symposium on Frontiers in Offshore Geotechnics (ISFOG III)*. Oslo, Norway, CRC Press.

Nibs (2003) National Institute of Building Sciences/Building Seismic Safety Council - National Earthquake Hazard Reduction Program - Recommended provisions for seismic regulations for new buildings and other structures (Federal Emergency Management Agency

450). *Part 1: Provisions*. Washington, DC, Building Seismic Safety Council.
Omar, M. N., Abbiss, C. P., Taha, M. R. & Nayan, K. A. M. (2011) Prediction of long-term settlement on soft clay using shear wave velocity and damping characteristics. *Engineering Geology*, 123, 259–270.
Paniagua, P., L'heureux, J.-S., Carroll, R., Kåsin, K., Sjursen, M. & Amundsen, H. A. (2017) Evaluation of sample disturbance of three Norwegian clays. *19th International Conference on Soil Mechanics and Geotechnical Engineering (ICSMGE)*. Seoul, South Korea.
Paoletti, L., Mouton, E. & Liposcak, I. (2013) Comparison of underwater MASW, seismic CPT and downhole methods offshore Croatia IN Coutinho, R. Q. & Mayne, P. W. (Eds.) *Proceedings 4th. International Conference on Geotechnical and Geophysical Site Characterisation (ISC'4)* Recife, Brazil, September 2012.
Parasie, N., Franken, T. & Peuchen, J. (2022) Assessment of seismic cone penetration testing for small strain shear modulus. *5th International Symposium on Cone Penetration Testing (CPT'22)* Bologna, Italy.
Peuchen, J., De Ruijter, M. R., Hospers, B. & Assen, R. L. (2002) Shear wave velocity integrated in offshore geotechnical practice. *SUT-OSIG-02-379 Society of Underwater Technology Conference*. London, UK (2002), Society of Underwater Technology.
Peuchen, J. & Gomez Meyer, E. (2021) Geo-intelligence from databases of offshore in situ tests in public domain. *6th International Conference on Geotechnical and Geophysical Site Characterisation (ISC'6)*. Budapest, Hungarian Geotechnical Society.
Peuchen, J., Kaltekis, K., Klein, M., Murali, M., Van Erp, F. C. W. & Hicks, M. A. (2020) Characteristic values for geotechnical design of offshore monopiles in sandy soils – case study. In(ed.).: proceedings (pp. - 1892–1901). . IN Westgate, Z. (Ed. *4th International Symposium on Frontiers in Offshore Geotechnics (ISFOG 2020)*. Deep Foundations Institute.
Pidlisecky, A. & Haines, S. S. (2011) Bayesian approach for determining velocity and uncertainty estimates from seismic cone penetrometer testing or vertical seismic profiling data. *Canadian Geotechnical Journal*, 48, 1061–1069.
Piriyakul, K. & Haegeman, W. (2008) Anisotropic shear wave velocity of overconsolidated Boom clay. IN Huang, A. B. & Mayne, P. W. (Eds.) *3rd International Conference on Geotechnical and Geophysical Site Characterization - ISC'3*. Taipei, Taylor and Francis, London.
Poulos, H. G. (2021) Use of shear wave velocity for foundation design. *Geotechnical and Geological Engineering*, Published online October 2021.
Powell, J. J. M. (2005) Session report: Technical session 1c: In situ testing. *Proc. 16th International Conference on Soil Mechanics and Geotechnical Engineering*. Osaka, Balkema, Rotterdam.
Powell, J. J. M. & Butcher, A. P. (1991) Assessment of ground stiffness from field and laboratory tests *In Proceedings Xth ECSMFE*. Fierenze, Rotterdam, Balkema.
Powell, J. J. M. & Butcher, A. P. (2004) Small strain stiffness assessments from in situ tests. IN Viana Da Fonseca, A. & Mayne, P. W. (Eds.) *2nd International Conference on Geotechnical and Geophysical Site Characterisation, ISC'2*. Porto, Millpress.
Powell, J. J. M., Dhimitri, L., Ward, D. & Butcher, A. P. (2016) Small strain stiffness assessment from in situ tests - revisited. IN Lehane, B. M., Acosta-Martínez, H. E. & Kelly, R. B. (Eds.) *5th International Conference on Geotechnical and Geophysical Site Characterisation (ISC'5)* Gold Coast, Australia, Australian Geomechanics Society.
Ramsey, N. (2020) Optimising Geotechnical Engineering Models (GEMs). IN S., H., S., P. & R., G. (Eds.) *Advances in Offshore Geotechnics. Lecture Notes in Civil Engineering, Vol 92*. Singapore, Springer.
Rice, A. H. (1984) The seismic cone penetrometer. Master of Science, Civil Engineering, University of British Columbia, Vancouver, B.C.
Robertson, P. K. (1990) Soil classification using the cone penetration test. *Canadian Geotechnical Journal*, 27, 151–158.
Robertson, P. K. (2009) Interpretation of cone penetration tests - a unified approach. *Canadian Geotechnical Journal*, 46, 1337–1355.
Robertson, P. K. (2015) Comparing CPT and V_s liquefaction triggering methods. *ASCE Journal of Geotechnical and Geoenvironmental Engineering*, 141, 04015037–1—04015037–10.
Robertson, P. K., Campanella, R. G., Gillespie, D. & Rice, A. (1986) Seismic CPT to measure in situ shear wave velocity. *Journal of Geotechnical Engineering ASCE*, 112, 791–803.
Robertson, P. K., Sasitharan, S., Cunning, J. C. & Sego, D. C. (1995) Shear wave velocity to evaluate flow liquefaction. *ASCE Journal Geotechnical Engineering*, 121, 262–273.
Salsabili, M., Saeidi, A., Rouleau, A. & Nastev, M. (2022) Development of empirical CPTu-V_s correlations for post-glacial sediments in Southern Quebec, Canada, in consideration of soil type and geological setting. *Soil Dynamics and Earthquake Engineering - Elsevier Press*, 154.
Shahri, A. A. & Naderi, S. (2016) Modified correlations to predict the shear wave velocity using piezocone penetration test data and geotechnical parameters: a case study in the southwest of Sweden. *Innov. Infrastruct. Solut., Springer*, 1:13.
Simonini, P., Ricceri, G. & Cola,S. (2007) Geotechnical characterization and properties of Venice lagoon heterogeneous silts. IN Tan, T. S., Phoon, K. K., Hight, D. W. & Leroueil, S. (Eds.) *Proceedings 2nd International Workshop on Characterisation and Engineering Properties of Natural Soils ("Natural Soils 2006")*. Taylor and Francis Group, London.
Soccodato, F. M. (2003) Geotechnical properties of Fucino clayey soil. IN Tan, T. S., Phoon, K. K., Hight, D. W. & Leroueil, S. (Eds.) *In Characterization and Engineering Properties of Natural Soils: Proceedings of the International Workshop*. Singapore A.A. Balkema, Rotterdam.
Stolte, A. C. & Cox, B. R. (2020) Towards consideration of epistemic uncertainty in shear-wave velocity measurements obtained via seismic cone penetration testing (SCPT). *Canadian Geotechnical Journal*, 57, 48–60.
Styler, M. A. & Weemees, I. (2016) Evaluating and reporting uncertainty in downhole shear wave velocities. *3rd International Conference on Performance-based Design in Earthquake Geotechnical Engineering*.

Vancouver, Canada, International Society for Soil Mechanics and Geotechnical Engineering, London, UK.

Tanaka, H., Hamouche, K. K., Tanaka, M., Watabe, Y., Leroueil, S. & Fournier, I. (1998) Use of Japanese sampler in Champlain Sea clay. IN Robertson, P. K. & Mayne, P. W. (Eds.) *Proceedings 1st. International Conference on Geotechnical Site Characterisation*. Atlanta, Georgia, Balkema, Rotterdam.

Tanaka, H., Locat, J., Shibuya, S., Soon, T. T. & Shiwakoti, D. R. (2001) Characterisation of Singapore, Bangkok and Ariake clays. *Canadian Geotechnical Journal*, 38, 378–400.

Teachavorasinskun, S. & Lukkunaprasit, P. (2004) A simple correlation for shear wave velocity of soft Bangkok clays. *Géotechnique*, 54, 323–326.

Teague, D. P. & Cox, B. R. (2016) Site response implications associated with using non-unique V_s profiles from surface wave inversion in comparison with other commonly used methods of accounting for V_s uncertainty. *Soil Dynamics and Earthquake Engineering - Elsevier Press*, 91, 87–103.

Tong, L., Che, H., Zhang, M. & Li, H. (2018) Review of shear-wave velocity prediction equations based on piezocone penetration test data: example from Yangtze River floodplain deposits at Nanjing, Jiangsu Province, China. *Quarterly Journal of Engineering Geology and Hydrogeology*, 51, 229–246.

Trafford, A., Donohue, S., Ellwood, R., Godfrey, A. & Wacquier, L. (2021) The application of distributed acoustic sensing for shallow marine investigations - an intertidal case study. *Near Surface Geoscience 2021 (NSG'21): The 27th European Meeting of Environmental and Engineering Geophysics*. Bordeaux, France (and Online), European Association of Geoscientists & Engineers (EAGE).

Trafford, A. & Long, M. (2020) Relationship between shear wave velocity and undrained shear strength of peat *ASCE Journal of Geotechnical and Geoenvironmental Engineering*, 146, 04020057–1 - 04020057–10.

Trak, B., La Rochelle, P., Tavenas, F., Leroueil, S. & Roy, M. (1980) A new appeoach to the stability analysis of embankments on sensitive clays. *Canadian Geotechnical Journal*, 17, 526–544.

Valsson, S. M., Dahl, M., Haugen, E. & Degago, S. A. (2020) Estimating shear wave velocity with the SCPTU and bender element. *18th Nordic Geotechnical Meeting*. Helsinki, Finland (Conferece held virtually January 2021), IOP Conf. Series: Earth and Environmental Science 710 (2021) 012017.

Wang, H., Lehane, B. M., Bransby, M. F., Askarinejad, A., Wang, L. Z. & Y., H. (2022) A simple rotational spring model for laterally loaded rigid piles in sand. *Marine and Structures*, Inpress.

Wood, T. (2016) On the small strain stiffness of some Scandinavian soft clays and impact on deep excavations, PhD thesis Department of Civil and Environmental Engineering, Chalmers University of Technology, Göteborg, Sweden. *Department of Civil and Environmental Engineering, Chalmers University of Technology, Göteborg, Sweden*. Göteborg, Sweden, Chalmers University of Technology.

Zhang, M. & Tong, L. (2017) New statistical and graphical assessment of CPT-based empirical correlations for the shear wave velocity of soils. *Engineering Geology*, 226, 184–191.

Zou, H., Liu, S., Cai, G., Puppala, A. J. & Bheemasetti, T. V. (2017) Multivariate correlation analysis of seismic piezocone penetration (SCPTU) parameters and design properties of Jiangsu quaternary cohesive soils. *Engineering Geology*, 228, 11–38.

Invited papers

Cone Penetration Testing 2022 – Gottardi & Tonni (eds)
© 2022 the Author(s), ISBN 978-1-032-35624-2

Uncertainties associated with CPT data acquisition

R. Soage Santos
Ørsted, London, UK

ABSTRACT: Cone penetration test (CPT) and piezocone penetration test (CPTU) are the dominant in situ testing techniques for soil investigations. CPT/CPTU data is used directly in design or used to derive geotechnical properties for design of infrastructure. Occasionally, acquired CPT/CPTU data exhibit anomalies which can cause uncertainties about use in geotechnical design. Consequently, it is very important to understand, and when possible, minimize the uncertainty associated to CPT/CPTU measurements so these can be considered in planning, design, and installation of infrastructure. The practical challenges associated to the quantification of uncertainty and the difficulties of its standardization are discussed. A comprehensive review of uncertainty contributors associated to CPT/CPTU measurements is provided. This document discusses and proposes a standardization approach that combines performance-based cone classification system based on calibration data, with a method-based approach to control and minimize the uncertainty contributions rising from execution of CPT/CPTU.

1 INTRODUCTION

The cone penetration test and piezocone penetration test (CPT/CPTU) are the most dominant in situ testing technique for soil investigations. CPT/CPTU provide valuable input for safe and economical design of many structures and therefore, assessing the representativeness of the CPT/CPTU results is essential.

This document broadly reviews the challenges related to the assessment of CPT/CPTU results quality. Most often, the CPT provides reliable data with depth, particularly the independent parameters q_c (cone resistance), f_s (sleeve friction), and u_2 (pore pressure). Occasionally, acquired CPT/CPTU data exhibit anomalies which can cause uncertainties about applicability in geotechnical design of offshore structures. Peuchen *et al.* 2020 provides detailed examples of data anomalies and discusses how to avoid them.

CPT/CPTU are among the most fundamental datasets used to define soil layers and infer parameters for safe geotechnical design of infrastructure and is often used directly in design of foundations (for instance pile capacity calculations). Thus, it is very important to understand the quality and therefore the uncertainties associated to CPT/CPTU results that are to be applied in design. The term uncertainty of measurement intends to express the doubt regarding the validity/representativeness of the results. Quantification of uncertainty is very important, because ignoring uncertainty may result in the choice of non-optimal geotechnical design and decision making which could lead to costly and unsafe infrastructure development. Understanding uncertainty is even more important when using the CPT/CPTU measurements directly in design. Note that additional uncertainty is introduced when inferring geotechnical parameters from CPT/CPTU data. Additional uncertainties associated to data interpretation is out of the scope of this document.

The objective of this document is to propose a way forward for maximizing the quality and representativeness of the CPT/CPTU results by controlling and reducing of the uncertainties.

Before proposing a new approach and methodology, it is necessary to understand and review potential sources of uncertainty (uncertainty contributors) and previous standardization efforts aimed to control and minimize uncertainty. The following sections will review standardization efforts and discusses the challenges associated to standardization.

2 CPT/CPTU STANDARDS

2.1 *Development in requirements*

There have been several documents describing requirements and recommendations for the cone penetration test. Before 1980s only cone penetrometers without pore pressure measurements were included. But, in later years the CPTU, or piezocone, has also been covered. In 1977 the Society of Soil Mechanics and Foundation Engineering (ISSMFE) issued a document that included recommendations regarding cone geometry and provided some requirement to precision: *The precision to be obtained should not be worse than the larger of the following values: 5 % of the measured value or 1 % of the measuring range of the sensor used.* Later documents have left out such references to the measuring range.

The Swedish Geotechnical Society Guideline in 1993 (SGF, 1993) introduced the concept of accuracy classes where the requirements were related to what the results were to be used for. ASTM released, in 1979, the first CPT standard publish by a national standardization institute. ASTM did not embrace the accuracy class approach but just included reference to estimates of precision of cone resistance (q_c) and sleeve friction (f_s). Later updates in ASTM 5778-12 (ASTM, 2012) and ASTM 5778-20 (ASTM, 2020) follows the same approach providing estimates of expected precision for q_c, f_s, and u_2 and providing requirements for the calibration errors expressed as maximum allowable errors as percentage of full-scale output (FSO).

In 2012, the first CPT/CPTU standard was published by an international standardization institution. ISO 22476-1:2012 (ISO 2012) opted for an application class scheme, defining four classes for given soil profiles and the use of CPT results. Each application class specifies minimum accuracy requirements for q_c, f_s and u_2 with an associated degree of uncertainty. In the ISO standard for Marine Soil Investigations 19908-1:2014 (ISO 2014) a simplification was made to accommodate offshore practice resulting in three application classes.

2.2 *Present status*

Two principal CPT/CPTU standards widely used are ISO 2012 and ASTM 2020. While the overall procedural approach is rather aligned between these two standards, there are still some fundamental differences regarding the uncertainty requirements of the CPT/CPTU results.

ISO 2012 and ISO 2014 propose a performance-based approach to classify CPT/CPTU results into application classes defined by limiting values of accuracy. Thus, these standards expresses the uncertainty requirements in terms of accuracy: *"If all possible sources of errors are added, the minimum accuracy of the recorded measurements should be better than the largest of the values given in Table 1. The uncertainty analysis should include internal friction, errors in the data acquisition, eccentric loading, temperature (ambient and transient) effects, pore pressure effects in gaps below and above the friction sleeve, and dimensional errors.* "

While the statement has the right intention it is challenging to implement in practice. Using the term accuracy implies the knowledge of the true value. It is not possible, to prove and document, with 100% confidence, how accurate the CPTU measurements are, since the true values are not known. The standard provides little guidance as to how errors and accuracy can be calculated. In practice, identifying and quantifying '*all possible sources of error*' is near impossible and this presents a fundamental problem with the demonstration of compliance.

The ASTM, 2020, approach avoids the intrinsic difficulties associated with expressing requirements in terms of uncertainty, by expressing requirements in terms of repeatability or precision. ASTM, 2020 states: *"There are little direct data on the precision of this test method, in particular because the natural variability of the ground"* and merely provides informative and generic estimates of the precision of q_c, f_s and u_2 from experience in uniform deposits. ASTM, 2020 clearly attempts to mitigate or diminish the uncertainty associated with the CPT/CPTU measurements by providing procedural requirements regarding the calibration of cone penetrometer in a laboratory environment and sets minimum requirements for the precision of q_c, f_s and u_2 in relation to a series of potential sources of uncertainty, although omits some important contributors. The most obvious omission is the requirement for the minimum uncertainty (or precision) of the reference source used in calibration. ASTM, 2020 states: *"In cases where higher level of precision is required, stricter calibration requirements would be required"* however not further advice is provided regarding the cases where higher precision is required or how to make the calibration requirements stricter. In summary ASTM, 2020, focuses on providing requirements for the calibration of cone penetrometers and for field methodology but does not provide requirements for assessing performance in the field.

None of these standards provide unequivocal guidelines on how to evaluate data quality. They only discuss some of the factors that could influence the data quality. A number of methods are used throughout the industry, and the approaches vary from company to company. This makes it very difficult to compare the quality of CPT/CPTU results from various systems and providers and creates difficulty for geotechnical practitioners, who may often need to take conservative judgements for design. The practical challenge related to this topic is setting up robust and detailed guidelines for the assessment of data quality of a performed test more or less immediately upon completion of the CPTU, so it can be repeated if necessary.

2.3 *Present challenges using standards*

Recent publications discussed in depth the difficulties of using ISO 2012 and ISO 2014 to assess the quality of CPT/CPTU results (Peuchen and Terwindt 2014 and 2015, Lunne et al 2017). The main difficulty arises from the definition of application class and the lack of procedural detail for the classification of CPT/CPTU results in practise. The vague nature of the recommendations opens up for different interpretation and can cause contractual disputes. Some of the practical questions that CPT/CPTU data providers and users are facing are listed here:

- How to determine and document that a certain Application Class can be obtained before the start of field work, for instance in the tender phase?
- Regarding the CPTU parameter accuracy requirements: is it an absolute requirement that all three parameters (q_c, f_s and u_2) shall fulfill the limiting values given in Table 1? Should the main emphasis be given to the measured parameters, which are most crucial for the interpretation methods?
- ISO 2012 Annex D, clause D2.1 refers to zero readings before and after a test: For CPTU's, a first attempt at recommendations for limiting values is to follow the minimum accuracy according to the applications given in Table 1. How reliable are the zero readings before and after for this evaluation? How does cleaning of the cones etc. play a role?
- What is the overall best approach for the evaluation of the resulting Application Class after a test has been completed?
- If a test is evaluated to be in a lower Application Class than required in the project specifications, how shall the results be treated in the definition of representative parameters for design?
- From a client's point of view: Which Application Classes and default values are reasonable to require in the project specifications? What is achievable in which conditions?

3 CHALLENGES ASSOCIATED WITH THE QUANTIFICATION OF UNCERTAINTY

The term uncertainty of measurement intends to express the doubt regarding the validity of the results. It follows, from the discussions in previous sections, that a detailed and robust methodology for the quantification of the uncertainty associated with CPT/CTU profiles is desired and necessary to resolve the challenges associated with the assessment of quality. Unfortunately, the quantification of measurement uncertainty for CPT/CPTU results is not trivial.

The formal definition of the term uncertainty of measurement is, in accordance with the International Vocabulary of Metrology (VIM:2012), as follows:" *non-negative parameter characterizing the dispersion of the quantity values being attributed to a measurand, based on the information used"*. ISO/IEC Guide 98-3:2008 adopt the use of uncertainty of measuring in the operational sense. This is to avoid the previous issues associated to terms like accuracy and error, which implies by definition the true value needs to be known in order to calculate accuracy.

ISO/IEC Guide 98-3:2008 and JCGM 100:2008 provides guidance for the estimation of uncertainty of measurement. Measurement uncertainty comprises, in general, many components or uncertainty contributors. The combined standard uncertainty of a measurement can be expressed as the positive square root of the sum of the uncertainty contributors. The expanded standard uncertainty can then be used to express the probability or level of confidence of the estimations of uncertainty by using a coverage factor.

Some of the uncertainty contributors may be evaluated by "Type A" evaluation of measurement uncertainty. Where the statistical distribution of the quantity values from series of repeated measurements under the same conditions are compared against a reference and can be characterized by standard deviations. Other uncertainty contributors may be evaluated by "Type B", which are evaluations based on scientific judgement considering probability density functions based on experience, theoretical modelling, manufacturers specifications or other information.

It is not possible to apply "Type A" approach to field measurements given the difficulties associated in finding a perfectly uniform site that removes soil variability from the uncertainty equation. In addition, the uncertainties may also depend on the soil type, and this make this approach unpractical.

Peuchen and Terwindt (2014, 2015) proposed a model that applies a "Type B" evaluation of uncertainty of CPT/CPTU profiles. Their work really represents an in-depth evaluation of the factors that can influence the various measurements during a CPTU. The analyses of uncertainties are based on input parameters found in literature, own investigations, some general knowledge on transducers, etc. The uncertainty analysis proposed becomes rather subjective and hazy when weighting coefficients are applied to the uncertainty calculation model for representing poor operational procedures and adverse soil conditions. Peuchen and Terwindt concluded that the approach proposed can be considerable conservative and that further studies were required.

In current practice the implementation of "Type B" evaluations for estimating the uncertainty of CPTU measurements is cumbersome and inevitably involves making conservative assumptions to fill the gap of knowledge regarding the uncertainty distributions of certain components. An added challenge is to reach consensus between all sectors, contractors, design engineers, consultants, certifiers, and researchers may have different interest from enterprises involve in manufacturing or those involve in the distribution of services.

Modelling uncertainty using a "Type B" is difficult to implement without adding excessive conservatism. It requires deep knowledge on the behaviour of each particular measurement system and this in turn requires substantial research from geotechnical practitioners, equipment operators and manufacturers. In particular research effort is required to gain sufficient understanding on:

- Development of standard methodologies and requirements to test and measure the behavior of

- a particular cone penetrometer design to minimize or eliminate subjective judgement when selecting uncertainty probability distributions and sensitivity coefficients.
- Development of an unambiguous methodology for reporting uncertainty probability distribution and sensitivity coefficients.

In summary, modelling uncertainty using a "Type B" evaluation can be a viable option for determining uncertainty of CPTU measurement, further maturation and consensus for calculation methodologies are required, before this approach can be incorporated into a standardization document.

4 CHALLENGES ASSOCIATED WITH STANDARDISATION

ISO Working Groups are responsible for the development of standardisation documents. Working groups are generally composed by technical experts and delegates appointed by the national standardisation participating in the standard. The main objective of a working group is delivering the minimum requirements to minimise the level of uncertainty associated to results and to deliver sufficient methodology descriptions to ensure consistent and repeatable results.

Working Groups are generally composed by diverse experts with different areas of competence and with specific experience on the local practise in their countries. Generally, the most important challenge is reaching agreement and consensus within the working group to ensure that the draft standard document is supported by the majority of countries during the balloting stages. Reaching consensus often requires long discussions, preparation and revision of formulations, datasets and evidence to demonstrate the scientific or operational rigor of a position or opinion.

Whenever possible, requirements should be expressed in terms of performance criteria, leaving maximum flexibility to technical development and innovation.

An important consideration for the Working Group is to use clear and concise language, well define terminology, clear unambiguous requirements to reduce the possibility of multiple interpretations. In this regard the provision of informative material in the form of examples, explanations and figures is very valuable to improve understanding and ease implementation.

Working Groups must consider the needs and the challenges from the industry and reach stakeholders to ensure key objectives, needs and challenges are served. During preparation of the standard the Working Group must take into consideration the cost of implementation of the standard in the industry against the value added and ensure, that the new requirements are practical and cost effective. Finally, the extent of the implementation period must be considered to ensure that the industry is able to react to requirement changes. In this regard Working Groups have often to consider the standard document as "work in progress" and consider a staged approach particularly when large changes in requirements are needed.

5 UNCERTAINTY CONTRIBUTORS

Whether a series of equations for quantifying the uncertainty associated with the CPT/CPTU measurements are formulated, or whether a series of effective testing methodology requirements are established to minimise the level of uncertainty, it is fundamental to start by mapping out the potential sources of uncertainty and classifying their relative weight on the overall CPT/CPTU quality.

The tables presented below shows the main uncertainty contributors associated to CPT/CPTU results grouped into three categories. Tables provide a subjective qualitative assessment of the impact on uncertainty, as function of the maximum cone resistance registered in the test. Additional commentary describing the contributors and the potential mitigation measures are included. Table 1A describes contributors associated with the measurement system; Table 1B lists the contributors linked to testing methodology (including procedure, equipment and operator dependency) and Table 1C presents the contributors intimately associated the effects, caused by the interaction between cone penetrometer and soil, during the penetration phase.

From the varied list of uncertainty contributors, it is clear, that the measuring system uncertainties are easier to measure with robust laboratory testing procedures. Soage Santos *et al.* (2022) discuss in detail the quantification of calibration uncertainty.

The uncertainties associated with testing methodology are described in detail in recent publications (Peuchen and Terwindt, 2014, Lunne *et al*, 2017, Kardan *et al.* 2016). Testing methodology related uncertainties can be minimised or fully removed by implementing best practise guidance and prescriptive procedures.

Uncertainties associated with the effects derived from cone penetrometer-soil interaction are more difficult to measure and control. The most important contribution to uncertainty during testing is temperature variations. The importance of temperature changes is widely discussed in literature (Boylan, 2008, McCallum, 2010, Peuchen, 2014). Transient temperature gradients resulting from frictional heat during cone penetration and from thermal flux through push rods causes heating of a cone penetrometer. Temperature-induced drift under in situ stress and axial loading is without a doubt the larger error contributor on CPTU results. Figure 1 shows an example of the magnitude of the drift and the behavior of the measurement system when a cone is

Table 1A. Uncertainties associated with the CPT/CPTU measurement system.

Uncertainty contributors	Potential impact to uncertainty of CPT/CPTU results			Comments	Potential mitigation
	Profiles with q_c<1MPa	Profiles with 1<q_c<3MPa	Profiles with q_c>3MPa		
Reference	High	Medium to high	Medium to high	Refers to the measurement of uncertainty or the reference used for the calibration of a sensor. The reference uncertainty should be several times lower than the instrument that is under calibration.	Introduction of a robust standard procedure for calibration of cone penetrometers. Definition of standard methodology for type A evaluation of uncertainty (in accordance with ISO/IEC Guide 98-3:2008 and JCGM 100:2008 recommendations) specific for the calibration of cone penetrometers under laboratory conditions
Reproducibility	High	Medium to high	Medium to high	Refers to the extent, to which consistent results are obtained when an experiment is repeated under a set of different conditions	
Repeatability	High	Medium to high	Medium to high	Refers to the closeness of successive measurements when carried out under as set repeatability conditions	
Resolution	High	Medium to high	Medium to high	Refers to the ability of a measurement system to detect small changes. The resolution indicates the smallest change in the measurand that a measurement system can detect.	
Zero drift	High	Medium to high	Medium to high	Refers to the difference in the measurement at zero load or pressure before and after applying a load or pressure increment	
Interpolation	High	Medium to high	Medium to high	Refers to the difference of the values given by the calibration interpolation equation and the actual measurements taken during the calibration	
Reversibility	High	Medium to high	Medium to high	Refers to the difference between two separate measurements taken at the same point, the first is taken during a series of increasing measurement values, and the other during a series of decreasing measurement values	
Apparent load transfer	High	Medium to high	Medium to high	Refers to the undesired influence experienced by a sensor caused by another sensor. For instance, when in compression type cone penetrometers sleeve friction strain gauge perceives changes, when the cone resistance sensor is loaded or vice versa.	
Normalised net area ratio	High	Medium to high	Medium to high	Applying a correct net area ratio for cone resistance corrections is of importance specially in clays. Best practise recommendations suggest the determination of net area ration in a in a pressure cell during calibration. It should be noted that the net area ratio varies with applied pressure and there is some variance along the measurement range.	
Dimensions and geometry	High	Medium to high	Medium	CPT/CPTU measurements are highly dependent on geometry and deviation from the standard recommended geometries will compromise the representativeness of the results. In addition, cone resistance and sleeve friction are reported in units of pressure thus the uncertainty associated to the determinations of cone and sleeve friction area will have an important impact on the CPTU reported measurements	

Table 1B. Uncertainties associated with testing methodology.

Uncertainty contributors	Potential impact to uncertainty of CPT/CPTU results			Comments	Potential mitigation
	Profiles with q_c<1MPa	Profiles with 1<q_c<3MPa	Profiles with q_c>3MPa		
Temperature	High	Medium to high	Medium	Changes in temperature will affect the results of the test due to sensors do not readily compensate for transient temperature changes. Prior to testing, the penetrometer should be brought to a stable temperature as close as possible to the expected ground temperature. This is particularly important when the penetrometer temperature is significantly higher or lower than the ground temperature.	Introduction of robust methodology, equipment and reporting requirements to prevent or minimise the error caused by these potential contributors.
Zeroing	High	Medium to high	Medium	Loads caused by trapped soil, or any other loads influencing the cone (handling, embedment in soil) can result in offsets being wrongly introduced while zeroing the measurement system and adding unnecessary uncertainty to CPTU datasets. In addition, unrepresentative zero reference values are likely to penalise the cone performance assessment after testing by comparing reference readings before and after the test	
Saturation	High	Medium to high	Medium	Gas or air trapped in the pressure chamber will cause delays in the sensor response (sluggish response) resulting in unrepresentative data. The pore pressure data will be invalid. This has a direct impact on the representativeness of the cone resistance, as cannot be corrected for pore pressure effects. Careful selection of filter elements, saturation fluid and saturation method are required to minimize the adverse effects on uncertainty.	
Rate of penetration	High	Medium to high	Medium	CPT/CPTU measurements are highly dependent on the penetration rate and deviations from the standardised rates will compromise the representativeness of the results.	
Placement of CPT rig at ground surface	Medium-high	Medium	Low-medium	The weight of the CPT rig can cause changes in the natural stresses in the ground. Placement of the weight need to be considered to avoid influencing the soil to be tested.	
Stops in penetration	Medium	Low to medium	Low	May cause adherence of soil to cone penetrometer body and thus false increase in cone resistance and sleeve friction. The acceleration and deceleration required for starting and stopping a test will inevitably cause local change in cone sensor readings	

Table 1C. Uncertainties associated with effects caused by cone penetrometer-soil interaction.

Uncertainty contributors	Potential impact to uncertainty of CPT/CPTU results uncertainty			Comments	Potential mitigation
	Profiles with q_c<1MPa	Profiles with 1<q_c<3MPa	Profiles with q_c>3MPa		
Temperature	High	Medium to high	Medium	Frictional heat developed during penetration can be significant and the consequence is large drifts will occur during testing. This is perhaps the most important source of uncertainty	Introduction of additional sensors to monitor temperature and bending changes and applied sound correction methods based on laboratory studies. Introduction of robust methodology, equipment and reporting requirements to prevent or minimise the error caused by these potential sources
Loss on saturation	High	Medium to high	Medium	It is sometimes difficult to maintain filter saturation during testing, particularly when pushing through a desiccated crust or through dilatant soils. Loss of saturation leads to sluggish pore pressure response and reliable fluid pressure data may not be consistently obtained. In instances testing approach need to be adjusted, for example by applying predrilling though a partially saturated soil to avoid losing saturation before reaching the saturated zone.	
Dimensions and geometry	Medium to high	Medium	Low to medium	As discussed, early variations dimensions and geometry of a cone can significantly contribute to uncertainty. Even when the dimensions and geometry are well known at the commencement of a test, the high frictional forces, and the abrasive nature of some soil types of my cause small variations on cone penetrometer dimensions and therefore affect the results. This is most likely to occur in sandy profiles and will have a significant effect on soft and loose sediments underlaying the sand layer.	
Bending	Medium	Low to medium	Low	During penetration the cone can be subjected to bending moments. This influence is generally small for q_c and f_s (error <10kPa and <1kPa respectively) for cone penetrometers using double bridge strain gauge.	
Inclination	Medium	Low to medium	Low	Control of inclination is of particular importance contributor to depth uncertainty. In addition, larger inclination will affect the cone resistance and sleeve friction readings. Cone resistance can be affected significantly (10%) in anisotropic soils when inclinations exceed 30 degrees	
Soil trapped or adhered	Low to medium	Low	Low	Trapped soil and dirt on gaps and seals can cause undesirable resistance that will impact the measurements specially when testing soft clays. Soils can also stick to the cone penetrometer surfaces effectively increasing the projected area and providing overestimates of q_c and f_s.	

submerged into a water bath at different temperatures in laboratory conditions.

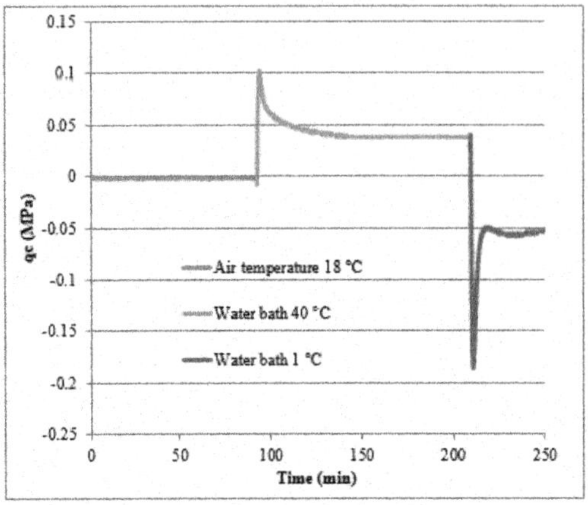

Figure 1. Example of verification data for temperature influence.

Additional uncertainty is introduced when inferring geotechnical parameters from CPT/CPTU data. These additional uncertainties associated with data interpretation is out of the scope of this document. Powell and Dhimitri, 2022 provide detailed discussion and advice for avoiding introduction of undesired uncertainty when deriving geotechnical parameters from CPT/CPTU results.

6 RECOMMENDED WAY FORWARD

The implementation of a rigid performance-based approach for quantifying CPT/CPTU uncertainties using modelling techniques ("Type B" uncertainty analysis) is complex and not fully developed. Modelling uncertainty requires maturation and large amount of research effort. Peuchen and Terwindt 2014 work has shown good potential for the application of this type of analysis, but currently it is not considered a feasible approach in practise as discussed in section 3. In this section a recommended methodology for standardising quality control is proposed.

There is ample opportunity to improve the overall quality of the CPT/CPTU results by implementing standard requirements with the objective of minimising the sources of uncertainties presented in Table 1. Different approaches can be followed to target the three main uncertainty groups:

1. Performance-based approach: Uncertainty analysis (Type A) to quantify uncertainty on measuring systems.
2. Method-based approach. Provide unequivocal procedures and equipment specifications to reduce uncertainty caused by operator and equipment bias.
3. Technological enhancement: Incorporating new technology that facilitates deeper understanding of the uncertainties rising from soil-equipment interaction.

6.1 *Application of uncertainty analysis to cone penetrometer calibration results*

The uncertainty contributions intrinsically associated to the CPT/CPTU measurement system can be better understood and quantified by conducting a series of well design and robust procedures for the calibration of cone penetrometers. Uncertainty analysis "Type A" is applied to the calibration results, allowing systematic quantification of the cone penetrometer ability to produce repeatable results under laboratory conditions.

Requirements for calibration of cone penetrometers and verification of cone penetrometers and uncertainty calculations are incorporated in (draft) standards published by ISO, particularly ISO/DIS 22476-1:2021 and ISO/DIS 19901-8:2022. The use of uncertainty analysis on calibration results implies the departure from the existing application class type approach and the inevitable implementation challenges discussed earlier. Instead, a series of "cone penetrometer classes" can be established using limiting values that defines the minimum uncertainty requirements to satisfy a given cone class. This provides a cone penetrometer classification framework that allows transparent comparison between systems, regardless of manufacturer or supplier. And facilitates the selection of cone penetrometers to match the project and design sensitivity requirements.

Calibration uncertainty analysis and performance indicators based on verification tests defines the best cone class that a particular system can obtain under laboratory conditions. Soage Santos *et al.* 2022 describes this scheme in some detail and provides example of the results.

6.2 *Improving and controlling quality by implementation of testing methodology requirements*

Most of the uncertainties associated to testing methodology (listed in Table 1B) can be greatly reduced by prescribing detailed methodologies and equipment requirements. For instance, zeroing the cone penetrometer acquisition system is critical. If zeroing is performed while the cone penetrometer is undergoing temperature changes or when the cone is embedded or covered in soil, the cone will experience significant drifts affecting the representativeness of the data. Thus, clear requirements regarding temperature stability and presence of undesired loads while zeroing the measurement system are necessary.

Significant uncertainty, up to 4% on cone resistance, can rise from using the default areas (cone design areas) in the conversion of force to pressure measurements rather than the actual measured areas. This can be readily prevented by ensuring that cone geometry is measured in a calibration laboratory and monitored in the field. Powell and Lunne 2005 and Powell et al. 2021 showed the impact of cone dimensions and geometry on CPTU results.

While CPT/CPTU measurement systems are generally reliable and robust, experience shows that every single component of the CPT/CPTU measurement system may deteriorate, suffer damage, or experience drifts during operations (transport, system set up or testing). Consequently, it is imperative to monitor the integrity of the system during field operations. A diagnostic check scheme can be introduced to classify the CPT/CPTU test results.

Reference readings and output stability are non-subjective diagnostic checks that can be measured and quantified. For instance, observed differences of reference readings taken before and after a test will indicate if the measuring system has experienced a drift during the penetration or retraction of the cone from the ground. The absolute difference of values observed can be used for classifying the CPTU results in accordance with prescribed limits to assign a "performance category" to the test.

CPT/CPTU "performance categories" does not provide quantification the uncertainty of a particular test but allows identification of results that may have been affected by equipment malfunction or procedural error. Certain CPT/CPTU applications, like the characterisation of soft clays for foundation design, may require highly representative and repeatable results and consequently better performance is required. The CPT/CPTU category scheme can accommodate this by defining allowable limits for a given soil profile or for a range of soil conditions. In order to avoid potential conflicts caused by interpretation, the soil categories can be linked to prescribed ranges of cone resistance, for example profiles showing q_c<1MPa can be use instead of using terms like "Very soft to firm clays".

It is important to highlight those changes such as ambient temperature, ambient pressure, the presence of dirt in the cone seals and adhesion of clayey soils can cause reference readings differences, that are not directly related to the changes that the measurement system may have experienced in the ground. For a performance category scheme to work as intended, it is fundamental, that reference readings are taken under similar conditions every time so that a meaningful comparison can be made. This challenge can be addressed by the incorporation of very clear testing methodology requirements.

Not all the uncertainty contributors associated to test methodology can be easily quantified and included in a CPT/CPTU performance classification scheme due to the impossibility to quantify them. However, qualitative assessment of test performance is very valuable and relevant information to the users and interpreters of the CPT/CPTU datasets. The introduction of rigid reporting requirements for capturing qualitative performance assessment can be a suitable approach to deal with this challenge. Some examples of qualitative performance observations are listed below:

- Information related to the responsiveness of sensor to soil changes
- Pore pressure sluggish responses
- Sleeve friction or cone resistance insensitivity caused by dirt trapped in cone seals and gaps.
- Pore water pressure cavitation effects experienced during testing
- Qualitative indication of the soil disturbance caused by the seabed frame or drilling activities

Draft ISO/DIS 22476-1:2021 and ISO/DIS 19901-8:2021capture quality monitoring by means of test categories, as discussed above, and provide highly detailed normative procedures ranging from field operations preparations to test termination and reporting activities.

6.3 *Improving and controlling uncertainty by encouraging technological development*

The uncertainty contributors resulting from the effects of the interaction between cone penetrometer and soil (see Table 1C) are especially challenging to quantify, control or correct. The introduction of additional technical requirements targeted to reduce and understand the uncertainty caused by the development of temperature variations and bending moment during testing will be greatly beneficial to improve the quality of CPT/CPTU results, especially in unfavourable soil profiles (i.e., dense sands, stiff to hard clays and interbedded soils profiles). Temperature changes of 10 degrees are not uncommon in unfavorable soils (i.e., Dense sands) and can cause drifts on q_c in the order of 300kPa, which is a very large error when trying to characterize a clay unit underlaying a dense sand unit.

Laboratory verification tests, designed to measure how temperature or bending moment variations affects the performance of q_c, f_s and u_2 sensors can be introduced. Performance indicator limiting values (expressed in terms of kPa/°C and kPa/N) can be included as part of the cone classification scheme described in section 6.2 such that cone penetrometer classes are defined by the combination of the uncertainty associated to the calibration and the test verification results.

Further, the inclusion of temperature sensors in cone penetrometers can additionally provide valuable information on sensor performance during field testing. As the effect of temperature changes on a particular cone design can be measured under laboratory conditions it is relatively straight forward to develop correction algorithms based on the variations of temperature recorded by the temperature sensor. The application of this technology can greatly reduce the uncertainty of CPT/CPTU results in unfavourable conditions where temperature changes dominate the uncertainty of the results.

Draft ISO/DIS 22476-1:2022 and ISO/DIS 19901-8:2022 introduces a cone penetrometer class that requires the inclusion of a temperature sensor. This clearly promotes technological development targeting the reduction of the uncertainties caused unavoidable temperature changes during testing in unfavourable conditions. It should also be noted that the required presentation of results of the ISO verifications covers the influences of temperature and bending on the parameters q_c, f_s, and u.

7 SUMMARY AND CONCLUSIONS

CPT/CPTU provide fundamental input for safe and economical design of many structures. Assessing the representativeness of the CPT/CPTU results is essential. Consequently, it is important to understand, and when possible, minimize the uncertainty associated to CPT/CPTU measurements, so these can be considered in planning, design, and installation of infrastructure.

The practical challenges associated to the quantification of uncertainty and the difficulties of its standardization have been discussed in depth in this document.

Uncertainty contributors associated to CPT/CPTU measurements have been categorized into three main groups: (1) Uncertainties associated with the measuring system, (2) Uncertainties resulting from testing methodology (including procedure, equipment and operator dependency) and (3) Uncertainties intimately associated with the effects caused by the interaction between cone penetrometer and soil. This classification of uncertainty contributors allows proposing a standardization approach that addresses each group with a combination of performance-based and method-based requirements, aiming to control and minimize the uncertainty contributions rising from execution of CPT/CPTU.

Uncertainties associated with the measurement system can be reduced by the introduction of a detailed calibration and verification procedures and prescribed methodology for uncertainty quantification. This also enables the development of a performance-based cone penetrometer class scheme. The approach allows for easy comparison of cone penetrometers, regardless of manufacturer or supplier, and assists the appropriate selection of cone penetrometer for specific applications.

Uncertainties resulting from test methodology can be addressed by implementing a quality monitoring scheme defining test categories based on quantitative performance indicators such as reference readings differences and output stabilities. In addition, detailed normative procedures ranging from field operations preparations to test termination and reporting activities will also improve repeatability and representativeness of the results.

Uncertainties intimately associated the effects caused by the interaction between cone penetrometer and soil are often unavoidable. The introduction of new technology that allows measurement of these uncertainties and enables the development of suitable corrections improving the quality and therefore the overall certainty of the CPT/CPTU measurements even in unfavourable conditions.

ACKNOWLEDGEMENTS

The author would like to thank his colleagues at ISO 19901-8 work group, Tom Lunne, Joek Peuchen, Tim Carrington for sharing their knowledge and for the challenging and fruitful technical discussions that greatly influenced the opinions expressed in this document. Special thanks go to Gülin Yetginer for her amazing work keeping the group working as a team with a focus on reaching consensus and delivering a much-improved draft standard document. Further, the author would like to acknowledge the valuable support and input received from all the members of the ISO 22476-1 workgroup, special thanks go to Håkan Garin, Kees-Jan van der Made, Marius Tremblay, Kristoffer Kåsin, Diederick Bouwmeester and Wouter Vervaele. Last but not least thanks to John Powell and Tom Lunne for providing moral and technical support and for sharing their vast experience and knowledge.

REFERENCES

ASTM International, 2012. *ASTM D5778-12 Standard Test-Method for Electronic Friction Cone and Piezocone Penetration Testing of Soils.* West Conshohocken: ASTM International.

ASTM International, 2020. *ASTM D5778-20 Standard Test-Method for Electronic Friction Cone and Piezocone Penetration Testing of Soils.* West Conshohocken: ASTM International.

Boylan, N., Mathijssen, F., Long, M. and Molenkamp, F. 2008. Cone penetration testing of organic soils. Proceedings, *Baltic Geotechnical Conference, September. Gdansk, Poland.*

International Organization for Standardization, 2008. *ISO/IEC Guide 98-3:2008 Uncertainty of Measurement – Part 3: Guide to the Expression of Uncertainty in Measurement (GU:1995).* Geneva: ISO.

International Organization for Standardization. 2012. *Geotechnical investigation and testing – Field testing – Part 1: Electrical cone and piezocone penetration tests, International Standard ISO 22476-1:2012. (With Technical Corrigendum 1, January 2013).* Geneva: International Organization for Standardization.

International Organization for Standardization, 2014. *ISO/DIS 19901-8:2021 Petroleum and Natural Gas Industries - Specific Requirements for Offshore Structures – Part 8: Marine Soil Investigations.* Geneva: ISO.

International Organization for Standardization, 2021. *ISO/DIS 22476-1:2021 Geotechnical Investigation and*

Testing – Field Testing – Part 1: Electrical Cone and Piezocone Penetration Test. Geneva: ISO.

International Organization for Standardization, 2022. *ISO/DIS 19901-8:2022 Petroleum and Natural Gas Industries - Specific Requirements for Offshore Structures – Part 8: Marine Soil Investigations*. Geneva: ISO.

ISSMFE International Society for Soil Mechanics and Foundation Engineering. 1977. Report of the subcommittee on standardization of penetration testing in Europe, Appendix A, recommended standard for the cone penetration test (CPT). *Proceedings of the ninth international conference on soil mechanics and foundation engineering*, Tokyo.

JCGM. 2008 Joint Committee for Guidelines in Metrology. *Evaluation of measurement data_ Guide to expression of uncertainty in measurements* JCGM 100: 2008.

JCGM. 2012 Joint Committee for Guidelines in Metrology. *International vocabulary of metrology – Basic and general concepts and associated terms (VIM)*. JCGM 200: 2012.

Kardan C., Viking k., Nik L., Larsson S., 2016. *Influence of operator performance on quality of CPTu results*. 17th Nordic Geotechnical Meeting. Challenges in Nordic Geotechnics. Reykjavik.

Lunne, T., Santos, R. Brink Clausen, J. & Powell, J.J.M. 2017. *Guidelines for Use of CPTU Application Classes According to ISO 19901-8: (2014)*. In: Society For Underwater Technology Offshore Site Investigation and Geotechnics Committee, ed. Offshore Site Investigation and Geotechnics: Smarter Solutions for Future Offshore Developments: Proceedings of the 8th International Conference 12-14 September 2017, Royal Geographical Society, London, UK: Volume 2. London: Society for Underwater Technology, pp. 300–307.

McCallum, A. Barwise, A. Soage Santos, R. (2010). *CPT in polar snow-equipment and procedures*, 2rd International Symposium on Cone Penetration Testing CPT10: 2010 – Huntington Beach, California.

Peuchen, J., Santos, R., Yetginer, A.G., Eckart, W.S., Carrington, T.M. & Lunne, T. 2020. *CPT data showing anomalies – assessment and potential postprocessing*. In Z. Westgate (ed.). *4th International Symposium on Frontiers in Offshore* Geotechnics *(ISFOG 2020): proceedings* (pp. 1026–1035). Deep Foundations Institute.

Peuchen, J. & Terwindt, J., 2014. *Introduction to CPT accuracy*. In 3rd International symposium on cone penetration testing CPT14: May 12-14 (2014). Las Vegas, Nevada.

Peuchen J and Terwindt J. (2015). *Measurement uncertainty of offshore Cone Penetration Tests*. Proc. ISFOG 15 Frontiers in Offshore Geotechnics III – Meyer (Ed.) © 2015 Taylor & Francis Group,

Powell JJM and Lunne T. (2005). *A comparison of different sized piezocones in UK clays*. Proc. XVIth ICSMGE, Osaka, September 2005. Vol. 2, pp729–734.

Powell JJM and Dhimitri L. and Ward, D. (2021). *The friction sleeve measurement in CPTU-Does size matters? – A new study*. 6th International Conference on geotechnical and geophysical site characterization. Budapest.

Powell JJM and Dhimitri L. (2022). *Watch out for the use of global correlations and "blackbox" interpretation of CPTU data*. 5th International Symposium on Cone Penetration Testing CPT'22. Bologna.

Soage Santos R., Gomez Meyer E, Peuchen J., Yetginer G., Lunne T., Carrington T., 2022. *Calibration of cone penetrometers according to International Organization for Standardization requirements*. 5th International Symposium on Cone Penetration Testing CPT'22. Bologna.

Swedish Geotechnical Society (1993) *"Recommended Standard for Cone Penetration testing"*. SGF Report 1:93, June 15 1992.

Cone Penetration Testing 2022 – Gottardi & Tonni (eds)
© 2022 the Author(s), ISBN 978-1-032-35624-2

CPT equipment: Recent advances and future perspectives

D.J. White
University of Southampton, UK

ABSTRACT: The basic configuration of the cone penetrometer test (CPT) has remained constant for half a century since the electric piezocone became widespread in the 1970s and the dimensions and test protocols were standardized soon after. The results of a conventional CPT performed today might be indistinguishable from a test conducted at the same location in 1972, aside from improved data logging and presentation. This discussion reflects on advances in CPT and other penetrometer equipment presented at this conference and in the recent literature, and presents discussion points to be considered during the event. Two general observations emerge: (i) advances in instrumentation, sensing and mechanical engineering have created a vast range of variations on the conventional CPT, providing excitement for technologists, researchers and a small subset of users; (ii) the conventional CPT is heavily standardized, commoditized and automated, and is progressively replacing the SPT in onshore practice. The result is a dispersion of practice: the majority of civil engineers base their ground characterization on a durable 1970s technology, updated with automation and improved data quality, while a minority are enjoying exploration of non-standard innovations – with mixed levels of impact. This conference provides a chance for us to (i) reflect where our activities sit in this spectrum, (ii) enjoy learning about recent advances, and (iii) debate how we will characterize the ground in another 50 years.

1 INTRODUCTION

1.1 *Purpose and scope of this discussion paper*

I was invited by the organisers of this conference to prepare a presentation for the start of the discussion session on the subject of CPT and other penetrometer equipment. This written contribution is to provide a record of the presentation in advance, and the presentation is intended to set the scene for a productive discussion.

The paper draws on the manuscripts presented at this event related to penetrometer equipment, as well as recent literature on the subject, and I have also tried to draw in wider context from the civil engineering industry, and other relevant industries, to give context to the drivers of CPT equipment developments.

The discussion that follows sets current developments in the context of the history of the CPT. This short contribution cannot comprehensively cover the full history and range of this technology, which is properly described by other keynote and review papers such as Broms & Flodin (1988), Lunne et al. (1997), Lunne (2012) and Massarch (2014).

Many of the advances described in the papers of CPT'22 are non-mainstream – relating to specialized modifications or applications of the CPT – so this review is a journey mainly through the exotic. However, there are contributions that provide incremental but useful improvements to traditional CPT equipment, which are suited to widespread adoption.

1.2 *Elements of penetrometer equipment*

To review advances in CPT equipment it is useful to define a baseline starting point, which is taken by defining a 'conventional' CPT. Also, it is useful to divide CPT equipment into the various functional elements of the test and examine the advances in each. Table 1 lists the adopted set of functional elements.

Table 1. Functional elements of CPT equipment.

Element	Description
Delivery	How the CPT equipment reaches the site location (at the soil surface)
Deployment	How the CPT advances into the subsoil, to the test location
Sensing	The instrument types and measurements taken during the test
Motion	The applied movement during the test
Scale	The size of the instrument

For the baseline case of the conventional CPT, these functional elements are shown in Figure 1. They are essentially unchanged since the 1970s, when standard dimensions and penetration rates emerged from early standardization work.

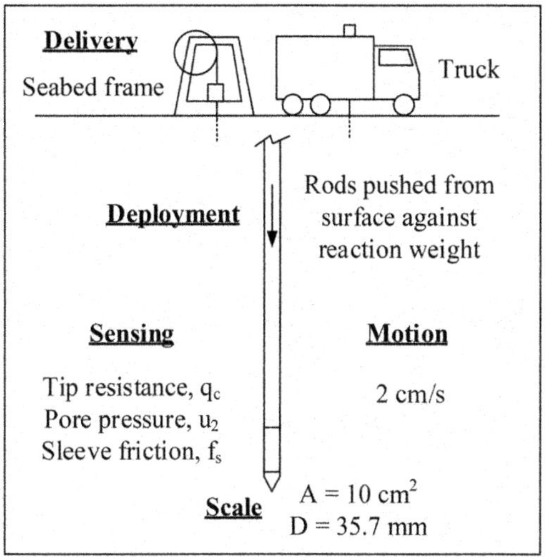

Figure 1. Conventional CPT systems – onshore and offshore.

2 CPT EQUIPMENT ADVANCES

2.1 *Introduction*

The following sub-sections discuss advances related to each functional element. The advances that feature in this review are summarized in Figure 2, which can be compared to the baseline of Figure 1. However, Figure 2 is not an exhaustive record of CPT equipment innovations, and many more can be found in the earlier CPT and ISOPT/ESOPT conferences.

2.2 *Delivery*

For *delivery* to onshore sites, early CPT rigs were mounted on rugged vehicles and today's fleets of commercial CPT rigs includes large and small wheeled or tracked vehicles, driven by people sat in a driving seat.

Recent onshore *delivery* advances have eliminated the wheels and tracks and in some cases the driver, to allow CPT testing in situations where these may not be used. This may, for example, be due to (i) extremely soft ground (e.g. tailings or marsh), (ii) hostile locals (e.g. for military trafficability assessments) or (iii) the lack of oxygen (e.g. on Mars) (Figure 3).

Uncrewed and remote-controlled onshore CPT delivery systems that have been trialled include aerial drones (Baez et al. 2020) and uncrewed wheeled or tracked vehicles (e.g. Scholz et al. 2014, Olmedo & Lipsett 2020). These technologies allow soil consistency assessments in wetlands, agricultural land and tailings ponds (Figure 3a-c). Robotic vehicles with advanced manipulators are widely used in agriculture for sowing, weeding and harvesting, and these carry soil condition sensors. It is a simple extension to add a penetrometer system to these vehicles. Interplanetary landers have also conducted penetrometry.

For *delivery* to offshore underwater sites, the remote-controlled CPT vehicle is not as esoteric as onshore. Trials of ROV-mounted CPT systems were reported in the early 1980s by Geise & Kolk (1983) and the GeoROV CPT system (Edmunds 2014) has found widespread use as a lightweight seabed penetrometer system over the past decade (Figure 3d).

During an ROV-based CPT test, the crew is above the apparatus, on the vessel that the unit is tethered to. A logical extension, aligned with trends in other offshore survey technologies, is for the crew to remain on land, and the vessel to be uncrewed. Autonomous (or uncrewed) surface vessels are now used for geophysical surveys, and an uncrewed geotechnical survey is a realistic prospect.

Compared to sampling operations, penetrometer testing is easier to perform in an uncrewed situation due to the complexity of sample handling and storage. Penetrometer testing will therefore be a central element of future uncrewed geotechnical surveys, both offshore and onshore. This likely increase in reliance on penetrometry is a driver of more advanced test types, as discussed in Section 2.5, to replace the reduction in data if sampling is eliminated or reduced.

2.3 *Deployment*

2.3.1 *Introduction*

For *deployment* of the cone penetrometer into the subsoil, a conventional onshore CPT rig usually has a hydraulic pushing system, with rods of 2.5--3.0 m length, that are connected after each push by the operator. Some offshore systems use coiled rods (e.g. Power & Geise 1994, Meunier et al. 2004), or a robotic drilling system (e.g. Kelleher & Randolph 2005). The vertical reaction force is provided by the vehicle weight, sometimes supplemented by ground anchors, or by the skirts of a suction-type foundation on the base of the unit (e.g. Boggess & Robertson 2010).

Advances in deployment of the penetrometer fall into two primary categories: (i) techniques to speed-up the deployment process and (ii) techniques to reduce the vertical required reaction force and therefore increase the potential depth of penetration.

2.3.2 *Speeding up deployment*

In this conference, Storteboom et al. (2022a) describe a new trademarked 'SingleTwist' rod connection system, which allows a long CPT string to be compactly stored and quickly assembled by robotics (Figure 4a). The rods are held together with a flexible ball joint, for storage on a reel, but require only 1/6th of a turn to become rigidly connected for deployment. The ball joint connection

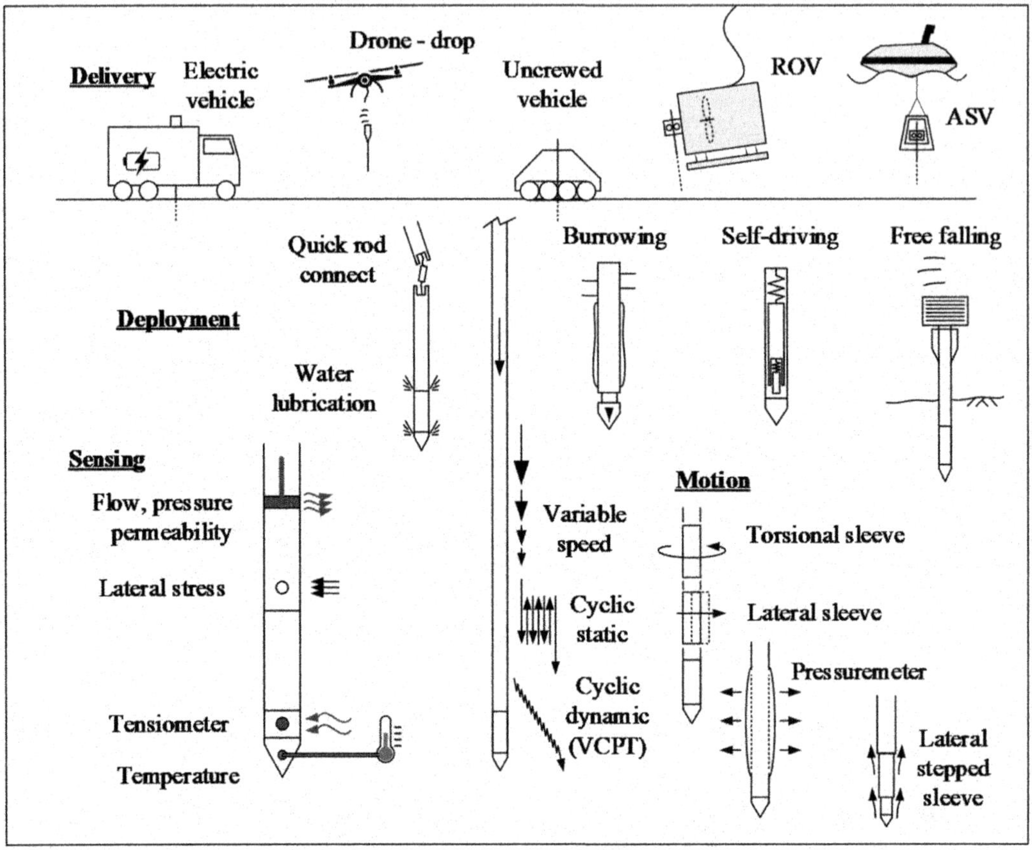

Figure 2. CPT equipment advances: selected from this conference and recent literature – as discussed in Section 2.

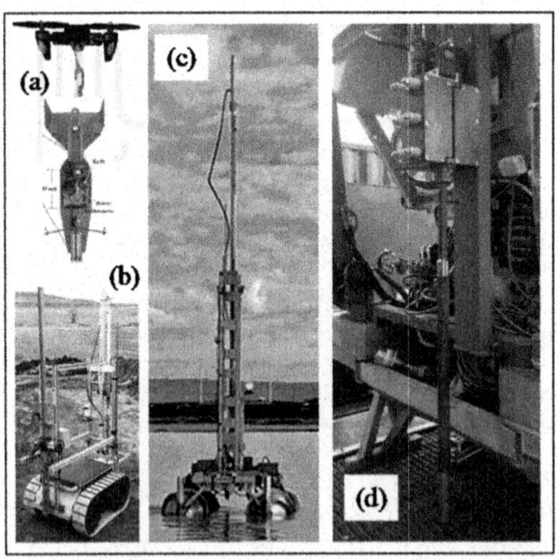

Figure 3. Advances in *delivery* of CPT equipment: uncrewed vehicles, (a) aerial (Baez et al. 2020), (b) land (Olmedo & Lipsett 2016), (c) land/water (HELIX25: www.copperstonetech.com) and (d) underwater (ROVcone www.bluefieldgeo.com).

features a protected data cable running through the joint.

An alternative speed-up technique, which also removes the need for any vertical reaction force, is to use a free-falling penetrometer (FFP). These systems are gaining wide use in research for shallow seafloor characterization (e.g. Stark et al. 2009, Steiner et al. 2014, Stark 2016) and their interpretation methods are evolving in parallel with free falling anchor research (Chow et al. 2017). Commercial FFP systems are also used for deepwater soft soils, sometimes with combined free-falling and static push test stages (e.g. Young et al. 2011).

In this conference, Kang et al. (2022) present sea trials in a relatively lightweight FFP, weighing 38 kg (Figure 4c). Studies such as this demonstrate how a self-contained low cost FFP data system (e.g. Randolph et al. 2018) allows multiple rapid CPTs to be performed.

However, these studies, and others (e.g. Stephan et al. 2015, Peuchen et al. 2017) show the challenge of achieving sufficient penetration in sands to reach the depth required for typical designs – such as assessing trenchability for cable route surveys. FFP penetration

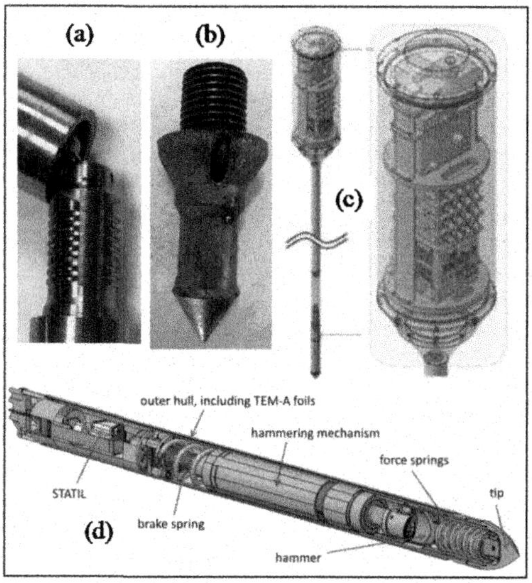

Figure 4. Advances in *deployment* of CPT equipment: (a) SingleTwist™ connectors (Storteboom et al. 2022a), (b) water lubrication (Yetginer-Tjelta et al. 2022), (c) free falling system (Kang et al. 2022) and (d) self-driving system (HP^3 Spohn et al. 2018).

of up to 15-20 m is reported in soft clays by the 'CPT stinger' (Jeanjean et al. 2012). However, the reported trials in sandy or stiff clay soils have achieved only 1-- 3 m penetration in some situations, even with a heavy 250-500 kg FFP unit. The challenge of achieving sufficient penetration may be exacerbated in dilatant sands by the undrained response. This effect also complicates the data interpretation, due to the change in drainage conditions relative to a standard CPT as well as the viscous rate effect (White et al. 2018). The depth range can be extended by a static push unit in the FFP (e.g. Jeanjean et al. 2012, Peuchen et al. 2017). This approach requires the embedment from the initial free fall stage to provide sufficient reaction.

2.3.3 *Reducing reaction force, raising depth range*
Twin advances of a lighter CPT system and an increased depth range can be achieved through methods that reduce or eliminate the external force required to push the rods through the ground.

A simple modification to reduce the pushing force that can form an addition to a conventional CPT system is a lubrication system to inject water or drilling mud around the rods (e.g. Jefferies & Funegard 1983.

In this conference, Yetginer-Tjelta et al. (2022) report a systematic investigation into different methods of water injection at the tip and rods of a seafloor CPT system (Figure 4b). This was part of a long term campaign on the Dogger Bank in the North Sea, with more than 500 CPTs. These improvement to the equipment have led to an approximate doubling of the depth achieved, including penetration through sands with $q_c > 100$ MPa.

2.3.4 *Bypassing reaction force: Self-burrowing*
An alternative to eliminating rod friction is to eliminate the rods themselves, and transform the CPT unit into a burrowing robot. Setting aside questions of navigation and power, this concept relies on the sleeve friction to provide a reaction force to overcome the cone tip resistance. Natural burrowers such as razor clams, earthworms and roots have been used as bio-inspiration for robots that could evolve into a rod-free CPT (e.g. Winter et al. 2014, Tao et al. 2020, Naclerio et al. 2021). Martinez et al. (2020) show that a roughened sleeve with a length-diameter ratio of L/D = 4-6.5 is required in dense sands – the most difficult conditions – to provide the required reaction.

An additional strategy to ease burrowing is to reduce the tip resistance by fluidizing the soil ahead of the robot, or in dry sand by aerating it. The approach has been successfully demonstrated in the laboratory by a steerable robot using an externally-provided air supply to fluidise the soil ahead (Naclerio et al. 2021).

The rise of robotics has led to many proposals for burrowing robots to be used across civil engineering applications. Perhaps self-burrowing CPTs will feature in future versions of this conference.

2.3.5 *Bypassing reaction force: Self-driving*
An alternative to self-burrowing is self-driving, in which the CPT is advanced by a driving mechanism that is incorporated within the machine itself. This concept was adopted by the Heat Flow and Physical Properties Package (HP^3) of the Insight Mission that took a lander to Mars during 2018.

The HP^3 system featured a CPT-like 'Mole' device, 400 mm long, 27 mm in diameter, with a mass of 860 g. The Mole was designed to pull a 5m-long Science Tether into the subsoil to a depth of 3 – 5m (Spohn et al. 2018, Kromer et al 2019) (Figure 4d).

The driving mechanism was designed to achieve 0.1-1mm of set per blow, at a rate of 0.28 blows/sec. Laboratory trials show this probe self-driving to 5 m depth in a sandy Mars-simulant soil of crushed basalt sand. Comparisons with conventional CPTs in the same bed showed that penetration at around 20 mm/min was possible in soil with $q_c \sim 1$ MPa (Spohn et al. 2018, Wippermann et al. 2020).

However, attempts to penetrate the surface of Mars were unsuccessful, despite the Mole operating correctly. The probe failed to overcome the initial penetration resistance of the shallowest few centimetres and created a wide hole around the probe. This prevented the shaft resistance on the Mole from providing resistance against the hammer recoil, contributing to the failure to penetrate. Despite various attempts over a period of two years, the Mole remained stuck in a shallow open hole on the surface of Mars (Spohn et al. 2021). Despite these difficulties on Mars, the project demonstrates the feasibility of self-burial by a CPT-like device using an internal hammer system.

2.4 *Sensing*

2.4.1 *Introduction*

The *sensing* on a conventional cone comprises primarily the three CPT parameters, tip and sleeve resistance, and pore pressure (i.e. q_c, f_s and u_2) with additional measurements of depth and tilt. The seismic piezocone test also provides measurements of shear wave velocity, V_s, at discrete depths.

A wide range of additional sensors have been incorporated within the cone, some of which are illustrated in Figure 2, based on contributions to this conference and other recent publications. These are discussed under the following headings of (i) mechanical, (ii) hydraulic and (iii) thermal.

2.4.2 *Advances in mechanical sensing*

It is well recognized that the fundamental strength and stiffness properties of a soil, in whatever constitutive framework is considered, do not uniquely correlate with the three basic cone parameters. This uncertainty can be reduced by taking one or more further mechanical readings, such as normal or shear stress values, by testing with a modified instrument equipped with additional sensors. These additional sensors are commonly mounted in further sleeves behind the conventional single friction sleeve.

For example, Isaev et al. (2022) describe in this conference a lateral stress cone, which is equipped with three pairs of lateral total stress and pore pressure sensors in successively larger diameter sleeves (Figure 5a). This type of 'lateral stress cone' provides data 'on-the-fly' (i.e. during steady penetration) that is analogous to a dilatometer or three points on a pressuremeter expansion curve. Correlations are explored to link these measured stresses, via their ratios or differences, to the in situ stress state and OCR.

A similar theme of multiple sleeves was proposed by DeJong & Frost (2002), using different values of roughness. In this case the multi-sleeve approach allows rough and smooth interface strengths to be measured, and the CPT sleeve to be tailored to the interface for which a design friction is required. By measuring sleeve friction at two positions, a new normalized friction parameter related to the degradation of friction can be calculated, and used as a fourth non-dimensional CPT parameter, to augment Q_t, F_r and B_q. This multi-friction parameter, MFP, allows the conventional range of CPT soil classification charts (e.g. Robertson 2009) to be supplemented by an extra chart using a derivative of MFP (Hebeler et al. 2018).

Friction sleeve data is sometimes treated with caution, due to potential unreliability from two sources: (i) errors in measuring the applied shear stress due to load leakage through the seals and/or low load cell sensitivity in subtraction cones and (ii) variations in the applied shear stress due to the sleeve roughness or minor differences in diameter or alignment of the sleeve. A review of these difficulties is presented in this conference by McConnell & Wassenaar (2022), who describe a new cone with improved load cell design to overcome these limitations and provide more reliable friction sleeve data in soft soils.

2.4.3 *Advances in hydraulic sensing*

Multiple contributions to this conference focus on the reliable measurement of hydraulic properties. Lunne et al. (2022) report on a revival of multiple pore pressure measurements – a review of early data that used the u_1 (tip and face) and u_3 (top of sleeve) sensors, alongside the conventional u_2 (shoulder). They show that normalized parameters using the difference or ratio between u_1 and u_2 can provide a reasonable correlation with OCR on a site-by-site basis, but the correlation with the conventional Q_t remains stronger.

One of the difficulties with CPT pore pressure data is the potential for unreliability due to poor saturation of the sensor filter element, air trapped elsewhere in the tip and cavitation. In this conference, Rocchi et al (2022) describe a new device that can be used to rapidly assure the saturation of a piezocone tip (Figure 5b). This device is used to test the saturation level by applying an impulse of pressure to the sensor, and then checking Skempton's B-value for the response.

Saturation is especially important when performing CPTs in unsaturated soils, where negative pore pressures are to be measured. In this conference, Tarantino et al. (2022) describe the development of a Tensiocone – a CPT attachment incorporating a high capacity tensiometer (HCD). They describe the laboratory validation of a new adapter that sits between the cone tip and friction sleeve, equipped with a surface-mounted tensiometer. Field trials did not show agreement between in situ tensiometers and the tensiocone measurements. This was traced to a cross-sensitivity between CPT axial load and tensiometer reading, which will be rectified in a new design of the sensor.

A further development of hydraulic measurement using the CPT is to impose a flow of water from the tip or shaft of the cone, and measure the associated pressure, in order to determine the local permeability. This adaptation is analogous to the conventional constant head borehole flow test (Hvorslev 1951). A similar adaptation has been applied to the pressuremeter, by withdrawing the device to create a test 'well' and applying controlled flow (Ratnam et al. 2001).

Gundersen et al. (2022), in this conference, describe trials using a 'flow cone' equipped to release water through a porous sleeve located

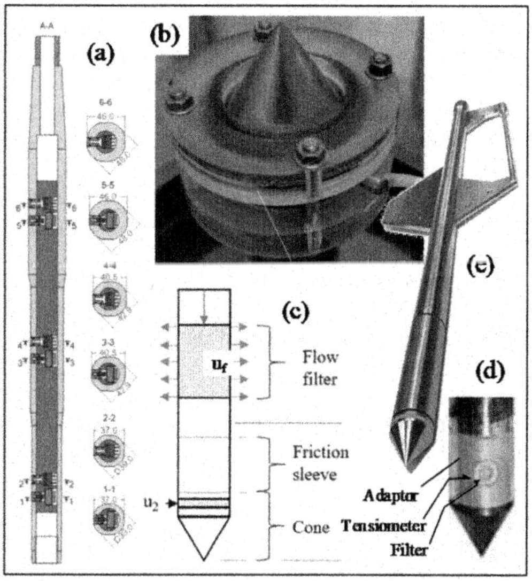

Figure 5. Advances in *sensing* on CPT equipment: (a) Tensiometer (Tarantino et al. 2022), (b) saturation verification (Rocchi et al. 2022), (c) the 'flow cone' (Gundersen et al. 2022), (d) Tensiometer adaptor (Tarantino et al. 2022) and (e) a parallel thermal probe element (www.datem.co.uk).

1.23 m above the shoulder of the cone, which is also equipped with pressure measurement (Figure 5c). Tests are reported from NGI's Halden test site, which comprises of silt. In these conditions, the sleeve measurement indicates an excess pore pressure remaining from the cone penetration process, and the dissipation response at the porous sleeve is used to infer the soil permeability. Prior studies using the same device at a sand site reported by Gundersen et al. (2019), show accurate direct measurement of permeability through constant head or falling head tests at the sleeve.

2.4.4 *Advances in thermal sensing*

CPTs are usually equipped with a temperature sensor within the tip, which can be used to check for temperature-induced drift of any sensors, and to also indicate the ground temperature – which may feature a step change at the depth of the water table.

More advanced use of the temperature data is described by Vardon et al. (2018, 2019). They describe the use of passive temperature responses to evaluate the thermal conductivity of the soil surrounding the CPT. The passive approach relies on the dissipation of heat generated by the penetrometer-soil friction during the installation process. This is a simplification of the active approach, which was developed earlier for geophysical research, which relies on an internal heating element to provide a pulse of heat that is recorded at other locations along the device. This type of 'thermal needle probe' can involve a thin extension from the front of a conventional CPT, or a parallel thinner element behind the friction sleeve, such as the example shown in Figure 5e produced by Datem.

These advances in the use of CPTs for thermal measurements reflect new markets related to ground heat energy systems and buried cables for offshore wind. Both of these applications relate to the energy transition and are rapidly growing markets.

2.5 *Motion*

2.5.1 *Introduction*

The *motion* in a conventional CPT is continuous penetration, usually at a rate of 2 cm/s, with intermittent pauses for rod changes or dissipation phases.

There have been many attempts to expand this motion through different speeds, patterns and directions of movement, some of which are the subject of papers in this conference. Changes in the pattern of vertical movement are made possible by advanced drive systems, which allow specific patterns of movement or load to be applied to the top of the cone rods. Meanwhile, modern miniature robotic technology has allowed miniature actuators to be incorporated in sleeve elements located behind the cone tip. This allows elements to be displaced relative to the main cone rod, with the corresponding soil response measured. The extensions can be summarized as follows:

1. Novel vertical movement: (i) different speeds, (ii) large amplitude cycles, (iii) small vibratory cycles;
2. Separate elements for other movements: (iv) expansion – i.e. the cone-pressuremeter, (v) rotation or (vi) lateral translation

Recent contributions to each of these test variations are briefly discussed below.

2.5.2 *Variations in the vertical movement*

Studies of large-amplitude cyclic CPTs – with vertical movement by several cone diameters – were prompted by similar testing of flow-round penetrometers (such as the T-bar or ball). For these devices, the cyclic penetration process is symmetric, with soil flowing in reverse around the protrusion. Cyclic T-bar or ball penetrometer tests provide a measure of soil sensitivity (e.g. Randolph 2004).

For a CPT, cyclic vertical movement may lead to a hole opening beneath the CPT, but if this does not occur the process can be similarly symmetric. Regardless of the flow at the cone tip, the cyclic movement on the friction sleeve also provides data that can support estimates of sensitivity or the degradation of friction on piles (e.g. Hebeler et al. 2005, Diambra et al. 2014, Shonberg et al. 2019).

Small amplitude high-frequency vertical cycles can also be applied concurrent with the standard installation push. This type of test and device is known as a Vibro-CPT (VCPT) or Vibropiezocone (e.g. McGillivary et al. 2000, Jorat et al. 2014, Al-Sammarraie et al. 2018). These vibrations can be imposed by an oscillating mass attached to the rods, or by control of the hydraulic actuator. VibroCPT installations typically feature 10-15 Hz vibrations on top of the standard 2 cm/s penetration, creating a vibratory amplitude on the order of 1 mm at the cone tip. Due to rod compression, accelerometer data from the tip is double integrated to find the tip motion.

The simplest interpretation of the VCPT is as the reduction ratio between q_c in a VCPT and an adjacent conventional CPT. This has been used as an indicator of liquefaction potential (e.g. McGillavary et al. 2000). In this conference, Al-Sammarraie et al. (2022) describe recent VCPT studies that have investigated more detailed interpretation of the q_c response within individual VCPT cycles. The aim is to link the measured response to pile-soil interaction models used for vibratory pile driving.

In softer soils, a further variation in vertical movement is to apply varying rates of penetration, altering the rate in steps – a so-called 'twitch' test (House et al. 2001, Randolph & Hope 2004, Silva et al. 2006). This test provides a basis to estimate the consolidation coefficient, albeit taking approximately the same time as a conventional dissipation test. However, a further benefit is that twitch results provide a ratio between the slow (drained) and fast (undrained) resistance, which can be linked to strength, stiffness or state (e.g. OCR) parameters (Schneider et al. 2007, Suzuki & Lehane 2015, Dienstmann et al. 2017, White et al. 2018, Mo et al. 2020).

Penetrometer tests that span a range of motion rates and also involve periods of cyclic movement and dissipation periods allow a wide range of soil strengths to be mapped out (e.g. Chow et al. 2019). These tests can mimic the loading history applied by infrastructure to the ground. The results may therefore be suitable for scaling to the changing capacity of a foundation system under 'whole life' loading (e.g. Zhou et al. 2020, Gourvenec 2020).

2.5.3 *Pressuremeter elements*

The earliest example of an additional actuated element combined with the CPT is the cone pressuremeter. Early studies using a pressuremeter incorporated above the cone led to interpretation solutions that focused on the cyclic and unloading stages, acknowledging that the initial penetration of the cone will have disturbed the in situ stress state (e.g. Houlsby & Withers 1988, Withers et al. 1989, Bellotti et al. 1989, Ferreira & Robertson 1992, Bolton & Whittle 1999, Whittle & Liu 2013).

Recently the cone pressuremeter has found new usage in relation to offshore wind, for the determination of soil stiffness for monopile foundations (e.g. Whittle et al. 2017). In this conference Vinco & Sachetto (2022) describe a new cone pressuremeter system designed particularly to verify the results of ground improvement activities (Figure 6a). For this application the system must be small, portable and suited to use within existing buildings, and requires only a low penetration force and depth capability. This system uses water as the control fluid, in common with another recently-developed pressuremeter described by Lopes dos Santos et al. (2019).

These new systems could support wider adoption of the cone-pressuremeter, taking advantage of the use of unload-reload loops and the unloading phase to gather strength and stress-strain or stiffness degradation curves from the measured data. It has also been demonstrated that pressuremeter tests can include large numbers of constant pressure-amplitude cycles, which may allow insights into the cyclic response, that can be translated into lateral p-y response models (e.g. Briaud et al. 1984, Lopes dos Santos 2020).

2.5.4 *Torsional or lateral motion elements*

Miniaturization of electromechanical systems allows actuators to be located within the cone sleeve. This allows concepts such as a torsional (t-θ) sleeve element (Frost & Martinez 2013, Figure 6b) and a laterally-moving p-y-type element (Diambra et al. 2022, in this conference, Figure 6c). These concepts offer the potential to provide more detailed soil characterization, by measuring load-

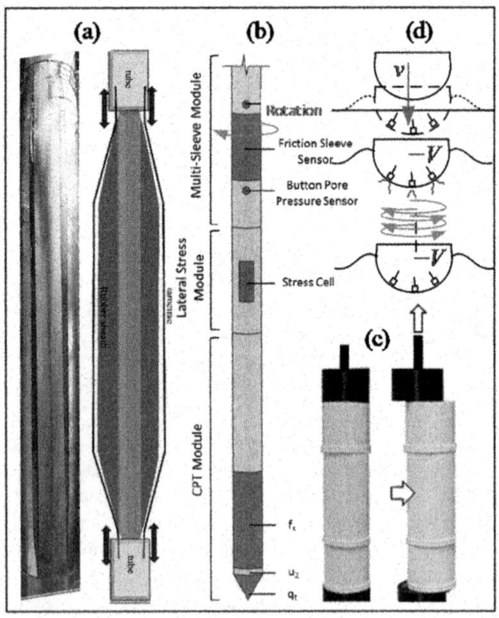

Figure 6. Advances in *motion* during CPT-type testing: (a) pressuremeter module (Vinco & Sachetto 2022), (b) Torsional sleeve (Frost & Martinez 2013), (c) lateral (p-y) sleeve (Diambra et al. 2022) and (d) rotational shallow penetrometer tests (White et al. 2022).

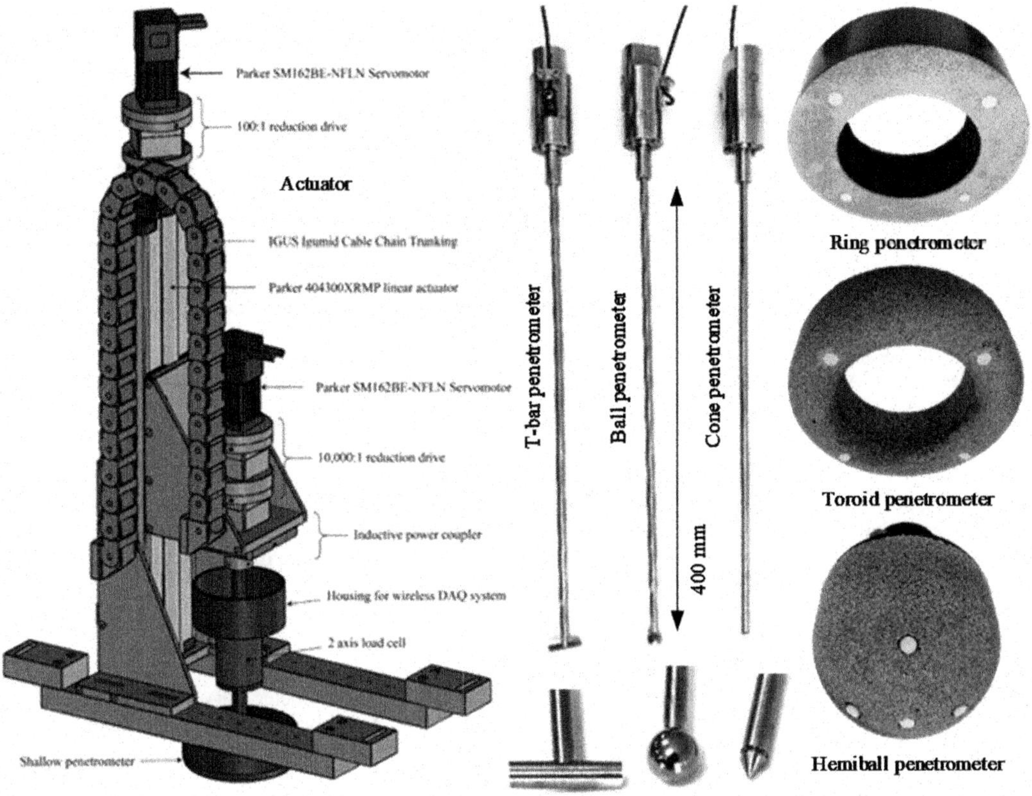

Figure 7. Shallow penetrometer system – a combination of CPT, T-bar and ball strength profiling tools and the shallow ring, toroid and hemiball devices for near-surface soil strength, interface strength and consolidation property measurement (White et al. 2022).

displacement responses for modes of soil deformation that are different to the cylindrical expansion of the pressuremeter.

For example, laboratory torsional shear tests indicate differences in the rate of pore pressure generation compared to axial shearing. Meanwhile, lateral deformation is more closely linked to pile p-y loading than cylindrical expansion. These locally-actuated sleeve concepts remain to be proven beyond trial prototype devices, but such robotic cones (or ROBOCONEs, to use the term coined by Diambra et al. 2022) could create exciting possibilities for complex in-situ tests.

The introduction of more complex dual-directional actuation in a penetrometer test has been successful for shallow devices, such as the ring, toroid and hemiball penetrometers. These tests involve penetration of the device by a fraction of a diameter, followed by phases of dissipation (on clay soils) and rotation (Figure 6d).

In this conference, we present test execution and interpretation procedures for a rotational penetrometry system that has been used for more than 300 tests over the past 6 years, including for several commercial projects (White et al. 2022). The system features electromechanical drives with feedback control, so a vertical load can be maintained while the device rotates and settles. Drained and undrained interface strength parameters, as well as consolidation properties, are derived from the tests.

3 HOW DOES CPT EQUIPMENT ADVANCE?

3.1 *Are equipment advances led by new ideas?*

This discussion and the wider CPT'22 conference feature numerous advances in CPT equipment, some of which will influence future practice, and many of which will not. So this closing part of the discussion offers some views on how the CPT equipment used in practice becomes more advanced. Hopefully this will prompt the sharing of wider views during the conference, and I look forward to learning from these.

The last equipment example discussed above – rotational testing using shallow penetrometers (Figure 7) – is an example that I have been heavily

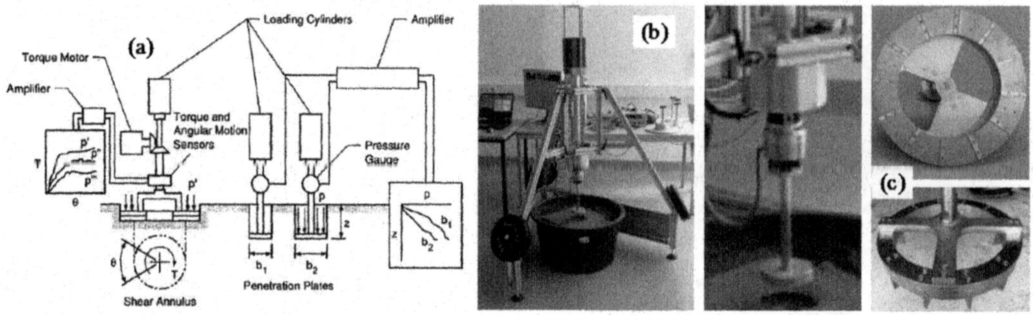

Figure 8. The bevameter – the original shallow penetrometer? (a) Bevameter principles (Bekker 1969), (b) Laboratory Bevameter system (Apfelbeck et al. 2011), (c) Shear plates for torsional Bevameter tests (top: Apfelbeck et al. 2011, bottom: Kim et al. 2021).

involved in, so I will use this as a living case study of an advance that might or might not have a future life.

These shallow penetrometers emerged from research at UWA at a time when we were heavily focused on seabed pipelines, and their friction on the seafloor. Centrifuge model testing of short pipeline elements was being used to assess design friction coefficients (White & Gaudin 2008). This led to the idea of a 'pipe-like' in situ test, but using rotation to allow longer distances of interface sliding. Prof. Mark Randolph and I proposed this concept as Dr Yue Yan's PhD topic and she developed the shallow penetrometer concept (Yan 2013). Dr Sam Stanier took the topic further in the RIGSS Joint Industry Project, with close support from industry, and co-supervising Dr Mark Schneider's PhD on this topic. The devices worked well, we devised rigorous interpretation methods, and the devices appeared to meet an industry need.

But was this type of test a new idea? It turns out not. After the PhD studies and the RIGSS JIP were completed, we discovered parallel research in the world of terramechanics and agriculture that involved a similar type of test – the bevameter (Bekker 1969, Wong 2010). This device was developed in the 1950s and resembles the ring penetrometer and the torsional plate penetrometer (Watson & Randolph 1998).

A bevameter test involves penetration and rotation of a ring-shaped device, with the pressure-sinkage and torque-rotation results being used to characterize the soil for assessment of trafficability (Figure 8). Bevameters have been developed for attachment to lunar landers and Mars rovers (Bekker 1969, Apfelbeck et al. 2011, Edwards et al. 2017). Their interpretation is rarely rooted in the same soil mechanics principles as are familiar for CPTs and our shallow penetrometers, but the test concept is similar.

So – shallow penetrometers might deserve to be counted as an advance, but they are not a new idea. The message is perhaps that (i) a new idea may create an advance but (ii) there may be plenty of old ideas that we are yet to develop into a suitable form that can be brought into the geotechnical market.

3.2 *Are equipment advances led by new markets?*

3.2.1 *Introduction*

The shallow penetrometer story leads to the question of markets for advances in penetrometer testing. Many of the advances reviewed in this paper target a particular market that requires additional information from a penetrometer test – through an additional *sensor* or *motion* stage. Or, the advances relate to a new market where conventional *delivery* and *deployment* methods are unsuited.

The offshore engineering examples discussed below illustrate three market pulls that have supported recent penetrometer equipment advances.

3.2.2 *High temperature high pressure pipelines*

The emergence of high temperature high pressure (HTHP) seabed pipelines around 20 years ago led to an increased focus on near-surface seabed strength. Engineered lateral buckling schemes for HTHP pipelines require reliable estimates of the embedment and the subsequent axial and lateral soil resistance (Bruton et al. 2007).

This market led to innovations such as miniature box-core-based CPT, T-bar and ball penetrometer systems (Low & Randolph 2010, Puech et al. 2010) and the shallow penetrometer systems (Yan et al. 2010, Boscardin & DeGroot 2015, White et al. 2017). These devices are designed to characterize only the shallowest half metre or so of the seafloor, by conducting tests in box core samples. There is a particular emphasis on remoulded properties, which control pipeline embedment, and interface strength, which controls axial pipeline friction.

3.2.3 *Subsea power cables*

The offshore wind industry has created a major market for cable route surveys, which has similarities with the on-bottom pipeline survey market. However, a particular feature of subsea cable design is the importance of thermal properties. The thermal insulation created by the trench backfill can have a strong influence on the cable rating and operating temperature (Dix et al. 2017).

A response to this market has been the development of new in situ thermal tests, that can be integrated within a conventional CPT-based route survey. Some of these devices (e.g. Figure 5e) use the violin bow-type design of thermal probe that has been used by the earth science community for studying heat flow from the deep earth into the oceans (e.g. Lister 1979). Once again – the idea is not new, but the concept has recently entered geotechnical survey practice in response to a market pull.

Another approach to a thermal measurement is the passive thermal CPT (T-CPT) system, which is easily added to a standard cone test. In this test the decay of the heat developed by the penetration process is used to infer the thermal conductivity. This simplicity has allowed the T-CPT to rapidly enter geotechnical survey practice.

A further driver for adoption of the T-CPT is the evidence that thermal conductivity measurements from samples are biased compared to in situ data. A review of hundreds of in situ and ex situ thermal conductivity measurements by Vardon et al. (2019) showed that the thermal conductivity measured in laboratory tests is consistently lower than in situ data. This is attributed to slight de-saturation and a reduction in density of the samples (Vardon et al. 2019).

3.2.4 *Project acceleration: Reliance on in situ data*

A further example of a market pull that is motivating advances in CPT equipment, is the need to accelerate the maturation of projects, from planning to operation. This is linked to the energy transition, with wind projects aiming for faster progression from planning or final investment decision (FID) to the beginning of energy production, compared to oil and gas projects.

As an illustration, offshore wind projects in Europe typically take 7-10 years to progress from initial planning to operation and in China the process is compressed into 2-4 years. In the past, major oil and gas projects often had a longer timeline than European offshore wind, and also a smaller footprint to survey.

The rest of the world needs to approach Chinese rates of project maturation if net zero targets are to be met. DNV (2021) predict a global wind capacity of 5.9 TW by 2050, which will require around a quarter of a million turbines to be installed in the next 30 years, with the annual rate of installation increasing by a factor of 10 compared to now.

One method to compress the project timeline is to increase the use of in situ geotechnical test data for determining design parameters. This can eliminate or reduce the importance of the period required for samples to reach the testing laboratories, be tested, and an interpreted set of design parameters be generated.

If in situ testing takes a larger role in determining design parameters, then there is a strong market pull for new types of in situ test, or extensions of the CPT, which can explore the detailed aspects of soil response that are conventionally determined from laboratory testing – such as cyclic behaviour, and stiffness parameters.

This market pull has created the driver for cyclic and small strain components of CPT-type tests such as the seismic CPT, the VCPT, the cone-pressuremeter and the locally-actuated lateral and torsional sleeves. The ability to determine thermal properties in situ, rather than relying on laboratory tests, also allows compression of the project timeline.

3.3 *Are equipment advances led by new technology?*

A final driver of equipment advances is the emergence of new technologies across mechanical and electronic engineering, as well as new types of vehicle. Over the past twenty years, the emergence of low cost sensors, data acquisition systems, actuators and communication technologies have permitted new types of sensor and testing protocol to be devised. Miniaturization has allowed additional equipment to be packed into the shaft of the CPT.

Also, the low cost and wide availability of mechatronic devices has opened up and democratized the process of equipment development. Small companies or individuals can now prototype sensors and devices that could previously only be produced and trialled by major organisations.

Massarch (2014), in his comprehensive review of the CPT, past and present, discussed a future in which sensors are so cheap that the device can be left in situ after the penetration test. This idea opens up the possibility that penetration tests are not a snapshot of the ground condition at a moment in time, but implant a sensor in the ground, that continues to operate, relaying back the condition of the ground through the life of the surrounding infrastructure. The boundary between in situ testing and field monitoring thus becomes blurred.

Automation technology also means that the execution of a CPT is becoming an 'office job'. In this conference, Storteboom et al. (2022b) describe a new hybrid CPT truck that operates using battery when on site. A carousel system automatically connects the CPT rods so the operator has minimal mechanical activities to perform. Storteboom et al. (2022b) report that *"the workplace of the CPT operator is turned into an office environment"*. They describe an improvement in working conditions, reduced absenteeism and higher productivity.

A final example of emerging technology is the digital environment used to hold and process CPT data, and the integration of this into the geotechnical design workflow and the digital model of the resulting construction. The geotechnical survey industry in the UK was an early adopter of standardized data formats, through the AGS standard, which is named after the Association of Geotechnical and Geoenvironmental Specialists, who are custodians of the standard. Version 1 of this data format was released in 1992, and version 4.1.1 is the current edition (AGS 2022).

The AGS format is widely used in the UK and worldwide, and improves the workflow and assurance of geotechnical data, as well as simplifying the sharing and public release of geotechnical data. For example, the Crown Estate (England and Wales) requires wind farm operators to release survey data to the Marine Data Exchange portal. This data commonly takes the form of reports, but for geotechnical in situ testing the data is often held in the AGS format, where it can be rapidly assimilated into other software for wider use.

A common software platform for storing and interpreting in situ data is gINT, which is owned by the Bentley software house. Bentley also own PLAXIS, and the two programs have growing interface capabilities. A consequence of this connectivity from raw CPT data to geotechnical numerical modelling software is that there is a single digital workflow from site measurements to design output.

This connectivity can be used to allow more rapid progression from survey to construction. It can also eliminate the human attention given to geotechnical data that is often held dear in our discipline. For example, the progression of data through factual reporting, interpretive reporting and ground model development involves detailed evaluation and thought processes that support geotechnical risk analysis, as well as design parameter choices.

On the other hand, we could take the counter view, which is that a smooth processing pipeline from in situ data towards design will liberate geotechnical specialists. We can spend more time assessing the information created by the data, rather than passing data between spreadsheets and different types of software.

3.4 *Equipment advances: Does standardisation help or hinder?*

Standardisation of the CPT emerged in the 1970s and has more recently been set out in various international standards produced by the ISO, CEN and others (e.g. ISO 2012, 2014).

Many advances in CPT equipment remain 'within standard', such as methods for efficient deployment of the rod string or improvements to load cells or data transfer. However, many advances in penetrometer equipment are 'non-standard', which creates difficulties when clients wish to assure data quality or set contract specifications and develop a competitive supplier market.

Two key challenges are therefore faced by developers of any new equipment that deviates from the standards:

1. *The need for a specification:* a test specification is required to define and allow assurance that any testing is conducted according to an appropriate procedure, agreed between client and contractor. A hierarchy of specification types, listed by the strength of their basis, might be: (i) a publication by the system developers, (ii) a guideline emerging from a Joint Industry Project, or developed with a verification agency, (iii) a guideline developed more openly by an industry technical committee or association (e.g. CFMS, CIRIA and the ISSMFE have supported guidelines on in situ testing) (iv) a recommended practice or standard produced by an international standards body (e.g. ISO, CEN, API).
   In practice, procedures generally evolve upwards through this hierarchy over a period of perhaps a decade – as seen by the procedures adopted for flow-round penetrometers that have now reached ISO 19901-8.
2. *The need for market competition.* A second challenge facing proponents of an advance in penetrometer equipment is that new test types will rarely be specified by clients if there is only one supplier. This places developers of new equipment in the difficult position of requiring a competitor to come into existence in order for demand to be created. This situation does not apply to equipment advances that are 'within standard', since the specification is unaffected.
   However, using the example of the shallow penetrometers for a final time, research funding for the RIGSS JIP from owners/operators was conditional on more than one contractor also being part of the project, in order to ensure that a competitive supply chain could emerge.

The heading for this final sub-section was perhaps a little frivolous. Standards are invariably beneficial for industry and should not hinder innovation. However, innovators of new 'non-standard' in situ test equipment need to be prepared to address the two challenges set out above. Developing reliable equipment that offers a technical benefit over standard test methods will not necessarily lead to industry adoption. Time and effort must be spent 'socialising' the technology among the community of practice, for example at events like this conference. It is then necessary to develop open guidance and specifications for execution and interpretation of non-standard tests. Finally, it might even be necessary to support open competition in the marketplace to avoid the technology being from a sole source.

4 CONCLUDING COMMENTS

The basic configuration of the cone penetrometer test (CPT) has remained constant for around 50 years. This discussion has reviewed recent advances in penetrometer equipment presented in this conference and elsewhere, considering separately the (i) delivery, (ii) deployment, (iii) sensing and (iv) motion aspects of the penetrometer test.

It is evident that new innovations in instrumentation, electronics and mechanical engineering have created a vast range of variations on the conventional CPT equipment. In some cases these advances provide a small but widespread improvement in penetrometer testing, such as improved rod connection systems, hybrid-powered CPT trucks and better saturation and calibration systems. In other cases the advances are more ambitious, aiming to incorporate new sensors and motion systems within a penetrometer, to allow new types of measurement to be incorporated in the testing process.

The result is a dispersion of practice: the majority of civil engineers base their ground characterization on a durable 1970s technology, updated with automation and improved data quality, while a minority are enjoying exploration of non-standard innovations – with mixed levels of impact.

This discussion, and the wider conference, is an opportunity to reflect, learn and debate how we will characterize the ground in another 50 years.

ACKNOWLEDGEMENTS

I am grateful for the kind invitation by the CPT'22 organising committee to present this discussion paper. In relation to the shallow penetrometer research, I acknowledge the support for that work from the RIGSS JIP (Fugro, Total, Woodside and Shell), as well as the collaboration with Dr Sam Stanier, Dr Henning Mohr, Prof. Mark Randolph and Dr Mark Schneider.

This review has also been supported by the EPSRC Offshore Renewable Energy Supergen Hub (grant EP/S000747/1) and by the EPSRC grant EP/W006235/1 (ROBOCONE – intelligent robotics for next generation ground investigation and design – PI Dr Andrea Diambra). All opinions and any mistakes or misrepresentations are my own.

REFERENCES

AGS (2022) Electronic transfer of geotechnical and geoenvironmental data, AGS4 Edition 4.1.1 – March 2022 Publ. Association of Geotechnical and Geoenvironmental Specialists

Al-Sammarraie D., Kreiter S., Kluger M.O., Goodarzi M. & Morz T. (2022) VCPT: an in situ soil investigation method to validate the vibratory pile-soil interaction models. Proc. 5th Int. Symposium on Cone Penetration Testing (CPT'22). Bologna. (this conference).

Al-Sammarraie D., Kreiter S., Stahler M.O., Goodarzi M. & Morz T. (2018) New vibratory cone penetration device for in-situ measurement of cyclic softening. International Symposium on Cone Penetration Testing, 2018.

Apfelbeck M., Kuß S., Rebele B., Schäfer B. (2011) A systematic approach to reliably characterize soils based on Bevameter testing. Journal of Terramechanics, 48(5):360–371

Baez V.M., Shah A., Akinwande S., Jafari N.H. & Becker A.T. (2020) Assessment of soil strength using a robotically deployed and retrieved penetrometer. IEEE/RSJ International Conference on Intelligent Robots and Systems (IROS) pp. 7324–7329, doi: 10.1109/IROS45743.2020.9341424.

Bekker, M.G. (1969) Introduction to Terrain-Vehicle Systems. University of Michigan Press, Ann Arbor, Mich.

Bellotti, R., Ghionna, V., Jamiolkowski, M., Robertson, P. K. & Peterson R.W. (1989) Interpretation of moduli from self-boring pressuremeter tests in sand. Géotechnique 39(2):269–292.

Boggess, R. and Robertson, P.K. (2010) CPT for soft sediments and deepwater investigations. International Symposium on Cone Penetration Testing, CPT'10. 2:127–136.

Bolton, M. D. & Whittle, R. W. (1999) A non-linear elastic/perfectly plastic analysis for plane strain undrained expansion tests. Géotechnique, 49(1):133–141.

Boscardin A.G. & Degroot D.J. (2015) Evaluation of a toroid for model pipeline testing of very soft offshore box core samples Frontiers in Offshore Geotechnics III – Proc. 3rd Int. Symp. on Frontiers in Offshore Geotech., ISFOG 2015. 363–368.

Briaud, J. L., Riner, K. B. & Ohya, S. (1984) Cyclic pressuremeter tests for cyclic lateral loads. Offshore Technology Conference, Houston. doi:10.4043/4678-MS

Broms, B.B. & Flodin, N. (1988) History of soil penetration testing. International symposium on penetration testing, ISOPT 1, 1:157–220.

Bruton D., Carr M. & White D.J. (2007) The influence of pipe-soil interaction on lateral buckling and walking of pipelines: the SAFEBUCK JIP. Proc. 6th Int. Conf. on Offshore Site Investigation and Geotechnics, London. 133–150.

Chow S.H., O'Loughlin C.D., White D.J. & Randolph M.F. (2017) An extended interpretation of the free-fall piezocone test in clay. Géotechnique. 67(12):1090–1103

Chow S.H., O'Loughlin C.D., Zhou Z., White D.J. & Randolph M.F. (2020). Penetrometer testing in a carbonate silt to explore changes in soil strength. Géotechnique 70(12):1160–1173

DeJong, J.T. and Frost, J.D. (2002). A multi-friction sleeve attachment for the cone penetrometer. ASTM Geotechnical Testing Journal, 25(2):111–127

Diambra A., Ciavaglia F., Harman A., Dimelow C., Carey J. & Nash D. F. T. (2014). Performance of cyclic cone penetration tests in chalk. Geotechnique Letters, 4 (3):230–237

Diambra A., Creasey J., Leonet J., Conn A., Ibraim E., Mylonakis G., White D.J., Cerfontaine B., Gourvenec S.M. & Igoe D. (2022) Concept design of a new CPT module for direct in situ measurement of p-y soil responses. Proc. 5th Int. Symposium on Cone Penetration Testing (CPT'22). Bologna. (this conference).

Dienstmann, G.; Schnaid, F.; Maghous, S.; Dejong, J.T. (2017). Piezocone penetration rate effects in transient gold tailings. Journal of Geotechnical and Geoenvironmental Engineering, ASCE. 04017116-1

Dix, J.K., Hughes, T.J., Emeana, C.J., Pilgrim, J.A., Henstock, T.J., Gernon T.M., Thompson, C.E.L. & Vardy, M.E., (2017) Substrate controls on the life-time performance of marine HV cables. Proc. 8th Int. Conf. Offshore Site Investigations and Geotechnics. Society for Underwater Technology. 88–108.

DNV (2021) Energy transition outlook. DNV, Hovik.

Edmunds J. (2014) Gaining traction. Offshore Engineer Magazine. 3 February 2014.

Edwards M.B., Dewoolkar M.M., Huston D.R. & Creager C. (2017) Bevameter testing on simulant Fillite for planetary rover mobility applications. J. Terramechanics, 70:13–26

Ferreira, R.S. & Robertson, P.K. (1992) Interpretation of undrained self-boring pressuremeter test results incorporating unloading. Can. Geotechnical J., 29:918–928.

Frost J.D. & Martinez A. (2013) Multi-sleeve axial-torsional-piezo friction penetration system for subsurface characterization. Proc. 18th Int. Conf. on Soil Mech. and Geotech. Eng. (18th ICSMGE). Paris, France, 1:527–530.

Geise, J.M. & Kolk, H.J. (1983). The use of submersibles for geotechnical investigations. Proc. Submarine Technology, The Design and Operation of Underwater Vehicles, Paper 7.3, London, Society for Underwater Technology.

Gourvenec S.M. (2020) Whole-life geotechnical design: What is it? What's it for? So what? And what next?: Keynote. In Proc. 4th Int. Symp. on Frontiers in Offshore Geotechnics. Deep Foundations Institute. 206–246.

Gundersen A.S., Carotenuto P., Lunne T., Walta A. & Sparrevik P.M. (2019) Field verification tests of the newly developed flow cone tool—In-situ measurements of hydraulic soil properties. AIMS Geosciences, 5 (4):784–803.

Gundersen A.S., Lunne T., Stelzer R., Blaker O, Tucker G. W. & Krogh L. (2022) Flow cone – new CPTU add-on module trialed in Halden silt. Proc. 5th Int. Symposium on Cone Penetration Testing (CPT'22). Bologna. (this conference).

Hebeler, G.L., Frost, D.J., Schneider, J.A. & Lehane, B.M. (2005). Cyclic friction piezocone tests for offshore applications, Int. Symp. on Frontiers in Offshore Geotech. 967–972.

Hebeler, G.L., Martinez A. & Frost J.D. (2018) Interface response-based soil classification system. Canadian Geotech. J., 55(12):1795–1811.

Houlsby, G.T. & Withers, N.J. (1988) Analysis of the cone pressuremeter test in clay. Géotechnique, 38 (4):575–587.

House, A.R., Oliveira, J.R.M. and Randolph, M.F. (2001) Evaluating the coefficient of consolidation using penetration tests. Int. J. Physical Modelling in Geotechnics, 1 (3): 17–26.

Hvorslev, M. J. (1951) Time lag and soil permeability in ground-water observations. US Army Corps of Engineers, Waterways Experiment Station Bulletin No. 36, Mississippi, USA.

Isaev O.N., Sharafutdinov R.F. & Zakatov D.S. (2022) Determination of natural stress state parameters for clay soils by using 3LSU-CPTU penetrometer. Proc. 5th Int. Symposium on Cone Penetration Testing (CPT'22). Bologna. (this conference).

ISO (2012). ISO 22476-1:2012: Geotechnical investigation and testing - Field testing - Part 1: Electrical cone and piezocone penetration test. International Standardisation Organisation

ISO (2014). ISO 19901-8:2014: Petroleum and natural gas industries - Specific requirements for offshore structures - Part 8. Marine soil investigations. International Standardisation Organisation [a 2022 edition is currently under international ballot]

Jeanjean, P., Spikula, D. & Young, A. (2012). Technical vetting of free-fall cone penetrometer. Proc. Int. Conf. on Offshore Site Investigation and Geotechnics, London, UK. 179–186

Jefferies M.G. & Funegard E. (1983). Cone penetration testing in the Beaufort Sea. Proc Conference on Geotechnical Practice in Offshore Eng., Austin, Texas, ASCE, 220–243

Jorat, M. E., Mörz, T., Schunn, W., Kreiter, S., Moon, V., & de Lange, W. (2014). Geotechnical Offshore Seabed Tool (GOST): a new cone penetrometer. Proc. 3rd International Symposium on Cone Penetration testing. 207–215

Kang H., Kwon O., Shin C., Seo J. & Jang I. (2022). Development of free fall cone penetration testing system. Proc. 5th Int. Symposium on Cone Penetration Testing (CPT'22). Bologna. (this conference).

Kelleher, P. J. & Randolph, M. F. (2005) Seabed geotechnical characterisation with a ball penetrometer deployed from the Portable Remotely Operated Drill. Int. Symp. on Frontiers in Offshore Geotechnics I, Perth, Australia. 365–371

Kim, J.T., Im, D.U., Choi, H.J., Oh, J.W., Park, Y.J., (2021) Development and performance evaluation of a bevameter for measuring soil strength. Sensors 21, 1541

Krömer O., Scharringhausen M., Fittock M., Tsakyridis G., Wippermann T., Witte L., Grott M., Knollenberg J., Spohn T., Krause C., Hudson T.L., Lichtenheldt R., Grygorczuk J., & Wisniewski L. (2019) Design details of the HP^3 mole onboard the InSight mission. Acta Astronautica. 164:152–167

Lister, C.R.B., (1979) The pulse-probe method of conductivity measurement,Geophys. J. Royal Astr. Soc., 57:451–461

Lopes dos Santos, A. & Dupla, J-C & Canou, J & Droniuc, N. (2019). Laboratory validation of an innovative mono-cell pressuremeter probe: test procedures and first results. Proc. XVII European Conference on Soil Mechanics and Geotechnical Engineering. Reykjavik, Iceland. 8pp.

Lopes dos Santos A. (2020). Determination of soil shear modulus at low strain level using an innovative pressuremeter probe: Application to cyclic pile design. Dissertation presented for the degree of Doctor of Université Paris-Est.

Low, H.E. & Randolph, M.F. (2010). Strength measurement for near-seabed surface soft soil using manually operated miniature full-flow penetrometer. ASCE Journal of Geotechnical and Geoenvironmental Engineering 136:1565–1573.

Lunne, T., Robertson, P.K. & Powell, J.J.M. (1997). Cone Penetration Testing in Geotechnical Practice, Blackie Academic/London, Routledge, New York: 312 p.

Lunne, T. (2012). The CPT in offshore investigations – a historic perspective. Geomechanics and Geoengineering: An International Journal. 2(2):75–101.

Lunne T., Ghanekar R., Tucker G.W., Santos R. & Krogh L. (2022) The revival of multiple pore pressure

measurements in the Cone Penetration Test. Proc. 5th Int. Symposium on Cone Penetration Testing (CPT'22). Bologna. (this conference).
Martinez A., DeJong, J.T., Jaeger, R. & Khosravi A. (2020). Evaluation of self-penetration potential of a bio-inspired site characterization probe by cavity expansion analysis. Canadian Geotech. J., 57(5):706–716.
Massarsch K.R. (2014) Cone Penetration Testing – A Historic Perspective. Proc. 3rd International Symposium on Cone Penetration Testing. 97–134
McConnell A. & Wassenaar E. (2022) An innovative new 3 MPa CPT – to detect and measure very small f_s values. Proc. 5th Int. Symposium on Cone Penetration Testing (CPT'22). Bologna. (this conference).
McGillivary, A., Casey, T., Mayne, P. W. & Schneider, J. A. (2000) An electro-vibrocone for site-specific evaluation of soil liquefaction potential. Proc. Geo-Denver 2000, Innovations and Applications in Geotechnical Site Characterization.
Meunier, J., Sultan, N., Jegou, P. & Harmegnies, F. (2004) First tests of Penfeld: a new seabed penetrometer. Proc. 14th Int. Offshore & Polar Eng. Conf., Toulon, France, 338–345
Mo, P-Q, Gao, X-W, Yang, W, Yu, H-S. (2020) A cavity expansion–based solution for interpretation of CPTu data in soils under partially drained conditions. Int J Numer Anal Methods Geomech. 44:1053–1076.
Naclerio, N.D., Karsai, A., Murray-Cooper M., Ozkan-Aydin Y., Aydin E., Goldman D.I. & Hawkes E.W. (2021) Controlling subterranean forces enables a fast, steerable, burrowing soft robot. Science Robotics 6:55 eabe2922.
Olmedo, N.A. & Lipsett, M.G. (2016) Design and field experimentation of a robotic system for tailings characterization. J. Unmanned Vehicle Systems 4 (3):169–192.
Peuchen J., Looijens P. & Stark N. (2017) Offshore characterisation of extremely soft sediments by free fall penetrometers. Proc. Int. Conf. on Offshore Site Investigation and Geotechnics. Society for Underwater Technology. 370–377
Power P. & Geise J. (1994) Offshore soil investigation techniques and equipment for the next century. Proc. of the Int. Conf. on Behaviour of Offshore Structures, BOSS'94, Boston, 1:97–109
Puech, A., Orcozo-Calderón, M. and Foray, P. (2010). Mini T-bar testing at shallow penetration. Int. Symp. on Frontiers in Offshore Geotechnics II, Perth, Australia. 305–310.
Randolph, M.F. (2004). Characterisation of soft sediments for offshore applications. 2nd Int. Conf. on Site Characterization, Rotterdam, The Netherlands, 1:209–232.
Randolph, M.F., Stanier, S.A., O'Loughlin, C.D., Chow, S. H., Bienen, B., Doherty, J.P., Mohr, H., Ragni, R., Schneider M.A., White D.J., Schneider J.A. (2018) Penetrometer equipment and testing techniques for offshore design of foundations, anchors and pipelines. Invited keynote Lecture. Proc. 4th Int. Symp. on Cone Penetration Testing, CPT '18. Delft. 3–24.
Randolph, M. F., and Hope, S. N. (2004) Effect of cone velocity on cone resistance and excess pore pressure. Proc. Int. Symp. On Engineering Practice and Performance of Soft Deposits, Yodogawa Kogisha Co. Ltd., Osaka, Japan, 147–152.
Ratnam S., Soga K., Mair R.J., Whittle R.W. & Tedd P. (2001) Permeability measurement using the self-boring pressuremeter. Proc. Int. Conf. on In Situ Measurement of Soil Properties and Case Histories, Bali, Indonesia. 667–672.
Rocchi I., Tonni L., Gottardi G. & Marcolongo M. (2022) Experimental procedure for checking the saturation degree of piezocone tips. Proc. 5th Int. Symposium on Cone Penetration Testing (CPT'22). Bologna. (this conference).
Schneider J.A., Lehane B.M. & Schnaid F. (2007) Velocity effects on piezocone measurements in normally and over consolidated clays. International Journal of Physical Modelling in Geotechnics. 7:23–34
Scholz C., Göttinger M., Hinck S., Möller K. & Ruckelshausen A. (2014) Automatic soil penetrometer measurements and GIS-based documentation with the autonomous field robot platform BoniRob. 12th Int. Conf. of Precision Agriculture.
Shonberg A., Harte M., Tucker G. Rattley M. & Gibbs P. (2019) Innovative use of the CPT to assess the in situ degradation of an offshore overconsolidated clay. Proc. Offshore Technology Conference. Paper OTC29414.
Silva, M. F., White, D. J., and Bolton, M. D. (2006) An analytical study of the effect of penetration rate on piezocone tests in clays. Int. J. Numerical & Analytical Methods in Geomechanics 30(6):501–527.
Spohn T., Grott M., Smrekar S.E., Knollenberg J., Hudson T. L., Krause C., Müller N., Jänchen J., Börner A., Wippermann T., Krömer O., Lichtenheldt R., Wisniewski L., Grygorczuk J., Fittock M., Rheershemius S., Spröwitz T., Kopp, E., Walter I., Plesa A. C., Breuer D. & Morgan, P. (2018) The Heat Flow and Physical Properties Package (HP3) for the InSight mission. Space. Sci. Rev., 214, 96
Spohn T., Hudson T.L., Witte L., Wippermann T., Wisniewski L., Kediziora B., Vrettos C., Lorenz R.D., Golombek M., Lichtenfeld R., Grott M., Knollenberg J., Krause C., Fantinati C., Nagihara S., & Grygorczuk J. (2021) The InSight HP^3 mole on Mars: Lessons learned from attempts to penetrate to depth in the Martian soil. Submitted to Advances in Space Research. On arXiv, ref. arXiv:2112.03234
Stark N., Hanff H., & Kopf A. (2009) Nimrod: a tool for rapid geotechnical characterization of surface sediments. Sea Technology. 50(4):2230–2235.
Stark N. (2016) Geotechnical site investigation in energetic nearshore zones: opportunities & challenges. Keynote lecture. Int. Conference on Geotechnical and Geophysical Site Characterisation 5. 69–80
Steiner A., Kopf A.J., L'Hereux J-S., Kreiter S. Stegmann S., Haflidason H. & Moerz T. (2014). In situ dynamic piezocone tests in natural clayey soils – a reappraisal of strain-rate corrections. Canadian Geotechnical Journal. 51(3): 272–288.
Stephan S., Kaul N., & Villinger, H. (2015). Validation of impact penetrometer data by cone penetration testing and shallow seismic data within the regional geology of the southern north sea. Geo-Marine Letters. 35(3): 203–219.
Storteboom O., Woollard M. & Allen B. (2022a) Increased safety and productivity for seabed Cone Penetration Test using the SingleTwist™ with ROSON. Proc. 5th Int. Symposium on Cone Penetration Testing (CPT'22). Bologna. (this conference).
Storteboom O., Woollard M. & Ooms B. (2022b) Sustainability in CPT practice: hybrid CPT track-truck. Proc. 5th Int. Symposium on Cone Penetration Testing (CPT'22). Bologna. (this conference).

Suzuki Y., & Lehane B.M. (2015) Cone penetration at variable rates in kaolin–sand mixtures. International Journal of Physical Modelling in Geotechnics 15(4): 209–219.

Tao J., Huang S., Tang Y. (2020) SBOR: A minimalistic soft self-burrowing-out robot inspired by razor clams. Bioinspiration and Biomimetics. 15, 055003.

Tarantino A., Capotosto, Bottaro F., Bellio M. & Gallipoli D. (2022) Tensiocone: a cone penetrometer with the facility to measure negative pore-water pressure. Proc. 5th Int. Symposium on Cone Penetration Testing (CPT'22). Bologna. (this conference).

Vardon P.J., Baltoukas D. & Peuchen, J. (2018) Thermal Cone Penetration Test (T-CPT). In Proc. 4th International Symposium on Cone Penetration Testing. 649–655.

Vardon P.J., Baltoukas D. & Peuchen J. (2019) Interpreting and validating the thermal cone penetration test (T-CPT). Géotechnique, 69(7): 580–592.

Vinco G. & Sachetto M. (2022) New portable pressiocone system for carrying out CPT & FDP tests. Proc. 5th Int. Symposium on Cone Penetration Testing (CPT'22). Bologna. (this conference).

Watson, P.G. & Randolph, M.F. (1998). Torsional vane and plate load tests. Proc. International Conference Centrifuge '98, Rotterdam, Netherlands. 1: 167–173.

White D.J. & Gaudin C. (2008) Simulation of seabed pipe-soil interaction using geotechnical centrifuge modelling. Proc. 1st Asia-Pacific Deep Offshore Technology Conference, Perth, Dec 2008. 28pp.

White D.J., Stanier S.A., Schneider M., O'Loughlin C.D., Chow S.H., Randolph M.F., Draper S.D., Mohr H., Morton J.P., Peuchen, J., Chow F.C., Fearon R. & Roux A. (2017) Remote intelligent geotechnical seabed surveys – technology emerging from the RIGSS JIP. Proc. Int. Conf. Offshore Site Investigation & Geotechnics. SUT, London. 1214–1222

White D.J., O'Loughlin C.D., Chow S.H. & Stark N. (2018) Interpretation of free fall penetrometer tests in sands: An approach to determining the equivalent static resistance. Proc. 4th Int. Symp. on Cone Penetration Testing. 695–701.

White D.J., Stanier S.A. & Mohr H. (2022) Characterisation of near-surface sediments using a blend of vertical and shallow rotational penetrometers. Proc. 5th Int. Symposium on Cone Penetration Testing (CPT'22). Bologna. (this conference).

Whittle R., & Liu L. (2013). A method for describing the stress and strain dependency of stiffness in sand. Proc. 18th Int. Conf. on Soil Mechanics and Geotechnical Eng. Paris.

Whittle R., Palix E., & Donaghy D. (2017). The influence of insertion process on determining the stiffness characteristics of chalk, using pre-bored, self-bored and pushed pressuremeters. Proc. 8th Int. Conf. on Offshore Site Investigation & Geotech. London Soc. for Underwater Tech. 308–315.

Winter A.G., Deits R.L.H., Dorsch D.S., Slocum A.H. & Hosoi A.E. (2014) Razor clam to roboclam: Burrowing drag reduction mechanisms and their robotic adaptation. Bioinspiration and Biomimetics. 9, 036009.

Withers N.J., Howie J., Hughes J.M.O. & Robertson P.K. (1989) Performance and analysis of cone pressuremeter tests in sands. Géotechnique 39(3): 433–454

Wippermann T., Hudson T.L., Spohn T., Witte L., Scharringhausen M., Tsakyridis G., Fittock M., Krömer O., Hense S., Grott M., Knollenberg J. & Lichtenheldt R. (2020) Penetration and performance testing of the HP3 Mole for the InSight Mars mission. Planetary Space Science. 181, 104870

Wong J.Y. (2010) Theory of ground vehicles. 4th Edition. Wiley, New York.

Yan Y., White D.J. & Randolph M.F. (2010) Investigations into novel shallow penetrometers for fine-grained soils. Proc. 2nd Int. Symp. on Frontiers in Offshore Geotechnics. Perth. 321–326

Yan Y. (2013) Novel methods for characterising pipe-soil interaction forces in situ in deep water. PhD thesis, University of Western Australia

Yetginer T.I., Bodtker S., Rose M., Lunne T., Meyer V. & Duffy C. (2022) Development of an enhanced CPT system for Dogger Bank. Proc. 5th Int. Symposium on Cone Penetration Testing (CPT'22). Bologna. (this conference).

Zhou Z., O'Loughlin C.D. & White D.J. (2020) The changing strength of carbonate silt: Parallel penetrometer and foundation tests with cyclic loading and reconsolidation periods. Canadian Geotechnical Journal 57 (11):1664–1683.

Cone Penetration Testing 2022 – Gottardi & Tonni (eds)
© 2022 the Author(s), ISBN 978-1-032-35624-2

Non-deterministic interpretation and applications of CPT testing data

M. Uzielli
Department of Civil and Environmental Engineering, University of Florence, Italy
Georisk Engineering S.r.l., Florence, Italy

ABSTRACT: The evolution of geotechnical design codes towards reliability-based and probabilistic concepts, the growing global awareness towards risk management for engineered and natural systems, and the global digital transition are steering geotechnical research and practice towards non-deterministic and data-centric approaches to geotechnical characterization and design. The quasi-continuity and high repeatability of CPT measurements inherently endow this testing method with a central role in the momentous shift in paradigm. This paper contributes principles and selected examples which exemplify the nature and use of non-deterministic CPT-based methods for geotechnical site characterization, design, and geohazards risk management. Current cautions and limitations to the development, dissemination, and implementation of these methods are arguably surmountable and provide stimulating opportunities for a collective effort on the part of the CPT community to meet the evolving requirements and trends in geotechnical practice, research, and education.

...ἔοικα γοῦν τούτου γε σμικρῷ τινι αὐτῷ τούτῳ σοφώτερος εἶναι, ὅτι ἃ μὴ οἶδα οὐδὲ οἴομαι εἰδέναι.

"... I seem, then, in just this little thing to be wiser than this man at any rate, that what I do not know I do not think I know either."

- Plato, Apology

1 LEVERAGING THE UNAVOIDABLE

"As far as laws of mathematics refer to reality, they are not certain; and as far as they are certain, they do not refer to reality."

- Albert Einstein, "Geometry and experience" (1921)

The variability in testing data and derived design parameters is a distinctive trait of the geotechnical discipline with respect to other engineering disciplines which deal with artificial materials. Measured properties effectively vary due to the natural heterogeneity and complexity of soils, which are generated and continuously modified by natural geologic and geomorphologic processes. In-situ effects due to stress state and stress history lead to spatially variable measurements even for compositionally homogeneous deposits. The spatial and temporal variability of geotechnical properties can be especially appreciated in the results of geophysical testing and geotechnical in-situ testing, which involve larger volumes of soils than laboratory tests. Moreover, the same test conducted at the same spatial location may produce different results due to the temporal variability of the natural environment (e.g., seasonal fluctuations in water content, groundwater level, etc.) and to imperfect measurement capability. Some degree of inaccuracy and imprecision is always present in testing operations.

The variability in measured values thus stems from inherent soil variability and measurement uncertainty. When transposed in quantitative engineering analyses, these variabilities result in uncertainty. When representative values are calculated from measured data (e.g., the mean value over a depth interval), additional statistical estimation uncertainty arises from the limited number of values used to calculate these values. Moreover, when testing data are used to calculated engineering design parameters, a further source of uncertainty is introduced due to transformation models which are always approximate and imperfect models of the physical world.

Given the evident existence of considerable uncertainties in geotechnical engineering, the question arises as to "whether" and "how" to address them. The complete neglection of uncertainty is unacceptable from both the ethical and technical viewpoints. Failing to acknowledge the existence and relevance of uncertainty would imply that basic geotechnical notions and geotechnical code prescriptions are not adhered to, since these do address uncertainty (either implicitly or explicitly, as discussed later). This paper focuses on how the geotechnical discipline has evolved from the primordial "deterministic" paradigm, in which the modeling of uncertainty is implicit, to the currently trending "non-deterministic" paradigm, in which uncertainties are modelled, processed, and reported explicitly, and is continuing to evolve towards a "data-centric" mode which fosters the synergy between engineering judgment and the increased availability of data, facilitating the treatment of uncertainty in characterization and design.

Through a specific focus on cone penetration (CPT) testing, this paper attempts to provide examples from research and case-study applications of the pivotal role of CPT testing in accompanying the geotechnical discipline towards an emerging non-deterministic and data-centric paradigm. Far from aiming to provide a structured theoretical treatment of non-deterministic and data-centric approaches and from contributing an exhaustive review of available examples, this contribution wishes to spark both an interest and the willingness, on the part of researchers and practitioners, to contribute proactively to the continued development, refinement, promotion, and dissemination of uncertainty-based and data-centric methods relying on CPT. Lastly, it aims to stimulate a constructive discussion focusing on how some current aspects requiring attention actually represent opportunities for the further refinement of methods for interpretation and use of CPT results in characterization and design.

While this paper advocates the increased use of non-deterministic, data-centric methods, it wishes to convey the awareness that such methods largely rest on the broad shoulders of the ingenious, pioneering deterministic methods which have laid the foundations of the geotechnical discipline, and which have nurtured its growth and evolution.

2 A HISTORY OF CHANGE

"The quest for certainty blocks the search for meaning. Uncertainty is the very condition to impel man to unfold his powers."

- Erich Fromm

The concepts of "determinism" and "non-determinism" stem from the fundamental question regarding the cause behind the occurrence of events, which has engaged philosophers in the East and West since ancient times. The deterministic approach prevailed over the stochastic approach in the technical community for many centuries. In the past decades, led by the high-energy physics community and facilitated by advancements in statistical-probability theory, enhanced data analysis capabilities, and increased computational power, the non-deterministic approach has replaced the deterministic one in many scientific and technical disciplines, and is progressively gaining momentum in others, including geotechnical engineering. Moreover, improved data collection technologies and computational power increasingly advocate the adoption of approaches making use of big data, which can further boost the performance of non-deterministic approaches. While a brief account of the transition from the deterministic to the non-deterministic and data-centric paradigm is provided below, readers are referred to Baecher & Christian (2003), Phoon (2020) and Tang & Phoon (2021) for further insights into the philosophical, conceptual, and operational evolution of the geotechnical discipline.

2.1 *The past: The illusion of certainty*

Geotechnical engineering, as all engineering disciplines, originated and developed speaking the deterministic language, in which uncertainties are neither modelled, processed, nor reported explicitly, but are lumped together and represented implicitly in design methods, for instance through factors of safety. Numerous research contributions and real-world cases continue to provide new evidence that deterministic approaches are prone to overestimating or underestimating the performance of geotechnical systems because they are unable to parameterize the "real" complexity in soil-structure systems and the inevitable approximation in the models used to obtain design parameters from testing data. For instance, neglecting to model the spatial variability of design parameters prevents the identification of relevant failure modes (e.g., non-symmetric failure modes foreseeable even in presence of symmetric foundations and axial loading conditions, see Figure 1).

Overconservatism and unconservatism are both to be avoided: while the first entails the excessive use of resources to attain a target level of performance, the second, more seriously, fails to achieve performance and can result in failures which bring human and economic losses.

2.2 *The present: Embracing uncertainty*

In the context of engineering disciplines, the adjective "non-deterministic" refers to a broad range of methods and techniques relying most frequently – but not exclusively – on statistical and probability theory.

The development of ingenious approaches based on concepts such as reliability-based and performance-based analysis and design, along with increased computational power and efficiency, have allowed civil engineering to embark on the evolution from the primordial deterministic paradigm towards the non-deterministic one. The latter is characterized by the explicit consideration of uncertainty. The adoption of the non-deterministic approach allows to exploit powerful mathematical tools such as statistics and probability to model, process, and report uncertainties quantitatively.

While all engineering disciplines originated within the deterministic framework, geotechnical engineering has been delayed in initiating its transition, probably due to the greater inherent difficulty in quantifying and modelling its uncertainties. The geologic environment is made up of two- or even three-phase natural materials and is thus inherently

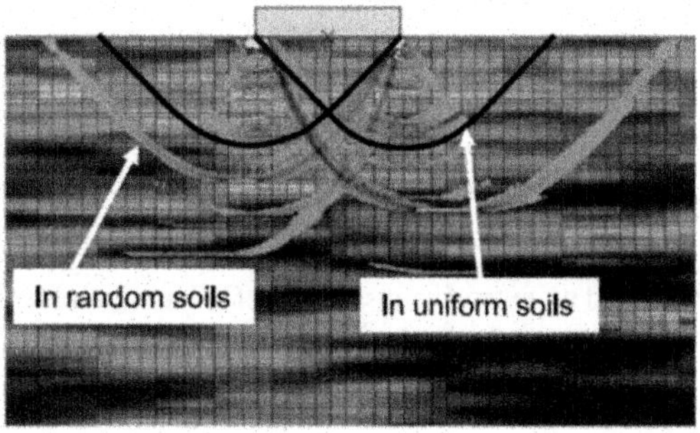

Figure 1. Comparison of failure modes for a surficial footing on uniform and spatially variable soils (from Li et al. 2014).

more complex and less easily characterized and modelled than structural materials such as steel, concrete, or timber. Nonetheless, in the past decades seminal research contributions and increased computational power have fostered and accelerated the development of non-deterministic methods.

The progressive shift towards non-deterministic methods is perhaps the most distinctive trait behind the development of evolutionary geotechnical design codes which are based on statistical, probabilistic, and performance-based concepts and which are gradually replacing the traditional deterministic design codes. Notable examples are Eurocode 7, Eurocode 8, the AASHTO (American Association of State Highway and Transportation Officials) LRFD Bridge Design Specifications, CHDBC (Canadian Highway Bridges Design Code), and JSHB (Japanese Specifications for Highway Bridges). The adoption of the non-deterministic paradigm enables the conceptual and formal harmonization between geotechnical and structural design, thus contributing to enhanced homogeneity and effectiveness in the design of soil-structure systems. Embracing non-determinism for geotechnical practice is no longer a choice in many parts of the world and is becoming less and less so at a global level. Despite the above, the transition to the non-deterministic paradigm is far from complete.

2.3 *The future: Data meets knowledge*

The momentous global process of digital transition which characterizes our time brings unprecedented possibilities. A new "data-centric" paradigm is emerging involving the synergy between the increased availability of data, evolutionary approaches such as machine learning, and expert engineering judgment (e.g., Phoon 2020). Such paradigm is extremely well-suited to synergize with non-deterministic approaches in defining the new identity of the geotechnical discipline.

Comprehensive compendia of literature estimates of uncertainties in soil parameters are available in the geotechnical literature (e.g., Uzielli et al. 2007; ISSMGE 2021). However, most estimates refer to total uncertainty. Prior to the introduction of the data-centric paradigm, soil property statistics determined from total uncertainty analyses could only be applied reliably to the specific set of site conditions, measurement techniques and procedures, and correlation models for which the design soil properties were derived. The site-specificity of variability parameters had thus constituted a long-standing limitation and an obstacle to the diffusion of non-deterministic approaches. A main strength of the data-centric paradigm thus lies in the provision of quantitative probabilistic methods to allow the integration between site-specific data collected for a given project and "big indirect data", i.e., existing data of any type collected from past stages of the same project or past projects at the same site, neighboring sites, or beyond.

The progressive surge of the data-centric paradigm is arguably entwined with the increasing centrality of the Bayesian approach in geotechnical engineering. Bayesian statistical analyses, differently from "frequentist" statistical analyses, provide analysts with the tools to update a "prior" probability model through the availability of new evidence (e.g., data observations) to obtain a "posterior" model. This approach is particularly suited for the geotechnical engineering discipline, in which it is often necessary to supplement site- and case-specific data (or aprioristic engineering judgment) with data from other sites, or to update characterization and design values as new data become available. The use of Bayesian methods in the data-centric approach promises to be useful in circumventing the limitations of site-specificity of geotechnical variability estimates, and in actually leveraging the availability of "indirect" data from other sites (see, e.g., the outcomes of Project DeepGeo at http://yo-1.ct.ntust.edu.tw/jge/files/articlefiles/v16i2202106031836319463.pdf) The data-centric paradigm arguably provides the natural evolution of the geotechnical discipline, provided that researchers and practitioners are provided with the necessary simple knowledge required to correctly interact with data (Tang & Phoon 2021).

"Data-centric" and "non-deterministic" are not synonymous. While the data-centric approach does not imply *per se* resorting to non-deterministic methods, the former clearly fosters the latter. Increased data numerosity increases the applicability, meaningfulness, and reliability of non-deterministic techniques. Many statistical and probabilistic analyses, for instance, rely on the availability of sufficient quantities of data, and the level of confidence of their outputs increases with increasing numerosity of data samples. As the quantity and quality of data are expected to continue to increase over time, failing to proactively confer further centrality to the data-centric approach in both geotechnical research and practice would be nothing short of a wasted huge opportunity.

2.4 *In medio stat virtus*

In their illuminating treatment of geotechnical uncertainty, Baecher & Christian (2003) opined that a hybrid approach involving both deterministic and non-deterministic methods is "proper to the geotechnical engineering discipline". This statement is increasingly and fully supported by a conspicuous, growing corpus of research and real-world applications, in which some aspects of a specific geotechnical analysis can be sufficiently well-described deterministically for practical purposes, while for others it is *convenient* to describe and model phenomena and variables as if they behaved, at least to some degree, in a random manner. A typical example is found in the quantitative modeling of spatial variability of soil properties, which is often pursued through the decomposition of data into a deterministic trend and a fluctuating, supposedly random component. The hybrid approach is fully compatible with the data-centric paradigm since the availability of more and better data is beneficial to both deterministic and non-deterministic analyses.

3 THE COMPETITIVE ADVANTAGES OF CPT TESTING

CPT testing has long been known to surpass other in-situ testing methods with respect to the high numerosity, quasi-continuity, and accuracy of its measurements. Moreover, CPT testing is highly standardized. Such features endow this testing method with several relevant "competitive advantages" in the light of the emerging data-centric, non-deterministic approach to geotechnical site characterization, geotechnical design, and geohazards risk analysis.

The small measurement interval which is unique to CPT testing among in-situ testing methods results in significantly higher data numerosity, i.e., in a higher number of measurements obtained for any given depth interval. Sample numerosity plays a decisive role in the applicability and significance of non-deterministic analyses. In most general terms, higher data numerosity is beneficial at least in terms of: (a) reduced statistical uncertainty; (b) improved modelling of inherent spatial variability and estimation of spatial averaging effects; and (c) improved estimation of transformation uncertainty. Ultimately, increased data numerosity contributes to the optimization of cost-performance optimization in geotechnical characterization and design by reducing the likelihood of unconservatism and the degree of overconservatism. Data-centric approaches relying on machine learning also enhance their performance (when not outright requiring) large data sets. Phoon et al. (2021) opined that CPT data is the only practical source of data for data-driven site characterization (DDSC).

The concepts of testing repeatability and accuracy is related to measurement uncertainty. The latter is due to equipment, procedural-operator, and random testing effects. Previous research (e.g., Kulhawy & Trautmann 1996) indicates that CPT testing repeatability is higher in comparison to that pertaining to other testing methods. Peuchen & Terwindt (2014) provided an extensive insight into CPT measurement uncertainty and affirmed that piezocone testing is closest to an ideal penetration tool than any other in-situ testing method. In principle, high accuracy is desirable in non-deterministic analyses because low epistemic measurement uncertainty contributes to a lower total uncertainty. High standardization allows more stringent control on testing data quality. Peuchen & Parasie (2019) highlighted the ongoing technical and normative activities aimed at the continued pursual of high CPT testing standardization.

The innumerable possible ramifications stemming from these virtuous attributes of CPT testing do not allow their exhaustive enumeration. Selected examples are provided in the following among the many ingenious available in the geotechnical literature.

4 NON-DETERMINISTIC CPT-BASED SITE CHARACTERIZATION

"Exploring the unknown requires tolerating uncertainty."

- Brian Greene

Non-deterministic approaches have been proposed for fundamental activities related to CPT-based geotechnical site characterization including soil behavior classification, stratigraphic profiling, and spatial variability modeling. Selected examples are presented in the following, and readers are referred to Cao et al. (2017) for a more comprehensive treatment of such topic.

4.1 *Soil behavior classification*

CPT testing is frequently used for soil behavior classification through approaches which link the quantitative measurements of the mechanical resistance to cone penetration to likely corresponding soil types. The association between mechanical resistance and soil type is pervaded by significant uncertainties stemming from in-situ state effects (e.g., cementation, overconsolidation), by the inter-site and intra-site variability in compositional and mechanical properties of soils, and by the complexity of soil behavior.

CPT-based soil behavior classification methods routinely used in geotechnical practice are overwhelmingly deterministic. While their use worldwide and in many different soil and site conditions attests to their high performance, these methods are inherently unable to quantify explicitly the degree of uncertainty in their outputs. Methods for the non-deterministic CPT-based soil classification have nonetheless been made available in the geotechnical literature since many years. Zhang & Tumay (2003) proposed the probabilistic region estimation method. Such approach provides a depth-wise profile of the probability that a set of CPT measurements (more specifically, cone resistance and friction ratio) correspond to cohesive-, intermediate-, or cohesionless-behavior soil type (see Figure 2).

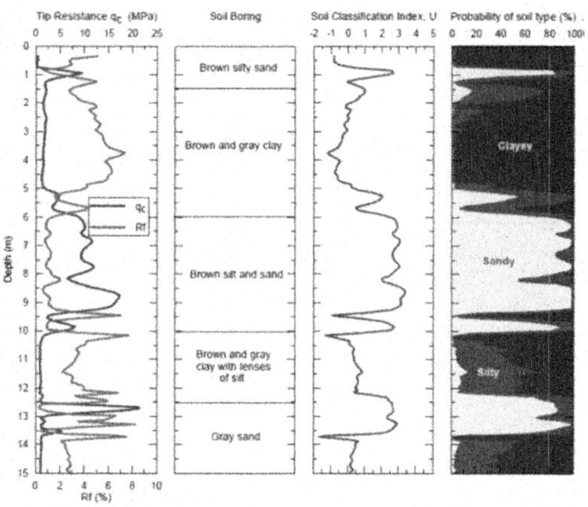

Figure 2. Non-deterministic depth-wise soil behavior classification (Zhang & Tumay 2003).

4.2 *Stratigraphic profiling*

Stratigraphic profiling is one of the main applications of CPT testing. Such process involves the identification of interfaces between soil units which can be considered sufficiently homogeneous and/or significant for specific purposes. Reliable stratigraphic profiling is also relevant to geotechnical design. For example, in CPT-based pile design, the identification of a stratigraphic interface between a layer with low bearing capacity and one with high bearing capacity and stiffness is fundamental in the definition of design pile length. Cone resistance in layered soil is affected not only by the layer currently being traversed but also by the underlying and, under certain conditions, overlying layers, depending on the position of the cone relative to such layers.

Several non-deterministic approaches for CPT-based stratigraphic profiling have been proposed. Wang et al. (2019), for example, developed a Bayesian unsupervised learning approach for automatic layer detection and soil classification with an explicit quantification of uncertainty, using CPT-based parameters (i.e., normalized sleeve friction F_R and normalized cone resistance Q_t) commonly used in the well-established soil behavior classification system by Robertson (1990). Figure 3 provides a visual representation of the approach, highlighting the capability of the method to simultaneously provide both soil behavior type and stratigraphic profiling from CPT testing data.

4.3 *Spatial variability modeling*

In-situ testing typically reveals that engineering properties of geomaterials vary spatially both in the horizontal and vertical directions (most often with higher variability in the latter case due to layering- and stress-related effects). In absence of geological, geomorphological, or geotechnical macro-discontinuities (e.g., layer interfaces) which should be accounted for by separating data by homogeneous soil units through stratigraphic profiling and soil behavior classification, soil properties (as parameterized, for instance, in CPT data) can be expected to vary gradually in conformity with the conceptual principle stated by Leibniz (1704) by which "*tout va par degrés dans la nature, et rien par saut*" ("everything occurs gradually in nature, and nothing by jumps"). Spatial variability can never be estimated with full precision and accuracy at any scale, from the micro-scale to the macro-scale, due to the finite number of tests and measurement uncertainty.

The importance of the quantitative modeling of spatial variability is increasingly acknowledged in research and practice as: (1) it is necessary for the conduction of simulation-based analyses based on techniques such as random field modeling; (2) it allows the spatial interpolation of in-situ testing measurements and of derived parameters and the quantification of uncertainties in interpolation estimates at unsounded locations, thereby providing support in the planning and integration of geotechnical investigation campaigns and optimizing their cost-performance ratio; (3) allows the quantitative estimation of the spatial averaging effect, by which the effects of variability on geotechnical design can be rationally calibrated with respect to the spatial extension of geotechnical structures (typically, pile foundations).

Multiple approaches and techniques have been used to the purpose of geotechnical spatial variability modeling. In many of these techniques, variability is often modeled most efficiently through the integration between deterministic and non-deterministic approaches, i.e., through the decomposition of data into a trend which is expressed deterministically (for instance through a polynomial function) and a fluctuating component, which is assumed to be a random variable, and which is investigated using statistical, probabilistic, geostatistical, and random field theory among others.

Regardless of the specific technique used, CPT is particularly suited for the modeling of spatial variability because of its small measurement interval, which allows, jointly with its high testing repeatability, more reliable estimation of small-scale variability in the vertical direction with respect to other in-situ testing methods. Moreover, higher data numerosity reduces statistical uncertainty in non-deterministic analyses such as the modeling of semivariograms for the modeling of spatial correlation.

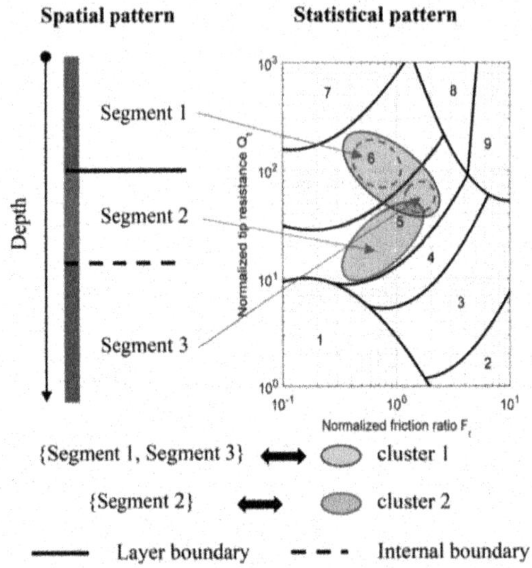

Figure 3. Statistical clustering for non-deterministic stratigraphic profiling (from Wang et al. 2021).

5 NON-DETERMINISTIC CPT-BASED GEOTECHNICAL DESIGN

Non-deterministic approaches are available for the utilization of CPT data in geotechnical design, both in the derivation of design parameters and in design methods. As shown in the following selected examples, such approaches allow improved and more rational calibration and assessment of the performance of geotechnical design. Readers are referred to Tang & Phoon (2021) for a more comprehensive insight into geotechnical transformation uncertainty.

Engineering models which relate in-situ and laboratory testing measurements to geotechnical design parameters, whether theoretical, empirical, experimental, or a combination thereof, are always approximate simplifications of physical soil-structure systems. Geotechnical models are thus invariably affected by bias (which relates to the model's precision) and/or dispersion (which refers to the model's accuracy). Empirical correlations have been made available to estimate numerous geotechnical design parameters from CPT testing results. Lunne et al. (1997) tabulated comparatively the degree of applicability of a wide range of in-situ testing methods to the estimation of geotechnical parameters, assessing that CPT testing overall outperformed other methods. Robertson & Cabal (2015) tabulated the perceived applicability of piezocone testing to the estimation of soil parameters for coarse-grained and fine-grained soils (Table 1), once again assessing the overall utility of CPT for geotechnical parameter estimation.

Table 1. Perceived applicability of CPTU for deriving soil parameters (from Robertson & Cabal 2015).

Parameter	Coarse	Fine
Relative density	2-3	-
State parameter	2-3	-
Young's and shear moduli	2-3	2-4
Overconsolidation ratio	5	1
Undrained strength	-	1-2
Coefficient of consolidation	3-4	2-3
Peak friction angle	2-3	4
In-situ stress ratio	5	2
Small-strain shear modulus	2-3	2-4
1-D Compressibility	2-3	2-3
Sensitivity	-	2
Permeability	3-4	2-3

Note: 1=high; 2=high to moderate; 3=moderate; 4=moderate to low; 5=low.

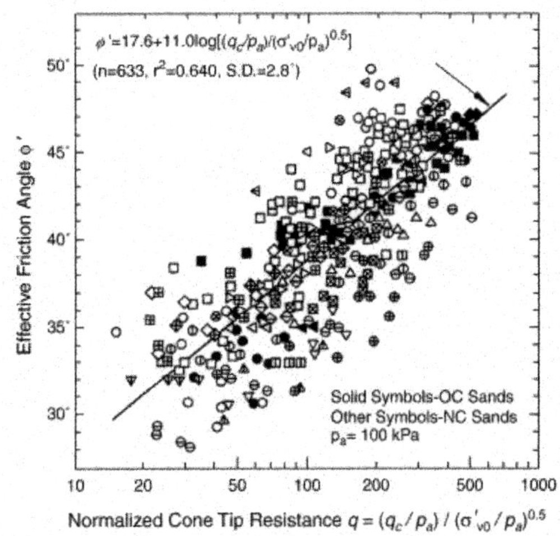

Figure 4. Source data and deterministic correlation for estimating friction angle from CPT (Kulhawy & Mayne 1990).

In the light of the non-deterministic insight into CPT testing, the "degree of perceived applicability" could be related conceptually to the reliability of the transformation models, i.e., to the magnitude of transformation uncertainty. However, this qualitative assessment can hardly be translated into quantitative probabilistic formulations which can be seamlessly used in new geotechnical design codes. Available CPT-based correlations are overwhelmingly deterministic. They are typically obtained as "best-fit", central tendency models from scattered data, and are thus amenable to an "intermediate" degree of conservatism.

An example of the transition to non-deterministic CPT-based estimation of geotechnical parameters is related to the estimation of effective friction angle of sands from stress-normalized cone resistance. Figure 5 shows the data used by Kulhawy & Mayne (1990) to develop their well-known deterministic model

$$\phi' = 11.0 \cdot \log_{10}(q_{t1}) + 17.6 \quad (1)$$

While goodness-of-fit statistics (typically, the coefficient of determination R^2) or descriptive second-moment dispersion statistics (e.g., standard deviation, variance) are provided together with the regression model in best-practice cases such as this one, these are in principle not sufficient to allow the model-based estimation of properties for different levels of conservatism. Moreover, the statistical distribution of data around the best-fit model cannot be characterized statistically beyond second-moment description.

Uzielli & Mayne (2019) proposed a probabilistic transformation model using a high-quality database comprising CPT and triaxial tests. The logarithmic formulation

$$\phi' = p_1 \cdot \ln(q_{t1}) + p_2 \quad (2)$$

has the same ease of practical applicability as the previous deterministic model while allowing the estimation of ϕ' for varying levels of conservatism (i.e., for different probabilities of exceedance P_{exc}) by varying model coefficients p_1 and p_2 as shown in Table 2. Figure 5 plots the curves corresponding to probabilities of exceedance 0.05, 0.50, 0.75, and 0.95 of effective friction angle in comparison with the deterministic model by Kulhawy & Mayne (1990). While the performance of the deterministic methods is overall comparable to the probabilistic method for a "median" probability of exceedance, it is apparent that the latter allows the fuller exploitation of the experimental dataset. It is also apparent that the deterministic model does not refer uniformly to a probability of exceedance for increasing values of q_{t1}. Thus, the degree of conservatism in the deterministic model is not established univocally.

The transition to a data-centric paradigm can further enhance the quality and utility of new-generation geotechnical transformation models. Site-specific data are typically sparse and incomplete. The geotechnical discipline has adapted to this fact by providing transformation models which are

Table 2. Coefficients of the logarithmic CPT-based model for the estimation of friction angle for various probabilities of exceedance (Uzielli & Mayne 2019).

P_{exc}	0.95	0.75	0.50	0.05
p_1	4.27	4.00	3.81	1.70
p_2	18.1	20.3	22.2	35.1

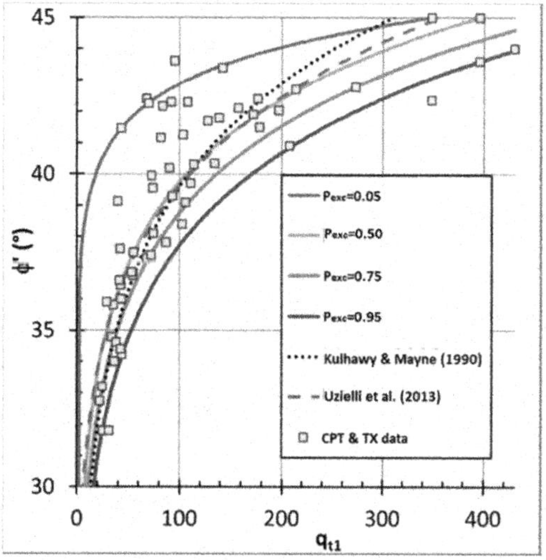

Figure 5. Comparison of probabilistic (Uzielli & Mayne 2019) and deterministic (Kulhawy & Mayne 1990) models for the estimation of effective friction angle from stress-normalized cone resistance.

founded on data from multiple sites. While geotechnical engineers are trained to operate the transition between "indirect" data (i.e., from other sites) to "direct" (i.e., local) data by applying transformation models and assessing the results critically using engineering judgment, non-deterministic approaches based on the data-centric paradigm allow an improved synergy between indirect and direct data. Through Bayesian approaches relying on the data-centric paradigm, it is possible to quantify the reliability of indirect data with respect to direct data and to provide a "hybrid" transformation model which relies on both datasets, and in which the weight of the "indirect" model is determined on the basis of objective criteria such as the "similarity" between indirect sites and the target site, e.g., in terms of soil types. For example, Ching et al. (2021) used a hierarchical Bayesian modelling framework to apply hybridization to the estimation of undrained strength of clays from CPT testing. Figure 6 illustrates indirect site-specific models as well as the hybrid model for the target site-specific model. Ellipses represent the uncertainties associated with the respective models, which are quantified explicitly.

An increasing set of non-deterministic CPT-based design methods are also available in the geotechnical literature. Cai et al. (2021) provided a framework for the CPT-based calculation of axial pile capacity in spatially variable soil. The study relied on the simulation of CPT profiles through random field modeling and on the statistical-based calculation of pile capacity for varying target levels of conservatism. Figure 7 provides a comparative visual representation of simulated profiles of cone resistance at preset spatial separation distances within the design site for smaller and larger horizontal correlation distances. Random field modeling also allows to refer explicitly to the scale of fluctuation, thus allowing the quantification of the spatial averaging effect by which the influence of the variability in soil properties is smoothed out along the shaft. The rigorous inclusion of the spatial averaging effect contributes to calibrating the effect of inherent soil variability and to reconduct rationally the degree of conservatism in design to its target level. CPT testing data are particularly suited for simulation purposes because of the higher reliability in the estimation of vertical spatial variability parameters stemming from the quasi-continuity of measurements in comparison to other in-situ testing metods.

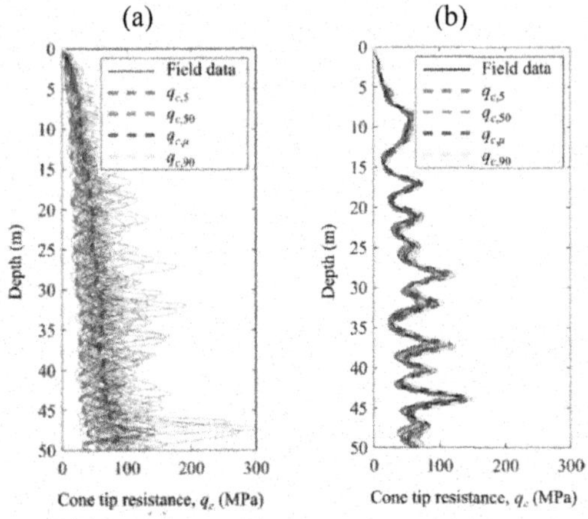

Figure 7. Simulation of horizontal site-scale variability of cone tip resistance for: (a) lower; and (b) higher horizontal correlation distance (adapted from Cai et al. 2021).

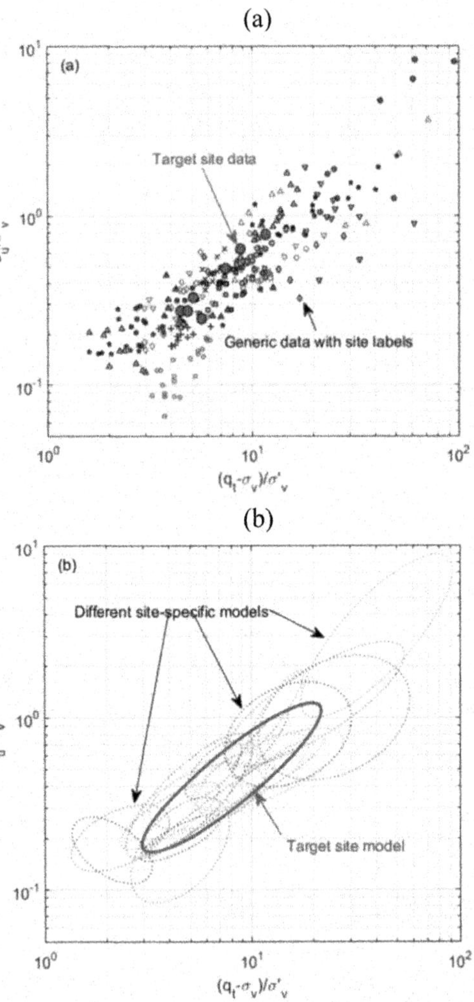

Figure 6. Hybridization of direct and indirect data-driven models for undrained strength from CPT: (a) target site and generic data; (b) resulting site-specific and target site models (Ching et al. 2021).

6 NON-DETERMINISTIC CPT-BASED GEORISK ANALYSIS

In-situ testing results are frequently used in the increasingly relevant contributions of geotechnical engineering to geohazards risk analysis. Methods have been made available for various hazards and related phenomena including the prediction of triggering of seismic liquefaction, seismically induced soil densification, lateral spreading, and slope stability, among others. The prediction of seismic liquefaction triggering has arguably received most attention. Numerous simplified liquefaction evaluation approaches based on the results of in-situ tests (SPT, CPT, DMT, seismic testing) have been made available over the past decades. Among these, CPT-based methods have been shown to provide significant advantages over others due to: (a) the capability to detect thin layers; (b) the repeatability of measurements; (c) the relative speed and economy.

As in all ramifications of the geotechnical discipline, deterministic methods for liquefaction assessment have preceded non-deterministic methods chronologically and are still today predominant even though literature findings increasingly indicate that non-deterministic methods provide multiple significant advantages. First, they allow users to define liquefaction boundary curves for target levels of conservatism, thereby contributing to cost-performance optimization with respect to the vulnerability and value of elements at risk in a specific area. Second, they explicitly include uncertainties in the estimation of the parameters and models used to estimate seismic demand and soil resistance parameters. Third, they allow a more realistic and physics-based modelling and treatment of geotechnical phenomena. The latter point can be well exemplified by the following.

A vast bulk of research and experimental evidence indicates that seismic liquefaction typically occurs in loose or medium-density, saturated cohesionless soils, while high-plasticity, fine grained soils are inherently "non-liquefiable" due to the relevance of the cohesive component of shear strength. A fundamental step in the implementation of simplified CPT-based methods lies in the assessment of inherent soil liquefiability. Such methods are developed from databases pertaining to soils which are expected to be inherently liquefiable based on prior knowledge, and are thus applicable solely to such soils, i.e., the calculation of the cyclic resistance ratio CRR, which parameterizes the cyclic resistance of soils to liquefaction, is pursued solely at depths corresponding to soils preliminarily assessed to be liquefiable. The applicability of simplified methods must be assessed depth-wise, as stratigraphic profiles can be very heterogeneous in terms of soil types and can include both liquefiable and non-liquefiable soil.

All simplified CPT-based liquefaction evaluation methods include quantitative approaches to the depth-wise assessment of the liquefiability or non-liquefiability of soils. Such assessment typically relies on the definition of deterministic threshold values of the soil behavior classification index I_c (typically varying between 2.4 to 2.6 among methods) which provide a strict boundary between liquefiable and non-liquefiable layers. The assessment resulting from the adoption of deterministic thresholds is invariably binary, i.e., a set of CPT measurements corresponds to either a "fully liquefiable" or "fully non-liquefiable" measured soil volume interval. The hypothesis of an abrupt discontinuity between non-liquefiability and full liquefiability is hardly compatible with well-established geotechnical knowledge and evidence by which liquefiability can be expected to decrease gradually with increasing percentage of fines in a soil. Incidentally, the non-binary effect of varying fines content is already included in simplified methods through the calculation of fine-sand equivalent stress-normalized cone resistance (usually denoted by $q_{c1N,cs}$), which takes continuous (albeit usually inferiorly and superiorly bounded), values, but does not intervene in the preliminary assessment of liquefiability. Moreover, calculated values of I_c are pervaded by uncertainty stemming from measurement depth offset effects, layering effects, and uncertainty in measurements of q_c and f_s. The progressive variation of the degree of liquefiability should be accounted for quantitatively.

Non-deterministic approaches to susceptibility evaluation can contribute to improving deterministic methods. Maurer et al. (2017) proposed deterministic and probabilistic correlations relating I_c to liquefaction susceptibility as defined by four well-established published criteria. Figure 8 shows the respective liquefaction susceptibility probability curves. Such curves, if duly integrated into existing deterministic methods, could provide a useful means to quantify liquefiability as a continuous (as opposed to binary) variable and to partially circumvent the limitations which are inherent to deterministic approaches.

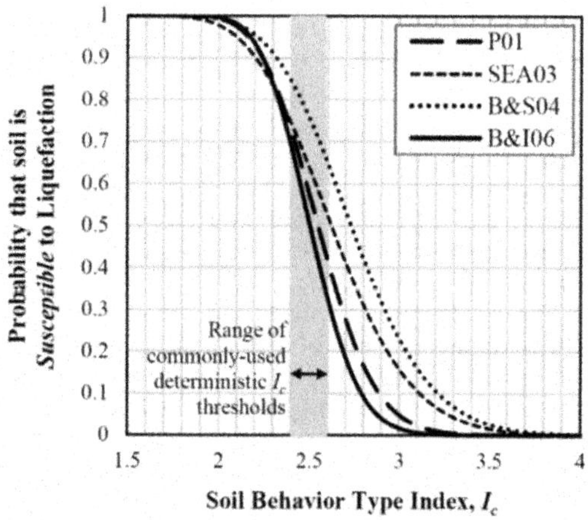

Figure 8. Probability of liquefaction susceptibility as a function of the soil behavior type index (from Maurer et al. 2017).

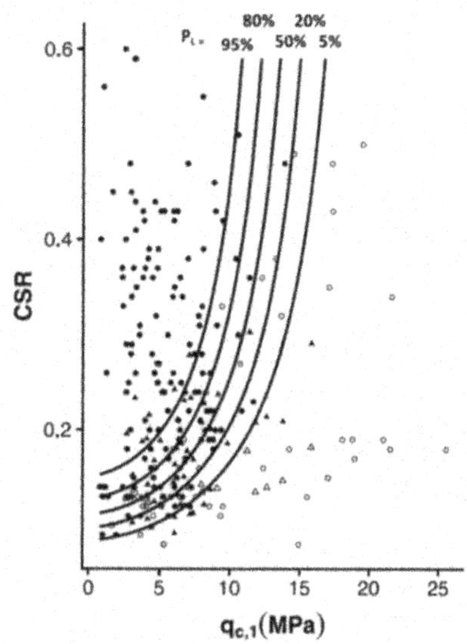

Figure 9. Probabilistic liquefaction boundaries (Schmidt & Moss 2021).

"Fully" non-deterministic CPT-based approaches, mostly relying on Bayesian methods for the prediction of liquefaction triggering, are available in the geotechnical literature and are being increasingly used in practice. In comparison with deterministic approaches which provide a single liquefaction boundary curve, these allow the selection of liquefaction boundaries for varying degrees of conservatism while accounting explicitly for uncertainties. Figure 9 provides an example of probabilistic liquefaction boundaries for varying probability levels as calculated from normalized cone resistance using the approach proposed by Schmidt & Moss (2021). This approach addresses and processes uncertainties in the parameters used in the estimation of cyclic resistance and cyclic stress, as well as uncertainties stemming from intra-event correlation in loading variables and differences in event sample sizes.

Non-deterministic methods can also be used to spatialize the outputs of deterministic simplified CPT-based methods. For example, Guan & Wang (2022) developed a probabilistic method for characterizing the horizontal spatial distribution of the deterministically calculated cyclic resistance ratio, cyclic stress ratio, and factor of safety against liquefaction in data from CPT soundings using Bayesian compressive sampling and Monte Carlo simulation. The method relies on the generation of random field samples and allows the modeling and quantification of spatial variability, interpolation uncertainty, and model uncertainty. Example outputs of the methods include spatial distributions of statistics of the factor of safety resulting from the interpolation of deterministic values obtained at 6 CPT sounding verticals. Figure 10a refers to the mean values of the factor of safety calculated from 10,000 realizations of random fields, while Figure 10b plots the spatial distribution of the coefficient of variation, which provides a quantitative measure of the level of uncertainty in the spatialized values of the factor of safety.

The availability of quantitative measures of uncertainty associated with interpolated values provides a fundamental advantage over deterministic characterization methods, at least because: (1) the parameter is defined probabilistically for direct use in reliability-based methods; and (2) it becomes possible to assess where estimation uncertainty exceeds thresholds of tolerability/acceptability, thus providing direct indications for the conduction of supplementary tests. The visual representation of the spatial variability of the interpolation estimates of the factor of safety allows rational decision-making regarding possible supplementary investigations.

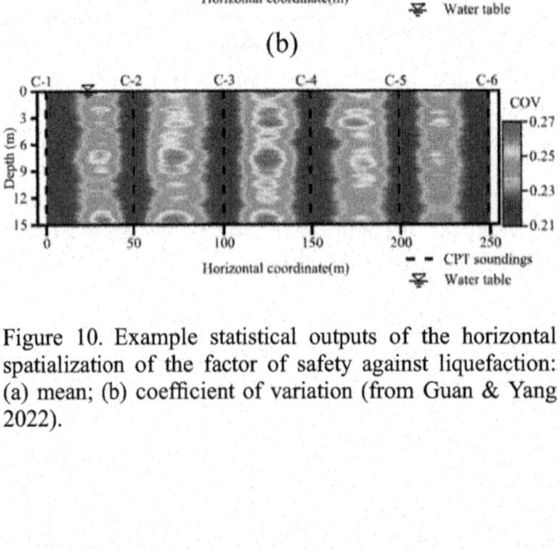

Figure 10. Example statistical outputs of the horizontal spatialization of the factor of safety against liquefaction: (a) mean; (b) coefficient of variation (from Guan & Yang 2022).

7 TURNING CAUTIONS INTO OPPORTUNITIES

The adoption and implementation of methods such as those presented in the previous sections should never occur in absence of engineering judgment and geotechnical knowledge. Cautions should be adopted at all stages of CPT-based analyses to ensure the geotechnical significance and formal correctness of the quantitative procedures relying on uncertainty modeling. This paper argues that current cautions can be successfully addressed through synergies between research and practice.

7.1 *Interpretation in layered soils.*

Previous studies explored the problem of cone penetration in layered soils analytically (e.g., Yu 2006) or experimentally, either in calibration chambers (e.g., Tehrani et al. 2017) or centrifuges (e.g., Mo et al. 2017). Some of these studies define "sensing distance" as the distance from a layer interface at which the cone resistance first starts changing because the cone is approaching it, and "development distance" the distance from a layer interface at which the cone resistance ceases to be affected by the interface as it moves away from it.

A further complication is given by the fact that the three fundamental CPT parameters pertain to different depths and to different volumes of soil. This fact stems from the physical geometry of the cone and by the inherent differences between the volumes of soils and the physical phenomena affecting cone resistance, sleeve friction, and pore pressure. These three measurements, involving different volumes of soil during penetration. Theoretical modeling of cone penetration has shown that such extent may depend on cone size, geometry, and soil type and stiffness. Thus, the volumes of soil involved in cone tip resistance and sleeve friction measurements are variable. The existence of a measurement depth offset between cone resistance, sleeve friction, and pore pressure is well-known. Approaches relying on statistical techniques such as cross-correlation have been proposed and are implemented in widely used commercial CPT interpretation software.

While CPT measurements ultimately reflect the true mechanical behavior of the in-situ soil volume in terms of its resistance to cone penetration, they do not provide fully reliable information if data are not corrected for layering effects, and measurement

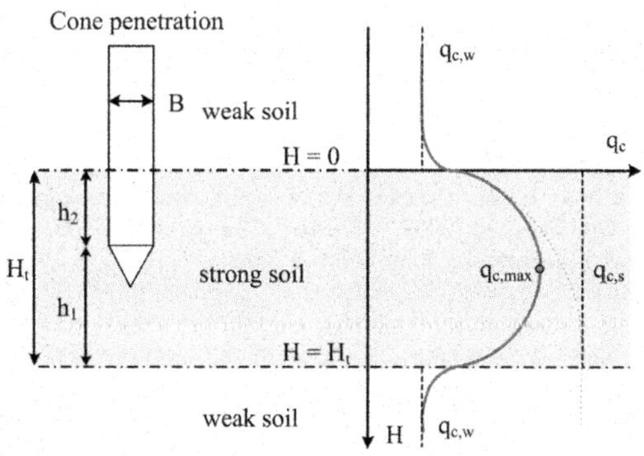

Figure 11. Schematic of cone penetration in multilayered soils: strong soil embedded in weak soils (Mo et al. 2017).

offset prior to being used for stratigraphic profiling, soil classification, and in the calculation of geotechnical design parameters. This is because depth-wise measured values do not correspond to the true values at the same depth. CPT-based approaches relying on the depth-wise processing of measurements should in principle be preceded by a pre-processing of data to account for layering effects and for the inherent depth offset between measurements. If such pre-processing is not conducted, these approaches will be pervaded by additional epistemic uncertainty, however sophisticated they may be. The magnitude of the epistemic uncertainty related to layering effects and depth offset tend to increase with stratigraphic complexity.

Despite the inherent complexity in physical phenomena involving cone penetration in layered soils, studies such as those mentioned above have provided ingenious analytical models for the estimation of sensing distance and development distance and for the correction of raw measurement data. The development of efficient algorithms for the translation of the results of these studies into software applications provides a stimulating challenge and an opportunity for considerable improvement in the significance of CPT data interpretation for routine implementation in non-deterministic methods for geotechnical site characterization and design. Operating in the light of the non-deterministic paradigm provides an ideal scenario as uncertainties in the "true" values of CPT measurements in layered profiles can be expressed explicitly.

7.2 *Improved quantification of measurement uncertainty*

The separation of the components of total uncertainty may prove to be a complex process depending on the quality and quantity of data from a specific case. The relative contribution of total uncertainty is also highly case-specific, and depends on site conditions and inherent spatial variability, and degree of equipment and procedural control in testing data. Geotechnical uncertainty models are typically additive (e.g., Phoon & Kulhawy 1999). The magnitude of individual uncertainty components can thus be estimated in reverse-mode, i.e., by subtracting known uncertainty terms from the total uncertainty which is observed. Consistent approaches to the quantification of measurement uncertainty would result in the availability of more reliable parameterization of measurement uncertainty, leading to more confident subtractive-mode implementation of uncertainty models to estimate aleatory and other epistemic uncertainties from total uncertainty. To this purpose, Peuchen & Terwindt (2014) advocated the development of a shared, less ambiguous glossary (e.g., "repeatability" is strictly not synonymous to "accuracy") and the compilation of a reference measurement uncertainty dataset from real cases (as opposed to "ideal" cases referring to unrealistically controlled conditions).

7.3 *Upgrading of geotechnical correlations*

While new, non-deterministic transformation models yielding design properties from CPT data such as the ones presented in this paper are progressively being made available, the vast majority of geotechnical correlations (including CPT-based correlations) remain deterministic. The full migration of the geotechnical discipline to the non-deterministic paradigm cannot prescind from the upgrading of geotechnical correlations to formats which are compatible with new paradigms. This could be aided by, for instance, through: (a) specific and systematic focus of new research on the development of non-deterministic correlations; and (b) a collective effort on the part geotechnical researchers and practitioners to contribute existing data for the non-deterministic reformulation of existing literature correlations.

8 CONCLUDING REMARKS

This paper has hopefully provided a meaningful insight into the possible benefits of non-deterministic methods for the interpretation of CPT data and its utilization in many and diverse aspects of geotechnical engineering. It has also aimed to highlight the inherent competitive advantage of CPT over other in-situ testing methods in meeting the demands of the new non-deterministic, data-centric paradigm of the geotechnical discipline.

The progressive migration of geotechnical design codes towards non-deterministic formats and the ever-strengthening global digital transition indicates that embracing and implementing such paradigm in practice is no longer a choice for the geotechnical community at large, but rather a beneficial, convenient obligation. Providing students, practitioners, and researchers with sufficient theoretical knowledge and practical tools to understand and implement non-deterministic methods and approaches is a necessity. ISSMGE TC304 "Engineering Practice of Risk Assessment & Management" recently exerted a formidable effort resulting in a state-of-the-art report containing a conspicuous bulk of quantitative estimates of inherent variability parameters as well as statistics for geotechnical design model factors, random field parameters of geotechnical properties, statistics for transformation uncertainties, and characteristic values of geotechnical parameters for use in non-deterministic design methods.

To further exploit the opportunities offered by the global digital revolution and improved measurement technologies, and to meet the requirements of evolutionary design codes, the geotechnical community must persist in the transition towards the data-centric approach, which is fully compatible with the non-deterministic paradigm. To facilitate and stimulate this transition, new geotechnical databases providing high-quality, validated information are made publicly available. An example database including CPT testing outputs is the ISSMGE TC304 database – 304dB at http://140.112.12.21/issmge/tc304.htm. This paper wishes to encourage the CPT community to actively contribute to this database.

The surge of data-driven approaches does not undermine the central role of geotechnical expertise and engineering judgment. Rather, new engineering paradigms and technological evolutions now require that geotechnical engineering judgment also rely on skills allowing the comprehension of uncertainties and the enhanced analysis of the data which geotechnical testing provides. A modern geotechnical engineering education would thus greatly benefit from structured training in statistical and probabilistic methods, specifically including Bayesian approaches, along with data analysis and machine learning methods. Current practitioners could receive these contents through high-quality professional development initiatives. On the academic side, geotechnical researchers focusing on non-deterministic and data-centric methods should continue striving to make their promising findings increasingly accessible through open-source sharing, free software, and on an improved communication skills to allow a more immediate comprehension of new findings on the part of the geotechnical community at large. ISSMGE-related initiatives focusing on these topics could be disseminated through structured protocols with academia and professional bodies.

The translation of non-deterministic mathematical frameworks of varying complexity and sophistication into practically utilizable executables for the interpretation of CPT data may further foster the transition of the geotechnical community towards cost-performance optimization in geotechnical site characterization, design, and in the analysis and management of georisks, in full accordance with current global trends aimed at sustainability, circularity, and community participation.

REFERENCES

Baecher, G.B. & Christian, J.T. 2003. *Reliability and statistics in geotechnical engineering*. New York: John Wiley & Sons.

Box, G. E. P., Luceño, A. & del Carmen Paniagua-Quiñones, M. 2009. *Statistical Control By Monitoring and Adjustment*. John Wiley & Sons.

Cai, Y., Bransby, F., Gaudin, C. & Uzielli, M. 2021. A framework for the design of vertically loaded piles in spatially variable soil. *Computers and Geotechnics* 134. DOI: 10.1016/j.compgeo.2021.104140

Cao, Z., Wang, Y. & Li, D. 2017. *Probabilistic approaches for geotechnical site characterization and slope stability analysis*. Springer. ISBN: 978-3-662-52914-0.

Ching, J., Wu, S. & Phoon, K.K. 2021. Constructing quasi-site-specific multivariate probability distribution using hierarchical Bayesian model. *Journal of Engineering Mechanics*, ASCE. DOI: 10.1061/(ASCE)EM.1943-7889.0001964.

Einstein, A. 1921. "Geometry and experience". Address to the Prussian Academy of Sciences in Berlin, January 27th, 1921.

Guan, Z. & Wang, Y. 2022. CPT-based probabilistic liquefaction assessment considering soil spatial variability, interpolation uncertainty and model uncertainty. *Computers & Geotechnics* 141. DOI: 10.1016/j.compgeo.2021.104504.

ISSMGE – Technical Committee TC304 "Engineering Practice. 2021. *State-of-the-art review of inherent variability and uncertainty in geotechnical properties and models*.

Kulhawy, F.H. & Mayne, P.W. 1990. *Manual on estimating soil properties for foundation design*. EPRI Report EL-6800, Electric Power Research Institute.

Kulhawy, F.H. & Trautmann, C.H. 1996. Estimation of in-situ test uncertainty. In C.D. Shackleford, P.P. Nelson and M.J.S. Roth (eds.), *Uncertainty in the Geologic Environment: From Theory to Practice*, ASCE Geotechnical Special Publication No. 58: 269–286.

Leibniz, G.W. 1704. *Nouveaux éssais* IV, 16, 12.

Li, J., Tian, Y. & Cassidy, M.J. 2014. Failure mechanism and bearing capacity of footings buried at various depths in spatially random soil *J. Geotech. Geoenviron. Eng*., ASCE. DOI: 10.1061/(ASCE)GT.1943-5606.0001219.

Maurer, B.W., Green, R.A., van Ballegooy, S. & Wotherspoon, L. 2017. Assessing liquefaction susceptibility using the CPT soil behavior type index. *Proceedings of the Third International Conference on Performance-Based Design in Earthquake Geotechnical* Engineering – *PBDIII*, Vancouver, July 16-19.

Mo, P.-Q., Marshall, A.M. & Yu, H.-S. 2017. Interpretation of cone penetration test data in layered soils using cavity expansion analysis. *J. Geotech. Geoenviron. Eng*., ASCE. DOI: 10.1061/(ASCE)GT.1943-5606.0001577.

Peuchen, J. & Terwindt, J. 2014. Introduction to CPT accuracy. *Proceedings of the 3rd International Symposium on Cone Penetration Testing*, Las Vegas.

Peuchen, J. & Parasie, N. 2019. Challenges for CPT accuracy classes. *Proceedings of the XVII ECSMGE-2019, Geotechnical Engineering foundation of the future*, Reykjavik. ISBN 978-9935-9436-1-3. DOI: 10.32075/17ECSMGE-2019-0049

Phoon, K.-K. & Kulhawy, F.W. 1999. Characterisation of geotechnical variability. *Canadian Geotechnical Journal* 36: 612–624.

Phoon, K.-K. 2020. The Story of Statistics in Geotechnical Engineering. *Georisk: Assessment and Management of Risk for Engineered Systems and Geohazards*. DOI: 10.1080/17499518.2019.1700423.

Phoon, K.-K., Ching, J. & Shuku, T. 2021. Challenges in data-driven site characterization. *Georisk: Assessment and Management of Risk for Engineered Systems and Geohazards*. DOI: 10.1080/17499518.2021.1896005.

Robertson, P.K. 1990. Soil classification using the cone penetration test. *Canadian Geotechnical Journal* 27:151–158.

Robertson, P.K. & Cabal, K.L. 2015. *Guide to Cone Penetration Testing for Geotechnical Engineering* – 6th Edition. Gregg Drilling & Testing, Inc.

Schmidt, J. & Moss, R. 2021. Bayesian hierarchical and measurement uncertainty model building for liquefaction triggering assessment. *Computers & Geotechnics* 132, 103963. DOI: 10.1016/j.compgeo.2020.103963.

Tang, C. & Phoon, K.-K. 2021. *Model uncertainties in foundation design*. CRC Press, Boca Raton. DOI: 10.1201/9780429024993.

Tehrani, F.S., Arshad, M.I., Prezzi, M. & Salgado, R. 2017. Physical modeling of cone penetration in layered sand. *J. Geotech. Geoenviron. Eng.*, ASCE. DOI: 10.1061/(ASCE)GT.1943-5606.0001809.

Uzielli, M., Lacasse, S., Nadim, F. & Phoon, K.-K. 2007. Soil variability analysis for geotechnical practice. Keynote paper, In T.S. Tan, K.K. Phoon, D. W. Hight & S. Leroueil (eds.), *Proceedings of the 2nd International Workshop on Characterisation and Engineering Properties of Natural Soils*. Singapore, November 29 – December 1, 2006. The Netherlands: Taylor & Francis.

Uzielli, M., and P. W. Mayne. 2019. “Probabilistic assignment of effective friction angles of sands and silty sands from CPT using quantile regression.” *Georisk: Assessment and Management of Risk for Engineered Systems and Geohazards* 13(4): 271–275. DOI:10.1080/17499518.2019. 1663388

Yu, H.-S. 2006. The first James K. Mitchell lecture in situ soil testing: From mechanics to interpretation. *Geomech. Geoeng.: An Int. J.*, 1(3),165–195.

Zhang, Z. & Tumay, M.T. 2003. Non-traditional approaches in soil classification derived from the cone penetration test. In G.A. Fenton & E. H. Vanmarcke (eds.), *Probabilistic Site Characterisation at the National Geotechnical Experimentation Sites*, ASCE Geotechnical Special Publication No. 121: 101–149.

Cone Penetration Testing 2022 – Gottardi & Tonni (eds)
© 2022 the Author(s), ISBN 978-1-032-35624-2

Combined use of CPT & DMT: background, current trends and ongoing developments

P. Monaco
University of L'Aquila, L'Aquila, Italy

ABSTRACT: This paper aims to introduce and stimulate the discussion on the use of CPT & DMT and on the potential of their mutual integration for enhanced site characterization and geotechnical design. It provides a brief overview of background information, current trends and ongoing developments on the combined use of CPT & DMT testing. Specific issues addressed in the paper include: (a) CPT-DMT comparisons and inter-relationships; (b) derivation of soil parameters from combined CPT & DMT data; (c) applications based on combined CPT & DMT data.

1 INTRODUCTION

The use of in-situ testing for geotechnical site characterization has continued to expand over the past few decades, gaining an increasing role over the traditional practice based on drilling and sampling for laboratory testing. Often today in-situ tests represent the major part of a site investigation, particularly in geomaterials that are difficult to sample and test using conventional methods (e.g., sands, tailings, semi-liquid soils). Mayne et al. (2009) summarized the key advantages of in-situ tests: (a) they can be done relatively quickly as compared with laboratory tests, (b) results are available immediately, (c) large numbers of data are obtained, and (d) vertical and lateral variability can be assessed over the site.

In this context, the cone/piezocone penetration test (CPT/CPTu) and the flat dilatometer test (DMT) have proven to be particularly efficient and expedient for routine site investigations. In fact: (a) both are direct-push probes, which do not need a borehole for advancement, (b) multiple measurements are collected from a single sounding, (c) the data are obtained digitally and can be accessed for use immediately after completion of the sounding or even in real-time during advancement of the probe, (d) their instrumental accuracy is high, unlike "older" techniques such as the standard penetration test (SPT), (e) technological upgrades, such as additional sensors or full automation (e.g., Medusa DMT, Marchetti et al. 2019), can be easily implemented.

The use of direct-push in-situ tests providing multi-measurements was emphasized by Mayne et al. (2009) as a more efficient approach to geotechnical site characterization. Particularly beneficial and versatile is their "hybrid" configuration that combines the advantages of full-displacement penetrometer probes with downhole geophysics (seismic piezocone SCPTu, seismic dilatometer SDMT), capable to provide information on soil behavior from the small- to the large-strain range.

A single SCPTu sounding provides up to five independent measurements: the basic CPTu measurements (cone resistance q_t, sleeve friction resistance f_s, pore pressure u_2), the shear wave velocity V_S, and information on soil consolidation / permeability parameters if dissipation tests are performed by monitoring the u_2 decay with time (e.g., time to reach 50% degree of consolidation, t_{50}). A single SDMT sounding provides a comparable number of independent measurements: the basic DMT pressure measurements (A, B and optional C corrected to p_0, p_1, p_2 respectively), the shear wave velocity V_S, as well as the compression wave velocity V_P when using the SPDMT configuration equipped with two pairs of receivers for S- and P-wave (Amoroso et al. 2020), and information on soil consolidation / permeability parameters if dissipation tests by monitoring the A-pressure decay with time are performed (e.g., reference time in the A-decay curve, t_{flex}).

For these reasons, Mayne et al. (2009) suggested that the SCPTu and SDMT direct-push tests should serve as the basis for the minimum required level of effort in routine and daily site investigation practices by the profession, in order that adequate amounts and sufficiently different types of data are collected for a given project. These data are complemented with sampling and laboratory testing.

The CPT and the DMT have in common the same equipment for insertion (push rig, rods) and a similar range of soil type applications (clays, silts, sands). Both tests are utilized in site investigations to obtain information on stratigraphy, soil types and parameters to be used with general or dedicated design methods. However, each test has advantages and limitations and may prove more adequate in specific applications. E.g., the CPT is commonly used for pile design and liquefaction assessment, the DMT generally provides good predictions of settlements of shallow foundations.

In common practice, the CPT is often the preferred primary in-situ test, supported by a much larger consolidated use. On the other hand, the DMT can provide distinctive contributions in a routine site investigation: (1) information on stress history, which has a dominant influence on soil behavior; (2) being an in-situ pressure-displacement test, the DMT results are more closely related to "working strain" soil stiffness than penetration tests.

Some questions that may arise are then: are the two in-situ tests equivalent? Is one test (CPT or DMT alone) self-standing and sufficient for exhaustive geotechnical site characterization? Are there any advantages in the combined use of CPT & DMT?

One notable emerging trend (Marchetti 2015, Marchetti & Monaco 2018) is the increasing diffusion in site investigation practice of a "multi-parameter / multi-test" approach, based on the combination of CPT & DMT. This approach benefits, besides from the multiple measurements provided by a single CPT or DMT sounding, also from the combination of the different measurements obtained from nearby CPT and DMT soundings. Examples of application of a multi-parameter / multi-test approach based on the combined use of CPT & DMT are illustrated in this paper, aiming to stimulate the discussion on the potential of the mutual integration of these two in-situ tests for enhanced site characterization and geotechnical design. The terms 'CPT' and 'DMT' are herein intended in a general sense, referring to any equipment configurations (i.e., CPT(u), SCPT(u); DMT, SDMT, Medusa (S)DMT).

2 CPT-DMT CORRELATIONS

Interrelationships between CPT & DMT parameters can be helpful to expand and improve correlations and applications by using existing experience and databases from one test and extrapolating to the other (Robertson 2012).

Robertson (2009a) reviewed published records of data from nearby CPT and DMT soundings, as well as existing correlations for geotechnical parameters, in an effort to identify possible intercorrelations. The key in this approach was the recognition that, since the main DMT interpreted parameters are normalized, they should be correlated with normalized CPT parameters. Correlations were tentatively established between the three "intermediate" DMT parameters, i.e., the material index I_D, the horizontal stress index K_D and the dilatometer modulus E_D (Marchetti 1980), and the CPT normalized cone resistance Q_{t1}, normalized friction ratio F_r and Soil Behavior Type (SBT) Index I_c, defined as follows:

$$I_D = \frac{p_1 - p_0}{p_0 - u_0} \tag{1}$$

$$K_D = \frac{p_0 - u_0}{\sigma'_{v0}} \tag{2}$$

$$E_D = 34.7(p_1 - p_0) \tag{3}$$

$$Q_{t1} = \frac{q_t - \sigma_{v0}}{\sigma'_{v0}} \tag{4}$$

$$F_r = \left[\frac{f_s}{(q_t - \sigma_{v0})}\right] 100(\%) \tag{5}$$

$$I_c = \left[(3.47 - \log Q_{t1})^2 + (\log F_r + 1.22)^2\right]^{0.5} \tag{6}$$

where p_0 = corrected first DMT pressure reading, p_1 = corrected second DMT pressure reading, q_t = corrected cone resistance, f_s = sleeve friction resistance, u_0 = pre-insertion in situ equilibrium pore pressure, σ_{v0} = in situ total vertical stress, σ'_{v0} = in situ effective vertical stress.

Robertson (2009a) explained that in most of the available published records the DMT K_D is normalized by σ'_{v0} as in the original formulation (Eq. 2) proposed by Marchetti (1980). Therefore, the CPT cone resistance Q_{t1} was normalized by σ'_{v0} in a consistent manner, i.e., assuming the exponent for stress normalization $n = 1$, as originally proposed by Robertson (1990), although currently used updated formulations (e.g., Robertson 2009b) refer to a normalized cone resistance Q_{tn} that uses a variable n:

$$Q_{tn} = \left[\frac{(q_t - \sigma_{v0})}{p_a}\right]\left(\frac{p_a}{\sigma'_{v0}}\right)^n \tag{7}$$

where p_a = atmospheric pressure in same units as q_t and σ_v, n = stress exponent that varies with SBT I_c (for $n = 1$, $Q_{tn} = Q_{t1}$).

In addition, stress normalization for K_D is a debated issue. Robertson (2009a, 2015) suggested that a more complex normalization for K_D, similar to Q_{tn}, would likely be more appropriate, especially in sands, and future CPT-DMT correlations could use more suitable normalized parameters. However, for typical stress levels in geotechnical engineering of about 65-200 kPa (i.e., ≈ 4-20 m) the normalization method has little influence on the normalized parameters, hence updated CPT-DMT correlations using a different normalization may not change significantly. On the other hand, Marchetti (2015) remarked that a linear increase with depth of the DMT p_0 (i.e., a nearly constant K_D calculated using $n = 1$) is observed to large depths at various test sites. In contrast, the increase of the CPT q_t with depth is generally less than linear, corresponding to $n < 1$. Such a different trend was attributed by Marchetti (2015) to the fact that the DMT blade, having a rectangular cross section with a width/thickness ratio ≈ 6, is less affected by arching than a conical probe, resulting in a more linear trend of p_0 with depth which justifies $n = 1$ for K_D.

The preliminary set of average CPT-DMT correlations proposed by Robertson (2009a) is:

$$I_D = 10^{(1.67-0.67I_c)} \tag{8}$$

When $I_c \leq 2.60$ (i.e., for sand-like soils):

$$K_D = \frac{0.144Q_{t1}}{\left[10^{(1.67-0.67I_c)}\right]} \tag{9a}$$

When $I_c > 2.60$ (i.e., for clay-like soils):

$$K_D = \beta(Q_{t1})^{0.95} + 1.05 \tag{9b}$$

$$E_D/\sigma'_{v0} = 5Q_{t1} \tag{10}$$

where the constant β varies with soil sensitivity (0.30 < β < 0.7), with an average value β = 0.3.

The correlations for I_D, K_D (Eqs 8, 9a, 9b) depend on soil type, reflected in the CPT index I_c, which is a function of both Q_{t1} and F_r. These correlations are shown in Figure 1 in the form of contours of DMT I_D, K_D on the CPT normalized SBT chart $Q_{t1} - F_r$.

The relationship between DMT I_D and CPT I_c (Eq. 8) was established considering that both parameters are used to identify soil type. It is noted that I_D, likewise I_c, is a parameter reflecting the mechanical soil behavior, not a soil classification index based on real grain size distribution (Marchetti 1980). Robertson & Wride (1998) suggested that I_c = 2.60 can be assumed as an approximate boundary between sand-like and clay-like soils. Based on Eq. 8, $I_c = 2.60$ corresponds roughly to $I_D \approx 1$. In a general sense, CPT and DMT results are drained in sand-like soils ($I_c \leq$ 2.60, $I_D > 1$) and undrained in clay-like soils ($I_c >$ 2.60, $I_D < 1$). The correlation for K_D (Eqs 9a, 9b) can be sensitive to the cut-off $I_c = 2.60$ when CPT data fall close to the boundary between clay-like and sand-like soils. DMT results in the transition region of silt-mixture soils may also be influenced by partial drainage.

Robertson (2015), based on additional pairs of CPT & DMT (e.g., Togliani et al. 2015), suggested a more simplified link between CPT Q_{t1} and DMT K_D, I_D for young uncemented soils (i.e., soils with little or no microstructure):

$$Q_{t1} = (1.5 \log I_D + 7.5) I_D K_D \tag{11}$$

The (smoother) contours of DMT K_D corresponding to Eq. 11 are also plotted in Figure 1 (red dashed lines) on the CPT normalized SBT $Q_{t1} - F_r$ chart.

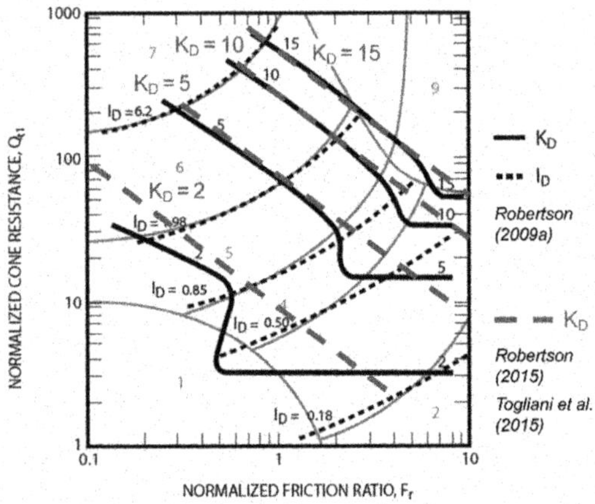

Figure 1. Contours of DMT K_D, I_D on the CPT normalized SBT $Q_{t1} - F_r$ chart for young uncemented soils (from Robertson 2017).

To stimulate a discussion on this topic, it is worth recalling what was observed by Robertson (2009a, 2015), i.e., the proposed CPT-DMT correlations are approximate and influenced by variations in in-situ stress state, soil density, stress and strain history, age, cementation, soil sensitivity. These correlations are unlikely to be unique for all soils, but they may form a framework for possible future refinements. Moreover, they may provide further insight into possible future correlations for the DMT with other geotechnical parameters and design applications, since the CPT has a somewhat more extensive theoretical background compared to the DMT, as well as a larger database of documented case histories for certain applications (e.g., liquefaction assessment). Marchetti (2011) recognized that procedures helpful for extracting as much information as possible from the field data represent a precious contribution since soil information is vital for the determination of reliable soil parameters for design, but costly to obtain. Correlations interconnecting CPT & DMT are useful in that translation formulas permit: (1) the use of interpretation methods or charts developed for one test with the results of the other test, (2) converting a database available for one test to a database for the other test. However, Marchetti (2011) stressed that intrinsic limits exist to the accuracy of the CPT-DMT translations, which, despite further refinements, are bound to remain of an approximate nature. These limits are mostly due to the different sensitivity to stress history of the representative normalized CPT and DMT parameters, as will be discussed in the following.

A different theoretical approach for establishing interrelationships between the CPT and the DMT in soft clays has been explored by Mayne (2016), Ouyang & Mayne (2017, 2018) and earlier studies, based on a CPT-DMT compiled database in clays that progressively expanded over the years.

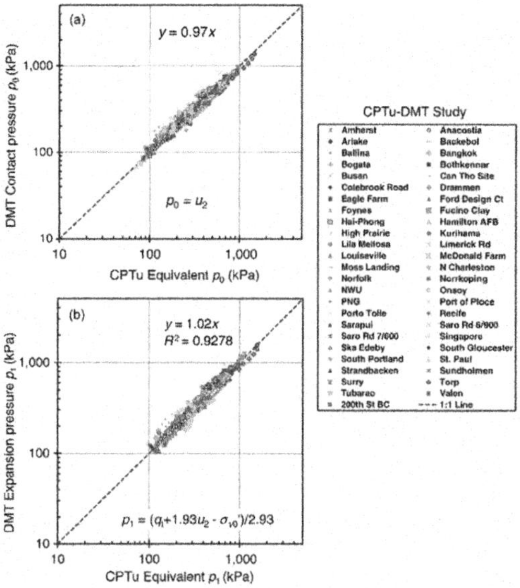

Figure 2. Comparison of DMT and CPTu-equivalent pressures p_0 (a) and p_1 (b) for 49 clays (adapted from Ouyang & Mayne 2018).

Mayne (2016) and Ouyang & Mayne (2017) established a theoretical nexus between CPTu readings (q_t and u_2) and DMT pressures (p_0 and p_1) based on spherical cavity expansion solutions for undrained penetration of both probes in soft to firm clays. This nexus would permit an exchange of interpretations between the two tests, offering a complementary extension of methodologies. Data from paired sets of CPTu-DMT results in a variety of clays were used to support and validate the theoretical links. The net cone resistance, q_{net}, defined as:

$$q_{net} = q_t - \sigma_{v0} \tag{12}$$

was linked to the DMT pressure readings p_0, p_1 by the following relationship:

$$q_{net} = 2.93p_1 - 1.93p_0 - u_0 \tag{13}$$

Ouyang & Mayne (2018) further expanded this concept and suggested that q_t and u_2 measured from the CPTu can be utilized to duplicate equivalent DMT pressures p_0 and p_1, and vice versa, according to the relationships:

$$p_0 = u_2 \tag{14}$$

$$p_1 = \frac{(q_t - 1.93u_2 - \sigma'_{v0})}{2.93} \tag{15}$$

Eq. 14 is in agreement with earlier studies (Mayne & Bachus 1989, Mayne 2006) that found the p_0 from DMT very similar to the pore pressures u_2 measured by CPTu in intact clays and clayey silts. Figure 2 shows a good agreement between the pressures p_0, p_1 measured by DMT and their CPTu-equivalent p_0, p_1 estimated by Eqs 14 and 15, based on 49 paired CPTu-DMT data sets in a variety of natural soft to firm clays and silts worldwide, having an overconsolidation ratio *OCR* between 1 to 2.5.

The approach proposed by Ouyang & Mayne (2017, 2018) is valid only for undrained penetration conditions, which are generally accepted to occur for both CPTu and DMT in low permeability clays at a standard penetration rate of 20 mm/s. The approach could be invalidated in intermediate permeability soils, such as silts, where drainage conditions may be different. The CPTu readings q_t and u_2 are taken during penetration (at time $t = 0$), whereas the DMT readings p_0 and p_1 are obtained at approximately $t = 15$ s and $t = 30$ s after penetration respectively, hence some pore pressure dissipation may have occurred prior to obtaining the DMT readings.

3 CPT VS. DMT DATA COMPARISONS

3.1 *Sensitivity to stress history of DMT vs. CPT*

The horizontal stress index K_D is a key parameter obtained from DMT interpretation. K_D reflects cumulatively various stress history effects (overconsolidation, aging, in-situ horizontal earth pressure, K_0), as summarized by Marchetti (2010).

Numerous researchers have observed that the DMT K_D is considerably more sensitive than the CPT cone resistance q_c in monitoring compaction in the field. The higher sensitivity of the DMT to stress history is confirmed by comparisons CPT-DMT in controlled calibration chamber (CC) testing conditions. Jamiolkowski & Lo Presti (1998), in CC tests in Ticino sand, found the DMT K_D much more sensitive to stress-strain history (including aging-like effects) than the CPT q_c. Lee et al. (2011) investigated the effects of stress history on CPT and DMT in CC testing on 40 large specimens of Busan sand having different relative density D_r and *OCR* in the range 1 to 8. The overconsolidation produced an almost negligible increase in the normalized cone resistance $q_c/(\sigma'_v)^{0.5}$ (Figure 3a), but a substantial increase in K_D (Figure 3b). It appears that, while the normalized q_c reflects essentially D_r and only to a minor extent OCR,

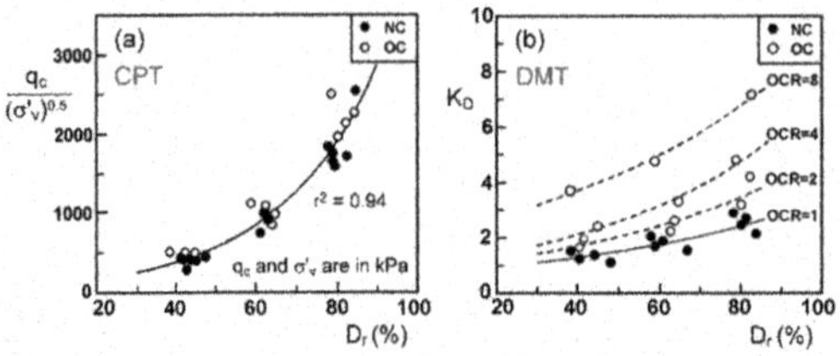

Figure 3. Effect of stress history on (a) normalized q_c from CPT, and (b) K_D from DMT in CC testing on Busan sand (modified from Lee et al. 2011).

K_D reflects not only D_r, but also stress history effects. The higher reactivity of K_D to OCR indicated by Figure 3, in agreement with other available experiences, implies that to the same normalized q_c may correspond different values of K_D (Marchetti 2016). In Figure 3a the coefficient of determination r^2 close to 1 for data points of any OCR suggests poor ability of the normalized q_c to distinguish OC from NC sands, hence estimating OCR from CPT alone appears problematic. This example supports the necessity of a multi-parameter/multi-test approach based on both DMT & CPT for estimating OCR in sand, as will be discussed in the following.

3.2 *Soil parameters estimated from DMT vs. CPT*

Many indirect correlations exist between DMT and CPT results, since both tests are used to estimate various geotechnical parameters.

The DMT interpretation is commonly based on the original correlations developed by Marchetti (1980) for "textbook" soils, although several updates, reviewed by Marchetti et al. (2001), have been proposed. Among the various parameters obtained from DMT, the 1-D constrained modulus M (M_{DMT}) is the most generally appreciated. According to Marchetti (1980) M_{DMT} is obtained from E_D (Eq. 3), which links soil stiffness to the p_0 and p_1 measured during membrane expansion (a "mini load test"), corrected as a function of I_D (soil type) and K_D (OCR). Several favorable comparisons in terms of both M_{DMT} vs. reference M and DMT-predicted vs. measured settlements have shown that, in general, M_{DMT} is a reasonable estimate of the "operative" or "working strain" modulus.

Numerous correlations have been proposed to estimate the constrained modulus M from CPT in various soil types. Typically, M is obtained by multiplying the cone resistance (q_c or q_t), or the net cone resistance ($q_t - \sigma_{v0}$), by an empirical coefficient α which depends on soil type and other soil properties, and may vary in a rather large range (e.g., 1 to 10).

Comparisons of moduli M estimated from DMT vs. CPT have been published by several authors. As an example, Amoroso et al. (2022a) compared the depth profiles of M estimated from DMT and CPTu before/after ground improvement with Rammed Aggregate Piers (RAP) in silty and sandy soils at the Bondeno test site, Italy. In natural soils they found a rather good agreement between M estimated from DMT (Marchetti 1980) and from CPTu using the correlations proposed by Robertson (2009b), as shown in Figure 4, while other CPT-based correlations (Lunne & Christophersen 1983, Senneset et al. 1988) provided M equal to about half the values of M obtained from DMT and from Robertson (2009b). The agreement between the latter two methods is not surprising. In fact, the development of the CPT correlations for M by Robertson (2009b) was partly "guided" by the author's correlations between normalized DMT and CPT parameters (Robertson 2009a), based on the recognition that M estimated from DMT has often been shown to provide excellent estimates of settlement (e.g., Monaco et al. 2006). However, Figure 4 also shows that in treated soils, using the same correlations, the increase in M from DMT was found much more evident than the increase in M from CPTu, suggesting higher ability of M_{DMT} to reflect the benefit of soil treatment.

As to the undrained shear strength (s_u) in clay, available comparisons of s_u from DMT vs. other tests at well-documented sites have often found that the depth profiles of s_u estimated from DMT (Marchetti 1980) plot in an intermediate position, in particular in between the s_u profiles obtained from CPT(u) assuming variable values of the cone bearing capacity factor (N_k or N_{kt}). An example is shown in Figure 5 (Marchetti et al. 2022), where the s_u values obtained in Fucino clay by Medusa DMT are compared with the s_u from CPT(u) and from other tests.

Comparisons CPT-DMT are also available in terms of pore pressure index and consolidation/drainage parameters inferred from dissipation tests. The DMT, though non provided with a pore pressure sensor as the CPTu, permits to determine the pre-insertion in-situ equilibrium pore pressure u_0 in sand and to discern layers of different permeability based on the closing pressure p_2. As described by Marchetti et al. (2001), in sand $p_2 \approx u_0$, while in clay $p_2 > u_0$ due to some excess pore pressure induced by penetration persisting at the time of the p_2 measurement. The pore pressure index U_D (Lutenegger & Kabir 1988) is defined as:

$$U_D = \frac{(p_2 - u_0)}{(p_0 - u_0)} \tag{16}$$

The parameter U_D is the DMT equivalent of the pore pressure ratio B_q from CPTu:

$$B_q = \frac{(u_2 - u_0)}{(q_t - \sigma_{v0})} \tag{17}$$

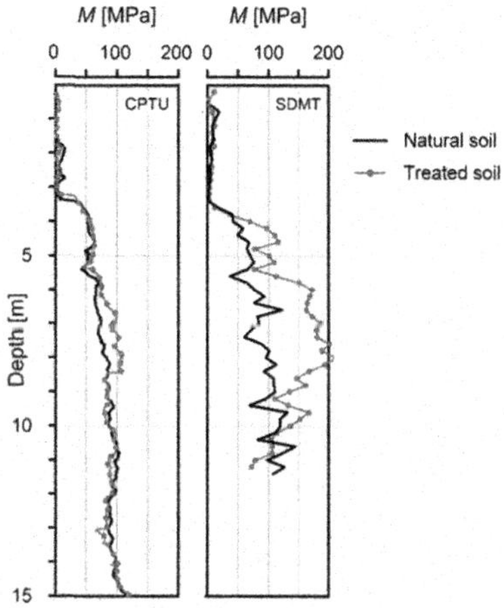

Figure 4. Comparison of constrained moduli M estimated from CPTu (Robertson 2009b) and SDMT (Marchetti 1980) in natural and treated soils at the Bondeno test site (modified from Amoroso et al. 2022a).

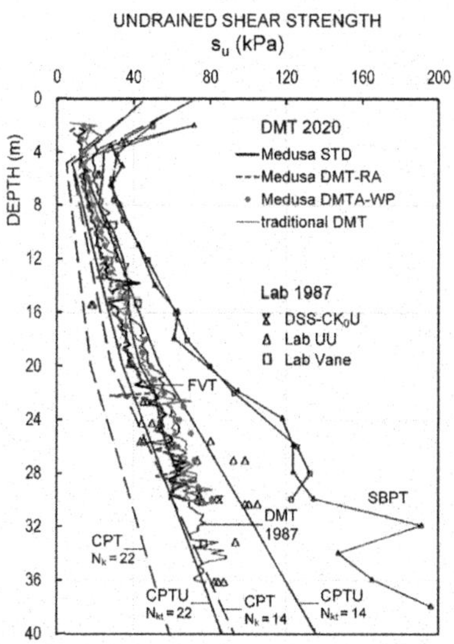

Figure 5. Comparison of profiles of the undrained shear strength s_u from standard/Medusa DMT, CPT/CPTu and other in-situ and laboratory tests (data from Burghignoli et al. 1991) at Fucino-Telespazio (Marchetti et al. 2022).

Similar to B_q, U_D may help delineate soil stratigraphy, particularly in highly stratified deposits. In free-draining soils $U_D \approx 0$, in non free-draining soils $U_D > 0$. The example in Figure 6 (Benoit 1989) shows a good agreement between DMT U_D and CPTu B_q.

The use of the DMT in intermediate soils (silty sands, silts, sandy silts) is being increasingly investigated (Schnaid et al. 2016, 2018). Monaco et al. (2021) presented preliminary results of tests carried out using the Medusa DMT adopting variable pressurization and penetration rates in intermediate soils in the Po river valley, Italy, combined with CPTus at variable penetration rates. In these soils, the highly accurate and repeatable time-for-reading facility of the Medusa DMT permitted to identify some trends in the variation of DMT pressure measurements in response to variable-rate testing conditions, i.e., a slower penetration/pressurization rate 'shifts' the interpretation towards drained behavior, while a faster penetration/pressurization rate 'shifts' the interpretation towards undrained behavior. In sand, a fully drained response was found for any penetration/pressurization rate. U_D appears as a suitable independent parameter to discern between drained, undrained or partially drained soil behavior and may be usefully combined with I_D to obtain some 'soil behavior type'. Ongoing research investigates the potential of combining variable-rate Medusa DMT and CPTu tests as an innovative approach for characterizing the in-situ behavior of intermediate soils.

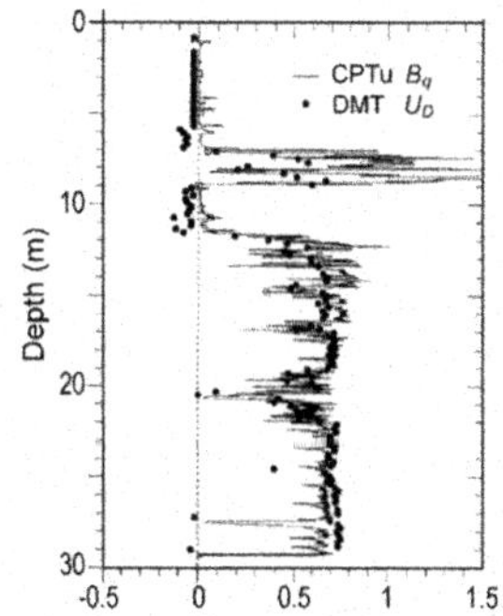

Figure 6. Comparison of U_D from DMT and B_q from CPTu (Benoit 1989, from Marchetti et al. 2001).

The DMT is also used to estimate in-situ consolidation/permeability parameters by means of dissipation tests. Differently from CPTu, the decay with time monitored by the DMT is not that of the pore pressure, but of the total contact A-pressure. This involves lack of worry over desaturation or poor saturation of a porous element, which may affect the u_2 measurement in CPTu dissipation tests. The Medusa DMT technology, which permits to acquire up to 3 repeated A-pressure readings per second, has extended the range of soil types in which A-dissipations are feasible from low permeability soils (clays, silty clays) to intermediate permeability soils (silts, silty sands).

4 DERIVATION OF SOIL PARAMETERS BY COMBINED USE OF CPT & DMT

4.1 *Estimating OCR in sand from DMT & CPT*

The combined use of DMT & CPT may provide estimates of the *OCR* in sand. Marchetti et al. (2001) suggested to use the ratio between the constrained modulus from DMT M_{DMT} and the cone resistance q_c from CPT as an indicator OCR in sand:

$$M_{DMT}/q_c = 5 - 10 \text{ in NC sands} \tag{18a}$$

$$M_{DMT}/q_c = 12 - 24 \text{ in OC sands} \tag{18b}$$

The above preliminary indications were based on available CPT-DMT data before/after compaction of sand fills (e.g., Jendeby 1992), where M_{DMT}/q_c was found to increase with the overconsolidation achieved by compaction, combined with data from CC research and from other tests.

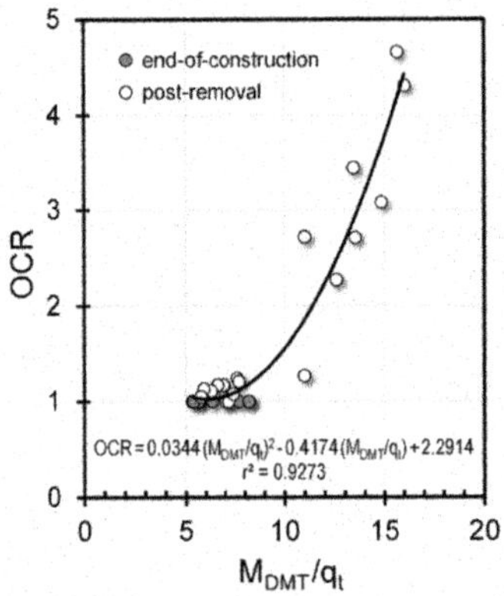

Figure 7. Correlation OCR-M_{DMT}/q_t in sands obtained from paired DMT-CPTu data at the Treporti trial embankment test site (modified from Monaco et al. 2014).

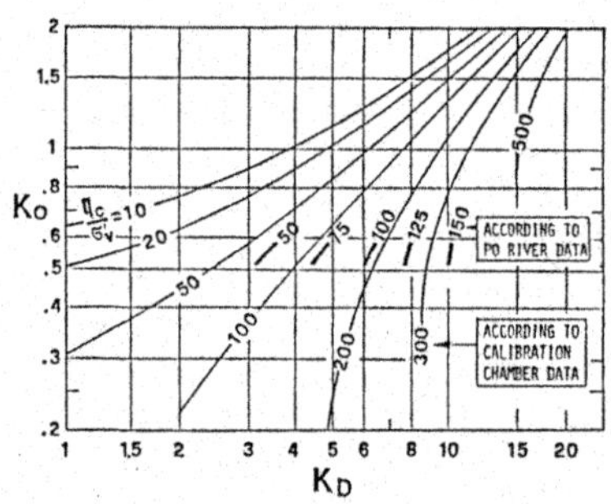

Figure 8. Chart for evaluating $K_0 = f(K_D, q_c/\sigma'_v)$ proposed by Marchetti (1985).

The possibility to estimate OCR in sand by the combined use of DMT and CPT was investigated by Monaco et al. (2014) as part of an extensive experimental study carried out at the Treporti Test Site, Venice, Italy. At this site, a full-scale cylindrical trial embankment (40 m diameter, 6.7 m height, applied pressure 106 kPa) was built and monitored from the beginning of its construction until its complete removal, four years later. Using same-depth values of M_{DMT} and q_t obtained in sand layers from paired DMT and CPTu soundings carried out at the times when reference "imparted OCR" were known, i.e. at end-of-construction and post-removal, Monaco et al. (2014) derived the correlation (Figure 7):

$$OCR = 0.0344(M_{DMT}/q_t)^2 - 0.4174(M_{DMT}/q_t) + 2.2914 \tag{19}$$

Using the same data set, Monaco et al. (2014) also constructed a correlation OCR-K_D based only on DMT. However the OCR-K_D correlation is not unique, but likely depending also on the relative density D_r. In fact, going back to the CC results shown in Figure 3b (Lee et al. 2011), it can be deduced that for a given D_r there is a one-to-one correspondence between K_D and OCR values (i.e., a vertical line having constant D_r in the diagram intersects each OCR curve for a unique K_D value), but if D_r is not constant the same OCR may result from a different combination of K_D and D_r. Hence a given K_D may be due either to a low D_r and a high OCR, or to a high D_r and a low OCR. In order to separate the D_r effect from the OCR effect, i.e., to pinpoint the right OCR, D_r pair and then to estimate OCR, the normalized q_c (Figure 3a) is also necessary to provide an indication of D_r on the horizontal axis. Hence both q_c and K_D would be needed to estimate OCR in sand, i.e., CPT alone or DMT alone are insufficient. The OCR-K_D correlation by Monaco et al. (2014) works well at the Treporti test site where it was calibrated because in these sands D_r is almost uniform, but it may not work at sites where D_r is variable. Based on CC tests on Busan sand samples having different D_r and OCR, Choo et al. (2015) proposed a correlation to estimate OCR in sand based on K_D and D_r, with D_r also estimated from DMT, hence using only DMT data. This OCR-K_D correlation needs to be calibrated for each specific sand.

The OCR-M_{DMT}/q_t correlation (Eq. 19) by Monaco et al. (2014), based on both DMT and CPTu data, appears to have more general validity and may provide broad OCR estimates in different sands. Additional research is encouraged to investigate the dependency of the correlation OCR-M_{DMT}/q_t on D_r, stress level and possibly sand type.

4.2 *Estimating K_0 in sand from DMT & CPT*

Similar considerations may be extended to the estimation of K_0 in sand. The $K_0 - q_c - K_D$ chart in Figure 8, elaborated by Marchetti (1985), is probably the first example of combined use of CPT & DMT. The chart permits to estimate K_0 once q_c and K_D are known. Figure 8 shows, besides the continuous curves obtained from CC tests, an additional scale for q_c/σ'_v based on 25 data points in Po river sand, where an average value $K_0 = 0.55$ was estimated.

Baldi et al. (1986) enriched such $K_0 - q_c - K_D$ chart with additional CC work and converted it into the following equations:

$$K_0 = 0.376 + 0.095K_D - 0.0017q_c/\sigma'_{v0} \quad (20a)$$

$$K_0 = 0.376 + 0.095K_D - 0.0046q_c/\sigma'_{v0} \quad (20b)$$

Eq. 20a was determined as the best fit of CC data, obtained on pluviated artificial sand, while Eq. 20b was obtained by modifying the last coefficient to predict "correctly" K_0 for the natural Po river sand.

Marchetti et al. (2001) recommended using Eqs 20a and 20b (Baldi et al. 1986) with the following values of the last coefficient: -0.002 in "freshly deposited" sand, -0.005 in "seasoned" sand. In this way, the influence of *OCR* on K_0 is incorporated by the choice of the last coefficient, which involves appreciable subjectivity (e.g., a sand could be assumed as "fresh" if K_D = 1-2, "seasoned" if K_D = 5-6).

Choo et al. (2015), based on CC testing on Busan sand, proposed a correlation to estimate K_0 in sand from DMT only, but similarly to *OCR* this correlation requires specific calibration for different sands.

Based on the CC data set by Baldi et al. (1986), Hossain & Andrus (2016) proposed to estimate K_0 in sand by use of the following correlation:

$$K_0 = 0.72 + 0.456 \log OCR + 0.035K_D - 0.194 \log(q_c/\sigma'_v) \quad (21)$$

in which the influence of *OCR* on K_0 is explicitly taken into account. Eq. 21 could be used in combination with *OCR* estimated by Eq. 19.

All the above considerations imply that the challenging task of estimating *OCR* and K_0 in sands requires a multi-parameter/multi-test approach, based on the combined use of DMT and CPT.

5 APPLICATIONS BASED ON COMBINED USE OF CPT & DMT DATA

5.1 *Use of DMT & CPT in monitoring ground improvement*

Several comparisons of pre-post CPT & DMT executed for monitoring ground improvement are available. Various studies have found the increase in M_{DMT} after soil treatment approximately twice the increase in q_c. For instance, Jendeby (1992) found an increase of the ratio M_{DMT}/q_c from ≈ 5-12 pre-compaction to ≈ 12-24 post-compaction of a loose sand fill (Figure 9a). Bałachowski & Kurek (2015) found the mean increase in M_{DMT} after vibroflotation of sand about 2.3 times higher than the corresponding increase in q_c and an increase of M_{DMT}/q_c from ≈ 2-10 pre-compaction to ≈ 10-24 post-compaction (Figure 9b). Amoroso et al. (2022a, b), using paired DMT and CPTu data before/after ground improvement with Rammed Aggregate Piers (RAP) at the Bondeno test site, found a similar trend of the ratio M_{DMT}/q_t (Figure 10), consistent with the observed increase in M_{DMT} higher than M from CPTu in treated soils (Figure 4). This evidence suggests that the DMT is more sensitive to stiffness variations as a consequence of the increase in horizontal stress produced by soil treatment, as reflected by the increase in K_0 (and *OCR*), also shown in Figure 10. Similar pre-post comparisons were reported by Amoroso et al. (2018). The finding that compaction (a sort of "imparted overconsolidation") increases both M_{DMT} and q_c, but M_{DMT} at a faster rate, inspired the use of the ratio M_{DMT}/q_c as a "proxy" of *OCR*, as previously discussed.

5.2 *Liquefaction assessment based on DMT & CPT*

The use of the DMT for liquefaction assessment is receiving increasing attention. Much of the interest on the development of simplified methods based on the DMT K_D derives from its high sensitivity to stress history, besides to other factors (e.g., D_r, K_0, aging) that influence liquefaction resistance. Numerous researchers (e.g., Pyke 2003, Leon et al. 2006 and many others) have pointed out that past stress-strain history (overconsolidation, aging) is likely to have a much greater effect on increasing liquefaction resistance than penetration resistance.

Simplified methods for estimating the cyclic resistance ratio (*CRR*) based on DMT K_D have been proposed over the years, including the most recent by Monaco et al. (2005), Tsai et al. (2009), Robertson (2012), Marchetti (2016), Chiaradonna & Monaco (2022) shown in Figure 11 (valid for magnitude M = 7.5 and clean uncemented sand).

All these methods have in common a strong link with CPT-based methods in their origin, in an effort to relate in some way to the field performance database that provides a vast experimental validation for current methods based on CPT, but is currently limited for DMT-based methods.

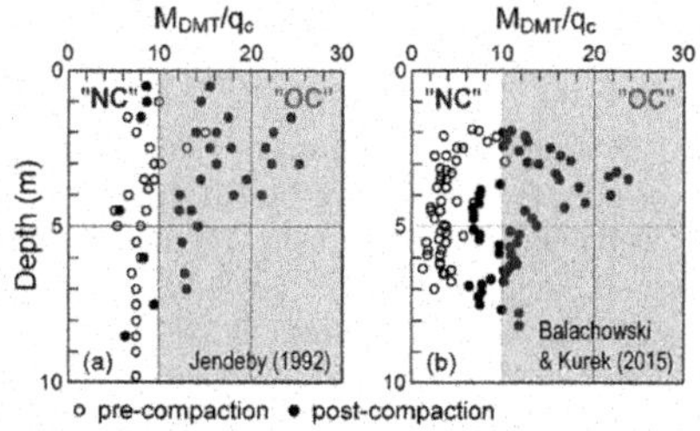

Figure 9. Ratio M_{DMT}/q_c before-after compaction of sand fills/deposits: data from (a) Jendeby (1992), (b) Bałachowski & Kurek (2015).

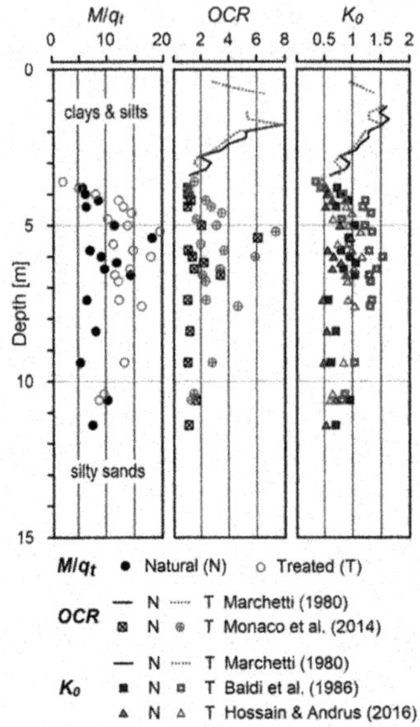

Figure 10. Profiles of M_{DMT}/q_t, OCR and K_0 obtained from combined DMT-CPTu data in natural and treated soils at the Bondeno test site (modified from Amoroso et al. 2022b).

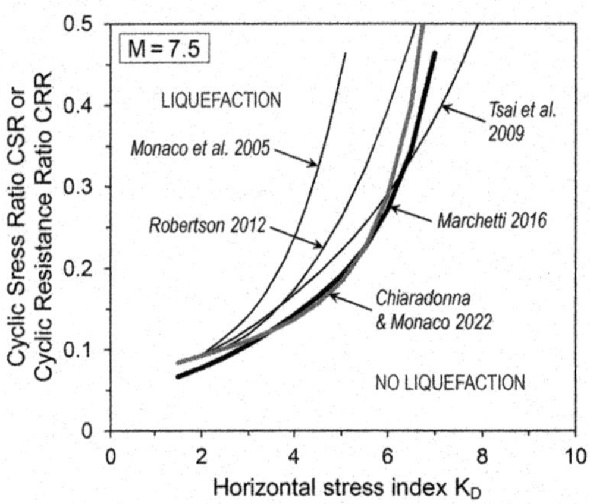

Figure 11. Recent CRR-K_D correlations for clean sand.

The CRR-K_D curve by Monaco et al. (2005) was obtained by translating "consensus" CPT (and SPT) curves, namely the CPT curve by Robertson & Wride (1998), using the relative density D_r as intermediate parameter. This approach is inevitably affected by some uncertainty in estimating D_r from the in-situ tests. Differently, Tsai et al. (2009) obtained a CRR-K_D curve by translating "consensus" CPT (and SPT) curves, again Robertson & Wride (1998) for CPT, by using direct correlations between the CPT normalized clean sand equivalent cone resistance and the DMT K_D, established using data from paired CPT-DMT soundings at five saturated loose sand sites in Taiwan that experienced liquefaction.

Using a similar approach, independently, Robertson (2012) re-elaborated the CPT-DMT data set collected by Tsai et al. (2009) and proposed, for young uncemented sands ($I_D > 1.2$ and $2 < K_D < 6$), an average relationship between the normalized clean sand equivalent cone resistance $Q_{tn,cs}$, defined as:

$$Q_{tn,cs} = K_c \cdot Q_{tn} \tag{22}$$

where K_c = correction factor for fines content estimated using I_c, and the DMT K_D:

$$Q_{tn,cs} = 25K_D \tag{23}$$

Based on Eq. 23, Robertson (2012) converted the Robertson & Wride (1998) CRR-$Q_{tn,cs}$ curve into a CRR-K_D curve. Later Marchetti (2016) obtained a new CRR-K_D curve by combining the same Eq. 23 with the CPT curve by Idriss & Boulanger (2006).

Chiaradonna & Monaco (2022) proposed an update of the CRR-K_D curve adopting the CPT-based framework by Boulanger & Idriss (2014). Using the CPT-DMT data set by Tsai et al. (2009), considering only CPT data having I_c between 1.5 and 2.6, a direct correlation was established between K_D and the clean sand equivalent cone resistance q_{c1Ncs} defined according to Boulanger & Idriss (2014):

$$q_{c1Ncs} = q_{c1N} + \Delta q_{c1N} \tag{24}$$

where q_{c1N} is the normalized cone resistance and Δq_{c1N} accounts for the effect of fines content.

Despite the dispersion, the data set (Figure 12) is well described by a linear trend with a slope of 20. Figure 12 also shows a second data set from the Scortichino site, Italy, where both CPT and SDMT data were available (Tonni et al. 2015), which is better interpreted by a linear trend with a slope of about 30. Based on this limited data, Chiaradonna & Monaco (2022) adopted an average coefficient of 25:

$$q_{c1Ncs} = 25K_D \tag{25}$$

which is compatible with Eq. 23 by Robertson (2012), also used by Marchetti (2016). By substituting Eq. 25 into the CRR-q_{c1Ncs} curve by Boulanger & Idriss (2014), a new CRR-K_D curve was obtained.

Average CPT-DMT correlations such as Eq. 23 (Robertson 2012) and Eq. 25 (Chiaradonna & Monaco 2022) may be useful to convert CPT- into DMT-based liquefaction triggering curves. On the other hand, the example in Figure 12 suggests that the $q_{c1Ncs} - K_D$ relationship varies from one site to another, presumably as a consequence of the higher sensitivity of K_D to stress history. Two sites may have similar q_{c1Ncs}, but different K_D depending on different stress history (Marchetti 2016). This implies that q_{c1Ncs} and K_D do not contain equivalent information, and may also explain in part the high dispersion of the CPT-DMT correlations (Marchetti 2015). Robertson (2015), recognizing the higher sensitivity of K_D to factors that cause soil microstructure (stress and strain history, age, cementation/ bonding), suggested using the $Q_{tn,cs} - K_D$ relationship to assess microstructure, noting that in young uncemented sands generally $Q_{tn,cs} > 16\ K_D$, while soils with significant microstructure tend to have $Q_{tn,cs} < 16\ K_D$.

In perspective, the DMT could offer a valuable integration to current methods for liquefaction assessment based on CPT, SPT or V_S, also given the general recommendation towards the use of "redundant" correlations based on different in-situ techniques / parameters. The main drawbacks of current DMT-based methods are: (1) lack of any correction factor for the fines content, and (2) limited experimental validation based on field performance data from real earthquakes. Ongoing research is attempting to fill these gaps. The implementation of an adequate case-history database for validation of the DMT-based approach could support the introduction of more consistent liquefaction triggering curves, considering also the fines content influence.

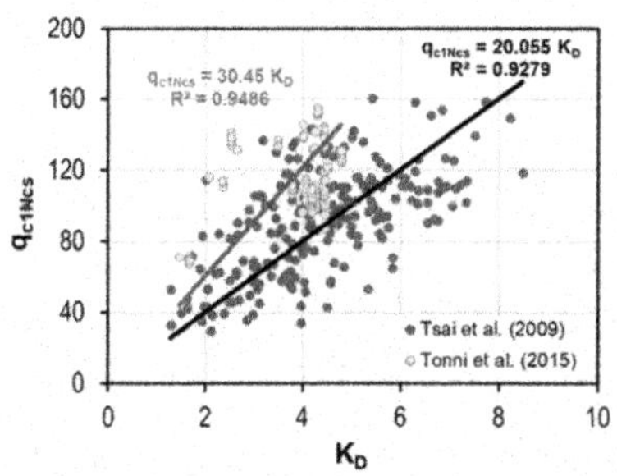

Figure 12. Relationship $q_{c1Ncs} - K_D$ from published CPT-DMT data records at two test sites (Chiaradonna & Monaco 2022).

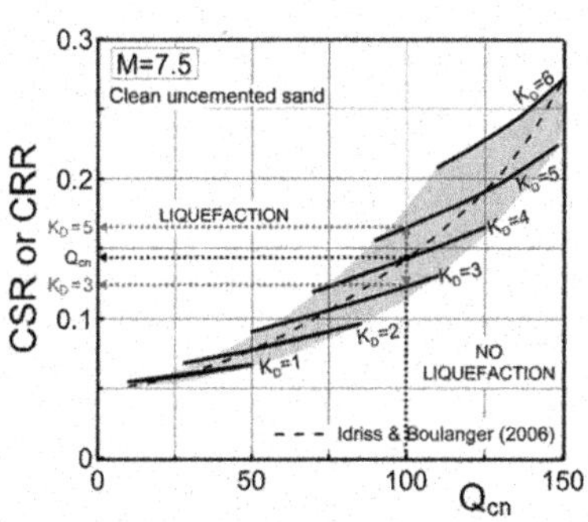

Figure 13. Correlation for estimating *CRR* based on both DMT & CPT for clean sand (modified from Marchetti 2016).

Besides the *CRR*-K_D curve, Marchetti (2016) also proposed a method to estimate *CRR* based on the combined use of DMT & CPT (Figure 13), aiming to complement the higher sensitivity of K_D to stress history with the much larger field performance support of CPT-based methods. The conceptual framework is the following: (1) estimate *CRR* using a "consensus" CPT liquefaction curve; (2) increase (shift upwards) the *CRR* predicted by CPT if K_D is high, reduce it (shift downwards) if K_D is low. As an example, Figure 13 for a normalized cone resistance Q_{cn} = 100 and K_D = 3 provides *CRR* = 0.125 (i.e., lower than predicted by the dashed CPT curve), while for the same Q_{cn} = 100 and K_D = 5 the chart provides *CRR* = 0.17 (i.e., higher than predicted by CPT). The method proposed by Marchetti (2016) for estimating *CRR* from both DMT & CPT is another example of multi-parameter / multi-test approach. It is expectable that an estimate based on two measured independent parameters could be more accurate than an estimate based on just one parameter.

6 PROPER MATCHING OF CPT & DMT DATA

One specific issue that arises when using any interpretation or design approach that requires the combination of CPT & DMT results is the proper matching of corresponding data. In fact, the CPT measurements are taken every 10-20 to 50 mm, whereas the DMT pressure readings are taken every 200 mm. Data from paired CPT & DMT soundings shall be matched at respective elevations. Care must be adopted even when nearby soundings are carried out at the same ground elevation, particularly in non uniform, highly stratified deposits.

Comparisons between individual same-depth values from nearby CPT & DMT profiles often show considerable scatter due to variations in soil stratigraphy and consistency, hence adjacent data from the same depth may not always represent the same soil. Sand deposits are often characterized by high variability in grain size distribution and D_r, then plots of individual data points from nearby CPT & DMT may show large scatter. The profiles of q_t may also differ slightly from those of adjacent DMT since the CPT senses soil slightly ahead and behind the cone tip (up to 15 cone diameters, depending on soil strength / stiffness and in-situ effective stresses, Ahmadi & Robertson 2005). The DMT appears to be less influenced by soil layers ahead and behind since the probe is stopped and the membrane expanded in horizontal direction. Hence, in interbedded soils the CPT may be influenced by adjacent soil layers somewhat more than the DMT (Robertson 2009a).

Any comparison between adjacent CPT & DMT should be done in terms of the near continuous profiles with depth, so that any variation in soil stratigraphy can be identified from the profiles. It is also common to compare values obtained at the same depth within relatively uniform sections of a deposit. As an example, at the highly stratified Treporti test site (Monaco et al. 2014) the paired CPTu-DMT data, including the data pairs $M_{DMT} - q_t$ shown in Figure 7, were carefully selected by retaining only pairs from uniform soil layers of significant thickness to avoid any possible mismatching of data.

In general, the following matching criteria may be adopted: (1) match paired CPT & DMT profiles accounting for respective ground elevations, if not coincident; (2) correct both CPT & DMT test depths for inclination; (3) select soil type (e.g., sand-like, $I_c \leq 2.60$ and $I_D > 1$, or clay-like, $I_c > 2.60$ and $I_D < 1$) and possibly depth range of interest; (4) identify "visually" or by cross-correlation of representative parameters (e.g., q_t and M_{DMT}) corresponding soil layers in CPT & DMT profiles, to detect any depth misalignment offset; (5) identify soil layers having a reasonable thickness (e.g., in which at least two DMT readings are present) and relatively uniform CPT & DMT parameters; (6) use average values of CPT & DMT parameters in each layer. It would be desirable that the matching algorithm is implemented in software tools, although some visual inspection and judgment may be still needed.

7 CONCLUDING REMARKS

Direct-push in-situ tests such as the CPT and the DMT, particularly in their powerful seismic configurations, have proven to be remarkably expedient in geotechnical site investigations. Recent experience highlights an increasing awareness of researchers and practitioners that an adequate site investigation for a given project cannot rely solely on one isolated testing technique, delineating the trend of expanding diffusion of a multi-parameter / multi-test approach in site investigation practice.

The CPT and the DMT can be effectively combined to obtain more accurate information on stratigraphy and stress history of soil deposits, more reliable estimates of soil parameters (e.g., *OCR* and K_0 in sand) and for various design applications (e.g., monitoring ground improvement, liquefaction assessment). Moving towards an in-situ multi-parameter / multi-test approach appears a logical trend. In this respect, the availability of the DMT stress history parameter K_D is important not only "per se", but also in combination with parameters obtained from CPT.

Each test has advantages and limitations. The CPT is often used more than the DMT, especially for smaller projects, being faster and less expensive (no pauses of penetration are required for obtaining the measurements, as for the DMT). However, the DMT has been shown to be more sensitive to stress history, which has a dominant influence on soil behavior. Hence, their mutual integration could helpfully merge the respective advantages and counterbalance their limitations. For instance, the CPT could be used for a "screening" of the site on a large scale, the DMT at locations of particular relevance for the project. Adjacent CPT & DMT at selected locations should be planned. The increase in time and cost of a site investigation program including paired CPT & DMT soundings can be efficiently optimized considering that both tests can share the same insertion rig and rods, and even the same cable when using the Medusa DMT (i.e., only the probe connected to the bottom rod needs to be changed). Such "redundancy" would be largely compensated by substantial increase in soil information.

Interrelationships between CPT & DMT can be used to expand and improve correlations and applications by using experience and databases from one test and extrapolating to the other test. The accumulation and sharing of paired data sets from companion series of CPT & DMT soundings at well-documented and benchmark test sites is a desirable address for future research and practice.

ACKNOWLEDGEMENTS

Special thanks to Peter K. Robertson for sharing his updates on CPT-DMT correlations. Thanks to Diego Marchetti, Sara Amoroso, Anna Chiaradonna for their helpful suggestions, to Gianfranco Totani and Giovanni Bosco for the shared experience on in-situ testing. Grateful thanks to Silvano Marchetti for his constant and bright guidance over many years.

REFERENCES

Ahmadi, M.M. & Robertson, P.K. 2005. Thin layer effects on the CPT q_c measurement. *Can. Geotech. J.* 42(9): 1302–1317.

Amoroso, S., Comina, C. & Marchetti, D. 2020. Combined P- and S-Wave Measurements by Seismic Dilatometer Test (SPDMT): A Case History in Bondeno (Emilia Romagna, Italy). *Geotech. Testing J.* 43(2): 383–393.

Amoroso, S., Garcia Martinez, M.F., Monaco, P., Tonni, L., Gottardi, G., Rollins, K.M., Minarelli, L., Marchetti, D. & Wissmann, K.J. 2022a. Comparative study of CPTU and SDMT in liquefaction prone silty sands with ground improvement. *J. Geotech. Geoenviron. Eng.* (accepted) http://dx.doi.org/10.1061/(ASCE)GT.1943-5606.0002801

Amoroso, S., Martínez, M.F., Tonni, L., Gottardi, G., Monaco, P., Rollins, K.M., Minarelli, L., Marchetti, D. & Wissmann, K.J. 2022b. Monitoring ground improvement by Rammed Aggregate Piers using a combined CPTU and SDMT approach at a silty sand liquefaction-prone site in Emilia-Romagna. *Proc. 5th Int. Symp. Cone Penetration Testing CPT'22, Bologna*, 8-10 June 2022.

Amoroso, S., Rollins, K.M., Monaco, P., Holtrigter, M. & Thorp, A. 2018. Monitoring Ground Improvement Using the Seismic Dilatometer in Christchurch, New Zealand. *Geotech. Testing J.* 41(5): 946–966.

Bałachowski, L. & Kurek, N. 2015. Vibroflotation Control of Sandy Soils. In S. Marchetti, P. Monaco & A. Viana da Fonseca (eds), *Proc. 3rd Int. Conf. Flat Dilatometer DMT'15, Rome*, 14-16 June 2015, 185–190.

Baldi, G., Bellotti, R., Ghionna, V., Jamiolkowski, M., Marchetti, S. & Pasqualini, E. 1986. Flat Dilatometer Tests in Calibration Chambers. *Proc. Spec. Conf. Use of*

In Situ Tests in Geotech. Eng. In Situ '86, *Blacksburg*, 431–446.

Benoit, J. 1989. Personal communication to S. Marchetti.

Boulanger, R.W. & Idriss, I.M. 2014. CPT and SPT liquefaction triggering procedures. *Report No UCD/GCM-14/01*, University of California at Davis.

Burghignoli, A., Cavalera, L., Chieppa, V., Jamiolkowski, M., Mancuso, C., Marchetti, S., Pane, V., Paoliani, P., Silvestri, F., Vinale, F. & Vittori, E. 1991. Geotechnical characterization of Fucino clay. *Proc. 10th Eur. Conf. Soil Mech. Found. Eng., Florence*, 26-30 May 1991, 1: 27–40. Rotterdam: Balkema.

Chiaradonna, A. & Monaco, P. 2022. Assessment of liquefaction triggering by seismic dilatometer tests: comparison between semi-empirical approaches and non-linear dynamic analyses. *Proc. 20th Int. Conf. Soil Mech. Geotech. Eng., Sydney*, 1-5 May 2022.

Choo, H., Lee, W., Lee, C., Kim, Y.-S. & Lee, M.-J. 2015. Perception of Overconsolidated States of Coarse-Grained Soils Using DMT Results. In S. Marchetti, P. Monaco & A. Viana da Fonseca (eds), *Proc. 3rd Int. Conf. Flat Dilatometer DMT'15, Rome*, 14-16 June 2015, 205–212.

Hossain, M.A. & Andrus, R.D. 2016. At-Rest Lateral Stress Coefficient in Sands from Common Field Methods. *J. Geotech. Geoenviron. Eng.* 142(12): 06016016.

Idriss, I.M. & Boulanger, R.W. 2006. Semi-empirical procedures for evaluating liquefaction potential during earthquakes. *Soil Dyn. Earthquake Eng.* 26(2-4): 115–130.

Jamiolkowski, M. & Lo Presti, D.C.F. 1998. DMT research in sand. What can be learned from calibration chamber tests. *1st Int. Conf. Site Characterization ISC '98, Atlanta*, 19-22 April 1998 (oral presentation).

Jendeby, L. 1992. Deep Compaction by Vibrowing. *Proc. Nordic Geotech. Meeting NGM-92, Aalborg*, 1: 19–24.

Lee, M.J., Choi, S.K., Kim, M.T. & Lee, W. 2011. Effect of stress history on CPT and DMT results in sand. *Eng. Geol.* 117(3-4): 259–265.

Leon, E., Gassman, S.L. & Talwani, P. 2006. Accounting for Soil Aging When Assessing Liquefaction Potential. *J. Geotech. Geoenviron. Eng.* 132(3): 363–377.

Lunne, T. & Christophersen, H.P. 1983. Interpretation of cone penetrometer data for offshore sands. *Proc. 15th Annual Offshore Technology Conf.*, 181–188.

Lutenegger, A.J. & Kabir, M.G. 1988. Dilatometer C-reading to help determine stratigraphy. *Proc. Int. Symp. Penetration Testing ISOPT-1, Orlando*, 1: 549–554. Rotterdam: Balkema.

Marchetti, D., Monaco, P., Amoroso, S. & Minarelli, L. 2019. In situ tests by Medusa DMT. *Proc. XVII Eur. Conf. Soil Mech. Geotech. Eng., Reykjavik*, 1-6 September 2019.

Marchetti, D., Monaco, P., Totani, G., Totani, F. & Amoroso, S. (2022). Comparisons CPT-DMT in soft clay at Fucino-Telespazio GeoTest site. *Proc. 5th Int. Symp. Cone Penetration Testing CPT'22, Bologna*, 8-10 June 2022.

Marchetti, S. 1980. In Situ Tests by Flat Dilatometer. *J. Geotech. Eng. Div.* 106(GT3): 299–321.

Marchetti, S. 1985. On the Field Determination of K_0 in Sand. *Proc. 11th Int. Conf. Soil Mech. Found. Eng., San Francisco*, 5: 2667–2673. Rotterdam: Balkema.

Marchetti, S. 2010. Sensitivity of CPT and DMT to stress history and aging in sands for liquefaction assessment. *Proc. 2nd Int. Symp. Cone Penetration Testing CPT'10, Huntington Beach*, 9-11 May 2010.

Marchetti, S. 2011. Discussion of "CPT-DMT Correlations" by P.K. Robertson. *J. Geotech. Geoenviron. Eng.* 137(4): 441–442.

Marchetti, S. 2015. Some 2015 Updates to the TC16 DMT Report 2001. In S. Marchetti, P. Monaco & A. Viana da Fonseca (eds), *Proc. 3rd Int. Conf. Flat Dilatometer DMT'15, Rome*, 14-16 June 2015, 43–65.

Marchetti, S. 2016. Incorporating the Stress History Parameter K_D of DMT into the Liquefaction Correlations in Clean Uncemented Sands. *J. Geotech. Geoenviron. Eng.* 142(2): 04015072.

Marchetti, S. & Monaco, P. 2018. Recent Improvements in the Use, Interpretation, and Applications of DMT and SDMT in Practice. *Geotech. Testing J.* 41(5): 837–850.

Marchetti, S., Monaco, P., Totani, G. & Calabrese, M. 2001. The Flat Dilatometer Test (DMT) in Soil Investigations – A Report by the ISSMGE Committee TC16. Official version approved by ISSMGE TC16 reprinted in R. A. Failmezger & J.B Anderson (eds), *Proc. 2nd Int. Conf. Flat Dilatometer, Washington D.C.*, 2-5 April 2006, 7–48.

Mayne, P.W. 2006. Interrelationships of DMT and CPT Readings in Soft Clays. In R.A. Failmezger & J. B Anderson (eds), *Proc. 2nd Int. Conf. Flat Dilatometer, Washington D.C.*, 2-5 April 2006, 231–236.

Mayne, P.W. 2016. Evaluating effective stress parameters and undrained shear strengths of soft-firm clays from CPTu and DMT. *Australian Geomech. J.* 51(4): 27–55.

Mayne, P.W. & Bachus, R.C. 1989. Penetration Pore Pressures in Clay by CPTU, DMT, and SBP. *Proc. 12th Int. Conf. Soil Mech. Geotech. Eng., Rio de Janeiro*, 291–294. Rotterdam: Balkema.

Mayne, P.W., Coop, M.R., Springman, S.M., Huang, A.B. & Zornberg, J.G. 2009. Geomaterial behavior and testing. *Proc. 17th Int. Conf. Soil Mech. Geotech. Eng., Alexandria*, 5-9 October 2009, 4: 2777–2872. Amsterdam: IOS Press.

Monaco, P., Amoroso, S., Marchetti, S., Marchetti, D., Totani, G., Cola, S. & Simonini, P. 2014. Overconsolidation and Stiffness of Venice Lagoon Sands and Silts from SDMT and CPTU. *J. Geotech. Geoenviron. Eng.* 140(1): 215–227.

Monaco, P., Marchetti, S., Totani, G. & Calabrese, M. 2005. Sand Liquefiability Assessment by Flat Dilatometer Test (DMT). *Proc. 16th Int. Conf. Soil Mech. Geotech. Eng., Osaka*, 4: 2693–2697. Rotterdam: Millpress.

Monaco, P., Tonni, L., Amoroso, S., Garcia Martinez, M.F., Gottardi, G., Marchetti, D. & Minarelli, L. 2021. Use of Medusa DMT in alluvial silty sediments of the Po river valley. *Proc. 6th Int. Conf. Geotech. Geophys. Site Characterisation ISC'6, Budapest*, 26-29 September 2021.

Monaco, P., Totani, G. & Calabrese, M. 2006. DMT-predicted vs observed settlements: a review of the available experience. In R.A. Failmezger & J.B Anderson (eds), *Proc. 2nd Int. Conf. Flat Dilatometer, Washington D.C.*, 2-5 April 2006, 244–252.

Ouyang, Z. & Mayne, P.W. 2017. Spherical cavity expansion nexus between CPTu and DMT in soft-firm clays. *Proc. 19th Int. Conf. Soil Mech. Geotech. Eng., Seoul, 17-22 September 2017*, 631–634.

Ouyang, Z. & Mayne, P.W. 2018. Effective stress parameters of clays from DMT. *Geotech. Testing J.* 41(5): 851–867.

Pyke, R. 2003. Discussion of "Liquefaction Resistance of Soils: Summary Report from the 1996 NCEER and 1998 NCEER/NSF" by T.L. Youd et al., *J. Geotech. Geoenviron. Eng.* 129(3): 283–284.

Robertson, P.K. 1990. Soil classification using the cone penetration test. *Can. Geotech. J.* 27(1): 151–158.

Robertson, P.K. 2009a. CPT-DMT Correlations. *J. Geotech. Geoenviron. Eng.* 135(11): 1762–1771.

Robertson, P.K. 2009b. Interpretation of cone penetration tests – a unified approach. *Can. Geotech. J.*, 46(11): 1337–1355.

Robertson, P.K. 2012. The James K. Mitchell Lecture: Interpretation of in-situ tests – some insights. In R. Coutinho & P.W. Mayne (eds), *Proc. 4th Int. Conf. Geotech. Geophys. Site Characterization, Porto de Galinhas*, 1: 3–24.

Robertson, P.K. 2015. Soil Behavior Type using the DMT. In S. Marchetti, P. Monaco & A. Viana da Fonseca (eds), *Proc. 3rd Int. Conf. Flat Dilatometer DMT'15, Rome*, 14-16 June 2015, 243–250.

Robertson, P.K. 2017. DMT-CPT Correlations. *Prof. Silvano Marchetti Memorial Session on In-Situ Testing, 19th Int. Conf. Soil Mech. Geotech. Eng., Seoul, 17-22 September 2017* (oral presentation).

Robertson, P.K. & Wride, C.E. 1998. Evaluating cyclic liquefaction potential using the cone penetration test. *Can. Geotech. J.* 35(3): 442–459.

Schnaid, F., Belloli, M.V.A., Odebrecht, E. & Marchetti, D. 2018. Interpretation of the DMT in silts. *Geotech. Testing J.* 41(5): 868–876.

Schnaid, F., Odebrecht, E., Sosnoski, J. & Robertson, P.K. 2016. Effects of test procedure on flat dilatometer test (DMT) results in intermediate soils. *Can. Geotech. J.* 53 (8): 1270–1280.

Senneset, K., Sandven, R., Lunne, T., By, T. & Amundsen, T. 1988. Piezocone Tests in Silty Soils. *Proc. Int. Symp. Penetration Testing ISOPT-1, Orlando*, 2: 955–966. Rotterdam: Balkema.

Togliani, G., Calzolari, L. & Menghini, A. 2015. Governolo & Gonzaga (Mantua, Italy) Experimental Sites: In situ Test comparison. In S. Marchetti, P. Monaco & A. Viana da Fonseca (eds), *Proc. 3rd Int. Conf. Flat Dilatometer DMT'15, Rome*, 14-16 June 2015, 259–270.

Tonni, L., Gottardi, G., Amoroso, S., Bardotti, R., Bonzi, L., Chiaradonna, A., d'Onofrio, A., Fioravante, V., Ghinelli, A., Giretti, D., Lanzo, G., Madiai, C., Marchi, M., Martelli, L., Monaco, P., Porcino, D., Razzano, R., Rosselli, S., Severi, P., Silvestri, F., Simeoni, L., Vannucchi, G. & Aversa, S. 2015. Interpreting the deformation phenomena triggered by the 2012 Emilia seismic sequence on the Canale Diversivo di Burana banks. *Rivista Italiana di Geotecnica* 49 (2):28–58 (in Italian).

Tsai, P.-H., Lee, D.-H., Kung, G.T.-C. & Juang, C.H. 2009. Simplified DMT-Based Methods for Evaluating Liquefaction Resistance of Soils. *Eng. Geol.* 103(1-2): 13–22.

Technical papers

Session 1: Equipment and procedures

Cone Penetration Testing 2022 – Gottardi & Tonni (eds)

Incorporation of SH source wave parameter "SH Polarization" within DST seismic trace characterization

Erick Baziw & Gerald Verbeek
Baziw Consulting Engineers Ltd., Vancouver, Canada

ABSTRACT: Downhole Seismic Testing (DST) is an important geotechnical testing technique for site characterization that provides low strain in-situ interval shear wave velocity estimates, which are fundamental design parameters for static and dynamic soil analysis. A challenging problem in DST is to obtain an accurate assessment or characterization of the quality of the acquired seismic data, which is then used to guide the analysis process to obtain the most accurate interval velocity values. The characterization process is referred to as Seismic Trace Characterization (STC). STC derives various independent parameters of the acquired seismic data at a particular depth, which are then fused together into a single classification. To date Baziw Consulting Engineers has identified five STC independent parameters. These five parameters are the linearity estimates from the polarization analysis, the cross correlation coefficient of the full waveforms at the particular depth and the preceding depth, a uniquely developed parameter referred to as the signal shape parameter, the signal-noise-ratio and the peak symmetry differential, which provides insight into the skewing or time shifting of the peak source wave response. This paper outlines a newly identified seismic trace feature that is independent of the parameters listed above. This new parameter is SH Polarization (SHP), which quantifies the desired polarization of the generated source on the horizontal plane compared with particle motion generated on the vertical plane.

Cone Penetration Testing 2022 – Gottardi & Tonni (eds)
© 2022 the Author(s), ISBN 978-1-032-35624-2

Methodology for obtaining true cone bearing estimates from blurred and noisy measurements

Erick Baziw & Gerald Verbeek
Baziw Consulting Engineers Ltd., Vancouver, Canada

ABSTRACT: Cone penetration testing (CPT) is an important and widely used geotechnical *in-situ* test for assessing soil properties and mapping soil profiles. CPT consists of pushing at a constant rate an electronic cone into penetrable soils and recording the resistance to the cone tip or cone bearing (q_m). These values (after correction for the pore water pressure to get q_t) are utilized to characterize the soil profile along with measured sleeve friction and pore pressure. The q_m measurements can have significant fluctuations when penetrating sandy, silty gravelly soils, resulting in "high" peaks due to interbedded gravels and stones and "low" peaks due to softer materials or local pore pressure build-up. Furthermore, the q_m values are blurred and/or averaged which results in the inability to identify thin layers and the distortion of the soil profile characterization. Baziw Consulting Engineers has invested considerable resources in addressing these two q_m measurement challenges. The q_mKF algorithm was developed to address the additive measurement noise. In this case the dynamics of q_m are modeled within a state-space mathematical formulation and a Kalman filter is then utilized to obtain optimal estimates of q_m. The q_mHMM algorithm implements a hidden Markov model smoother so that true cone bearing are obtained from the averaged/blurred q_m values. This paper outlines the integration of the q_mKF and q_mHMM algorithms and demonstrates the performance first with test bed simulations (to show the functionality of the algorithm) and then through the analysis of various actual q_m data sets.

Cone Penetration Testing 2022 – Gottardi & Tonni (eds)

Evaluation of parameters inducing desaturation of a piezocone: Saturation liquid viscosity and exposure to dry sand

G. De Backer, R.D. Verastegui-Flores, W. Vervaele, L. Vincke & K. Haelterman
Department of Mobility and Public Works, Flemish Government, Ghent, Belgium

ABSTRACT: Several CPTU tests were performed in both ideal and adverse conditions at a test site in Zwijnaarde. The influence of exposure to unsaturated soils was examined by holding well-saturated piezocones in dry sand for several minutes without a protective rubber membrane. After the exposure to dry sand, penetration was performed from the ground level through the unsaturated and saturated subsoil. These tests show suboptimal pore pressure measurements, compared to the tests under ideal circumstances, regardless of the duration of the exposure to dry sand. Results reveal that, in adverse conditions, the densely packed sandy top layer, rather than the exposure to dry sand, has contributed to the desaturation of the piezocones. The impact of the viscosity of the saturation liquid on pore pressure quality was studied by performing tests with a 50 and 100 cSt silicone oil. Under ideal circumstances, the viscosity does not play a role in terms of quality of the pore pressure measurements. However, in adverse conditions, results showed a better performance for a silicone oil with 100 cSt viscosity.

Cone Penetration Testing 2022 – Gottardi & Tonni (eds)
© 2022 the Author(s), ISBN 978-1-032-35624-2

Large diameter cone penetrometers: What is an appropriate location for the transition to the rod diameter?

D.A. de Lange, T.A. van Duinen & D.J. Peters
Deltares, Delft, The Netherlands

ABSTRACT: 15 cm^2 cone penetrometers are typically pushed to depth by means of 10 cm^2 rods. Therefore, a transition in diameter will be present at a certain distance from the cone shoulder. Although the standards give a requirement for the position of this transition, cone penetrometer configurations that don't meet this requirement are often preferred by CPT sounding companies and are being used in practice. This raises a dilemma. In order to fuel discussion, a first step was made by performing two series of 9 CPTs with different penetrometer geometries. The effects of the penetrometer designs on the measured quantities were investigated. From a preliminary analysis of the results no systematical differences between the different penetrometers were observed. When these preliminary findings are better understood and validated, the standards can be updated.

Cone Penetration Testing 2022 – Gottardi & Tonni (eds)

Issues related to piezocone sleeve friction measurement accuracy in soft sensitive clays

B. Di Buò & M. D'Ignazio
Tampere University / Ramboll Finland Oy, Tampere, Finland

T. Länsivaara & M. Haikola
Tampere University, Tampere, Finland

ABSTRACT: Over the past decades, the piezocone testing has been increasingly adopted for field investigation as it offers a quick and cost-effective methodology for subsoil profiling and geotechnical parameters estimation. Although the piezocone testing has revealed good applicability worldwide, difficulties are encountered in complex soil conditions, such as soft sensitive clays. One of the key issues in such soils is represented by the poor quality of the sleeve friction (f_s) measurement due to inaccuracy and poor resolution of the sleeve sensor. This paper investigates the influence of f_s data quality on soil parameters determination with particular emphasis on the soil behavior type (SBT) chart classification. The field investigation was conducted in a soft sensitive clay site located in Finland using two different penetrometers: a standard piezocone and an advanced piezocone characterized by enhance accuracy sleeve friction sensor. Results show that the use of high-resolution piezocone plays a key role in soft clays to avoid misleading soil type classification.

Cone Penetration Testing 2022 – Gottardi & Tonni (eds)
© 2022 the Author(s), ISBN 978-1-032-35624-2

Flow cone – new CPTU add-on module trialled in Halden silt

A.S. Gundersen, T. Lunne, R. Stelzer & Ø. Blaker
Norwegian Geotechnical Institute

G.W. Tucker & L. Krogh
Ørsted

ABSTRACT: Hydraulic conductivity is an important soil parameter for design of shallow foundation concepts for offshore wind, and there is a need for a new tool that can measure this parameter in a reliable way. A new module has been developed that can be mounted behind a standard piezocone test (CPTU) probe. Water can be injected in a controlled manner into the CPTU equipment and flow out into the surrounding soil through a filter an offset behind the friction sleeve, while water pressure (u_f) is measured by a transducer mounted in the filter itself. During penetration of the CPTU probe, water flows out at a constant rate, while u_f is measured in addition to CPTU parameters q_c, f_s and u_2. At desired depth intervals penetration can be stopped and either a dissipation test or constant flow rate test can be carried out to determine hydraulic conductivity.

Hydraulic fracture tests can be performed in low permeability soils, where water flow is used to induce a vertical crack in the soil. The vertical crack is then allowed to close while pore pressure is monitored, from which the closing pressure can be determined and subsequently used to determine the in-situ K_0-condition.

This paper describes a series of hydraulic conductivity tests and a hydraulic fracture test carried out at one of Norway's recently established geotechnical test sites, a silt dominated site in Halden. Several tests were successfully carried out and the results were benchmarked against hydraulic conductivity as measured by falling head tests in standpipes and laboratory tests. In general, the results compared well to credible benchmarking tests, showing a promising potential for this tool.

Cone Penetration Testing 2022 – Gottardi & Tonni (eds)
© 2022 the Author(s), ISBN 978-1-032-35624-2

Estimating in-situ frozen loamy soil viscosity from CPT

O.N. Isaev
Gersevanov Research Institute of Bases and Underground Structures (NIIOSP), "Research Center of Construction" JSC, Moscow, Russia

ABSTRACT: Viscosity is used for calculation of settlements and deformations of frozen loamy soils and ice caused by viscous flow at constant speed under action of long-term loads. A method for determining the viscosity coefficient of frozen loamy soil by CPT was developed. Experimental studies by CPT were carried out in a quasi-static mode (a series of short tests at low well-controlled speeds) and in a step-increasing relaxation-creeping mode (a series of penetrometer stabilizations in which the penetrometer and the speed of its penetration gradually decreased). Comparative data «unit cone resistance q_c – velocity of penetration v_c» were analyzed using mathematical statistics methods. For frozen loamy soil, a generalized regression equation $q_c = M(v_c)^m$ was established. Parameters M and m are calculated based on empiric equations, depending on the soil temperature. Viscosity values obtained by uniaxial compression CPT were compared. As a result of statistical analysis, an empirical transition coefficient between them was established.

Cone Penetration Testing 2022 – Gottardi & Tonni (eds)

Determination of natural stress state parameters for clay soils by using 3LSU-CPTU penetrometer

O.N. Isaev, R.F. Sharafutdinov & D.S. Zakatov

Gersevanov Research Institute of Bases and Underground Structures (NIIOSP), "Research Center of Construction" JSC, Moscow, Russia

ABSTRACT: The article presents the results of experimental studies on the use of lateral stress measurements by using the LS-module (includes full and pore pressure sensors) of 3LSU-CPTU penetrometer to determine the parameters of the natural stress state (NSS) of clay soils (the coefficient of lateral pressure at rest K_o, the effective pre-compaction stress *POP*, the over-compaction stress σ'_c, the overconsolidation ratio *OCR*, the natural total σ_{ho} and effective σ'_{ho} horizontal stress in the soil). In experiments, the LS-module with three pairs of sensors located in areas with different diameters (Ø37 mm, Ø41 mm and Ø45 mm) was used. The results of a statistical analysis of the relationships between LS-module (twenty-six direct and derived types were considered) and NSS parameters of clay soils are presented. It is shown that K_o, *OCR*, σ'_{ho} in the soil is determined with the greatest accuracy.

Cone Penetration Testing 2022 – Gottardi & Tonni (eds)
© 2022 the Author(s), ISBN 978-1-032-35624-2

Towards correlating seabed penetrometer and chirp sonar measurements

R. Jaber & N. Stark
Virginia Tech, Blacksburg, VA, USA

ABSTRACT: Cone Penetration Testing (CPT) and seabed coring represent key methods to determine geotechnical properties offshore. Physical testing is typically combined with geophysical surveying, i.e., chirp sonar sub-bottom profilers, and results are qualitatively related to extrapolate geotechnical properties between physical test locations. Literature suggests that more quantitative relationships can be established between geotechnical properties and geoacoustic responses. However, specific correlations are still rare, and correlating frameworks are often complex and in need of additional parameters. This study attempts to explore correlations between geotechnical properties measured using a portable free fall penetrometer (PFFP) and determined from soil samples and chirp sonar at a frequency of 8-12 kHz. Preliminary results focusing on fine-grained sediments using data collected in the York River, Virginia, USA suggest relations between sediment properties using PFFP measurements and geoacoustic properties using chirp measurements and laboratory testing.

Cone Penetration Testing 2022 – Gottardi & Tonni (eds)

Development of free fall cone penetration testing system

Hyoun Kang, Osoon Kwon, Changjoo Shin, Jungmin Seo, Insung Jang & Man Dong-Woo
Korea Institute of Ocean Science & Technology, Busan, Korea

ABSTRACT: For the rigorous design of offshore structures, the geotechnical characteristics of the seabed ground should be rigorously investigated, in the deep water, seabed ground surveys are complex equipment and costly. Free fall type ground survey equipment that can acquire a simple and comparatively accurate ground is developed when only the characteristic investigation of the shallow depth ground such as the submarine pipeline, the submarine cable burial is the main concern. In this study, FFCPT was developed. Based on behavior results of numerical model simulation for FFCPT, it was manufactured. Water depth was obtained using pressure sensor and penetrated depth was estimated by the double integration of acceleration signal. In addition, we can verify vertical penetration at real field test of FFCPT using 2axis of inclinometer. So, the results of the seabed type offshore CPTu experiment and the developed model were compared and the performance was evaluated.

Keywords: FFCPT (cone penetration testing system), tip resistance, sleeve friction, pore pressure

Cone Penetration Testing 2022 – Gottardi & Tonni (eds)
© 2022 the Author(s), ISBN 978-1-032-35624-2

Evaluation of statistical fluctuation of measured data from nuclear density cone penetrometer

Muthusamy Karthikeyan
Surbana Jurong Consultants Pte Ltd, Singapore

ABSTRACT: This paper describes research conducted to investigate the statistical fluctuation and interpret the density profiles obtained by a nuclear-density cone penetrometer (ND-CP). The laboratory results show that the amount of fluctuations present in the ND-CP measurements can be approximated by Normal or Gaussian distribution. The coefficient of variation for RI count and BG count data's is 2.4% and 15.1% respectively. The moving-average technique has been employed for the measured raw data at different time intervals in order to reduce the statistical fluctuation. It was found that the error due to the statistical fluctuation in ND-CP measurement is less than 1% if an averaging span of 10 cm is used. The raw data obtained from the field results when averaged over 10cm depth shows that the cone profile count distribution trend is not distorted or deformed compared to the original raw data, which means the loss of information about the density through the filtering is negligible. This averaging is necessary to smooth the measured wet density profile. Therefore, the ND-CP measurements are averaged over a span of 10cm depth to minimize the statistical fluctuations.

Cone Penetration Testing 2022 – Gottardi & Tonni (eds)
© 2022 the Author(s), ISBN 978-1-032-35624-2

On the accuracy and precision of the seismic cone penetration test – a field test study on the seismic source

O. Koreta, A.H. Augustesen & L. Krogh
Ørsted

K. Lundvig & S. Bøtker-Rasmussen
Geo

ABSTRACT: The seismic cone penetration test (SCPT) is one of the most frequently used *in situ* testing methods for deriving the small strain shear modulus. Offshore SCPT results often present large scatter, in particular when data are obtained with different operational setups. No standard or industry consensus exist in terms of requirements to system key characteristics, operational testing procedures, interpretation and the numerical quality control of the traces obtained. Seismic sources currently deployed as part of the offshore industry practice have been developed individually by the various suppliers. This paper presents the findings of an onshore experimental study with the objective to increase the understanding of the influence of the offshore seismic source setup on SCPT results under controlled operational procedures. Accuracy and precision of the SCPT are investigated as part of the study.

Cone Penetration Testing 2022 – Gottardi & Tonni (eds)
© 2022 the Author(s), ISBN 978-1-032-35624-2

The revival of multiple pore pressure measurements in the cone penetration test

T. Lunne
Norwegian Geotechnical Institute, Oslo, Norway

R.K. Ghanekar
Consultant, Panvel, India

G.W. Tucker, R. Santos & L. Krogh
Ørsted, Denmark

ABSTRACT: In common practice the most usual positioning of the filter location for the piezocone (CPTU) is at the cone shoulder (u_2) which is recommended by ISO 19901-8:2014 and ISO 22476-1:2012. However, these ISO standards allow additionally for pore pressure measurements to be taken at the cone face (u_1) or behind the friction sleeve (u_3). The triple element piezocone (CPTU3) offers a solution whereby measurements of pore pressure can be taken simultaneously at all three locations. By taking these three pore pressure measurements the soil behaviour classification may be enhanced by interpretation in terms of c_h from u_1, u_2 and u_3 and by correlations to overconsolidation ratio (OCR) and lateral stress ratio (K_0). In this paper the correlations to OCR are explored by establishing a database from historical data as presented in international geotechnical literature.

Cone Penetration Testing 2022 – Gottardi & Tonni (eds)

Comparisons CPT-DMT in soft clay at Fucino-Telespazio GeoTest site

D. Marchetti
Studio Prof. Marchetti, Rome, Italy

P. Monaco & G. Totani
University of L'Aquila, L'Aquila, Italy

F. Totani
Consultant Engineer, L'Aquila, Italy

S. Amoroso
University of Chieti-Pescara, Pescara, Italy
Istituto Nazionale di Geofisica e Vulcanologia, L'Aquila, Italy

ABSTRACT: This paper presents the comparison of results obtained from CPT/CPTU tests carried out in past investigations and from recent DMT tests carried out using the automated Medusa DMT, adopting different test procedures (standard, repeated *A*-readings, *A*-reading while penetrating), at the benchmark soft clay test site of Fucino-Telespazio, Italy. In particular, the depth profiles of the undrained shear strength s_u of the Fucino clay obtained from CPT/CPTU interpretation are compared with the s_u profiles obtained from standard DMT and from Medusa DMT using different test procedures, as well as with the s_u values obtained from field vane test (FVT), self boring pressuremeter test (SBPT) and laboratory tests available from past investigations, resulting generally in good agreement.

Cone Penetration Testing 2022 – Gottardi & Tonni (eds)

An innovative new 3MPa CPT – to detect and measure very small f_s values

A.J. McConnell
Founder of Insitu Geotech Services Pty Ltd (IGS)

E.J.C. Wassenaar
Regional Manager Asia-Pacific at Geomil Equipment B.V.

ABSTRACT: CPT testing of extremely soft soils and tailings materials requires the ability to detect and measure very low q_c values and extremely low f_s values. The former of these can and has been solved by use of low capacity cones with high quality well-calibrated load cells. The latter, detection and measurement of extremely low f_s values is an industry-wide problem, often treated as an "elephant in the room". This paper describes development of an innovative new CPT cone that the authors believe has largely solved this problem.

Cone Penetration Testing 2022 – Gottardi & Tonni (eds)
© 2022 the Author(s), ISBN 978-1-032-35624-2

Assessment of seismic cone penetration testing for small strain shear modulus

N. Parasie, T. Franken & J. Peuchen
Fugro, Nootdorp, The Netherlands

ABSTRACT: Reliable values of in situ shear modulus at small strain (G_{max}) are important for foundation design, particularly for large investments in offshore wind energy. The seismic cone penetration test (SCPT) method is commonly considered as primary reference for providing G_{max} of soil. The SCPT data acquisition process covers a combination of cone penetration testing and seismic downhole testing. Values of shear wave velocity are correlated with G_{max}. This paper presents a complete inventory of SCPT standard uncertainty components for G_{max}. These components are ranked as Low, Medium or High, indicating relative impact on a combined standard uncertainty budget for derived values of G_{max}. Supplementary discussion is provided based on theory and practice, with suggestions on uncertainty reduction. The focus of the assessment is on conventional SCPT systems, with a hammer impact source at ground surface and a dual-receiver seismic cone penetrometer.

Cone Penetration Testing 2022 – Gottardi & Tonni (eds)
© 2022 the Author(s), ISBN 978-1-032-35624-2

Experimental procedure for checking the saturation degree of piezocone tips

I. Rocchi
Department Environmental Engineering, Technical University of Denmark, Denmark

L. Tonni, G. Gottardi & M. Marcolongo
Department DICAM, University of Bologna, Italy

ABSTRACT: A correct measurement of the water pore pressure is essential to obtain reliable CPT results. Although this has always been a major concern, it still remains a possible relevant source of error in common engineering practice. In addition to usual reasons, such as malfunctioning of the equipment, its poor calibration and/or maintenance, a lack of complete saturation of the piezocone tip can play a major role. In fact, a correct and timely pore pressure measurement relies entirely on the full saturation of the pore pressure system. The variety of saturation fluids and methods that is encountered in practice is a clear evidence of the still existing uncertainties on the most suitable procedure. Furthermore, unless the saturation process is explicitly detailed, its choice is typically left to the contractor and the measurement quality can only be established a posteriori, after the test has been completed. This paper describes a novel experimental equipment that enables quantifying the degree of saturation of the piezocone tip prior to testing. The methodology proposed is inspired to the B-value check, which is routinely used in geotechnical laboratory testing to assess the degree of saturation, e.g. prior to triaxial testing. In addition, it could also be applied after a test, so that it would be possible to assess whether saturation was retained while testing. This promising approach, when extended to different saturation procedures and validated by engineering practice, has the potential to become a relevant benchmark for reliable CPT testing.

Cone Penetration Testing 2022 – Gottardi & Tonni (eds)
© 2022 the Author(s), ISBN 978-1-032-35624-2

Using penetrometer in situ and in box-core testing to obtain design information for lazy wave riser-soil interaction

O. Safaqah
ExxonMobil, Houston, Texas, USA

H.E. Low, S. Pant & S. Ingarfield
Fugro

M.F. Bransby & M.F. Randolph
Centre for Offshore Foundation Systems, Oceans Graduate School, The University of Western Australia, Australia

Z.J. Westgate
Department of Civil and Environmental Engineering, University of Massachusetts Amherst, USA
Formerly Fugro

ABSTRACT: Geotechnical assessment of vertical seabed stiffness during small vertical cycles of riser movement is important for riser fatigue analysis. These stiffness values change both with the amplitude of the cycles (because of non-linear stiffness) but also as the seabed consolidates under the small amplitude loading during long-term operation. To encapsulate both effects in design, a series of site-specific small displacement ball penetrometer testing was conducted both at seabed and in box-core samples on deck during a geotechnical survey for a recent deep water project. The paper describes the cyclic load and displacement controlled in situ testing, the results obtained, and how this data aided design. Of particular note were the long duration cyclic tests which were carried out for sufficiently long so as to not only capture secant stiffness reduction due to remoulding, but also subsequent stiffness increases due to consolidation. This novel test program performed on intact soil (as opposed to reconstituted samples in the laboratory or geotechnical centrifuge) confirms that consolidation hardening can occurs in naturally structured near-seabed soils.

Cone Penetration Testing 2022 – Gottardi & Tonni (eds)
© 2022 the Author(s), ISBN 978-1-032-35624-2

Calibration of cone penetrometers according to International Organization for Standardization requirements

R. Soage Santos
Ørsted, London, UK

E. Gómez Meyer & J. Peuchen
Fugro, Nootdorp, The Netherlands

G. Yetginer
Equinor, Stavanger, Norway

T. Lunne
Norwegian Geotechnical Institute, Oslo, Norway

T. Carrington
Fugro, Wallingford, UK

ABSTRACT: Detailed requirements for calibration of piezocone penetrometers are incorporated in standards published by International Organization for Standardization (ISO). The use of these standards can provide input for comparisons of cone penetration test (CPT) systems deployed in practice by means of cone penetrometer classes. It is important that parties specifying or supplying CPT data take note of the implications of the new requirements and the opportunities and benefits of appropriate selection of cone penetrometer classes. In addition, the information available from calibration and verification of a particular cone penetrometer can provide input into estimation of uncertainties of data points in CPT profiles used for design of structures.

This paper focusses on background information about topics considered for development of the ISO requirements. These topics included (1) practical and economical test methods achievable in a calibration laboratory, (2) assessment of differences in exposure conditions applied in the calibration process and site conditions likely to be encountered during actual cone penetration testing, and (3) cone penetrometers that incorporate ancillary sensors and algorithms for reducing the influence of temperature on CPT results.

Cone Penetration Testing 2022 – Gottardi & Tonni (eds)
© 2022 the Author(s), ISBN 978-1-032-35624-2

Long-term strength determination of frozen soils by CPT

Ivan Sokolov & N.G. Volkov
Fugro, Russian Federation

ABSTRACT: Frozen soils show rheological behavior which results in changing mechanical properties in time under applied load. It is highly complicated to determine the long-term strength of a frozen soil. Rheological behavior of frozen soils appears in creep or stress relaxation. This paper presents a new method to measure long-term strength of frozen soils using CPT equipment named Stress Relaxation Test (SRT). Conventional tests are focused on the creep behavior by creating constant loads and measuring strains. The SRT is based on stress relaxation approach by creating constant deformations and measuring stress by a cone. Cone resistance (q_c) would give compression long-term strength, as for sleeve friction (f_s) – shear long-term strength. SRT long-term strength results can be applied directly to pile bearing capacity estimation. A comparison between such estimation and results from a full scale static pile load test is provided in the paper.

Cone Penetration Testing 2022 – Gottardi & Tonni (eds)

Efficiency examined of hands-free Cone Penetration Testing using the SingleTwist™ with COSON

O. Storteboom & M. Woollard
A.P. van den Berg, Heerenveen, The Netherlands

J. Verhagen
GWR Ingenieursbureau, Rotterdam, The Netherlands

ABSTRACT: The Cone Penetration Test (CPT) cabin as working environment, built on trucks, crawlers or Track-Trucks®, will more and more develop from a workshop into an office. Before, operators were mainly dealing with manual operations to keep the production going. Nowadays and in the future, it can be increasingly expected that time "on board" will be spend on design or other office-related work next to performing CPTs. This paper describes a system that has been designed to support this market development. By integrating the patented SingleTwist™ technology in a CPT cabin with the COSON continuous pushing system, an automatic and hands-free CPT machine is created. The assembly and disassembly of the CPT string takes place fully automatically. This paper describes the efficiency of the SingleTwist™ technology. The City of Rotterdam, the first owner of a Track-Truck with COSON-ST, deployed their system in different projects. One project is chosen to further elaborate on its advantages and experience in practice.

Cone Penetration Testing 2022 – Gottardi & Tonni (eds)

Sustainability in CPT practice: Hybrid CPT Track-Truck

O. Storteboom & M. Woollard
A.P. van den Berg, Heerenveen, The Netherlands

B. Ooms
BAM Infra Nederland, Gouda, The Netherlands

ABSTRACT: Sustainability in construction is of great importance in government projects, but also of growing concern in the market in general. In this way, earth can be passed on to future generations in a healthy way. A sustainable realization of construction projects, within strict environmental standards, is the reason for companies to invest in electrification of equipment, so without carbon and nitrogen emissions. In terms of CPT, BAM Infra Nederland has taken the world's first hybrid truck with tracks into use early 2021. The heavy 4x4 truck is, while at the project location, electrically powered without emitting harmful substances such as carbon and nitrogen emissions. Both the CPT soil investigation and driving on the tracks are executed completely electrically. On the road, the truck is powered by the usual diesel engine. BAM Infra Nederland has now used the hybrid CPT Track-Truck for various projects. In the paper, experiences from practice are further elaborated and explained.

Cone Penetration Testing 2022 – Gottardi & Tonni (eds)

Elastic soil properties investigated using seismic tests to complement the CPT

O. Storteboom & M. Woollard
A.P. van den Berg, Heerenveen, The Netherlands

J.L. Rangel-Núñez
UAM-Azc and Ingeum Ingeniería, Mexico

ABSTRACT: Seismic tests are performed to investigate the elastic properties of the soil. Elastic soil properties are essential input for the prediction of soil-surface motions related to earthquake excitations and for the design of foundations for vibrating equipment, for the assessment of offshore structure behavior during wave loading and to predict deformations around excavations. To investigate these elastic properties of the soil, seismic tests can be performed. The digital Icone is easily extendable by click-on modules to measure additional CPT parameters. The Icone Seismic module is the instrument to investigate the elastic soil properties. This click-on module is automatically recognized, creating a true plug & play system. It contains three accelerometers to receive left and right shear waves as well as compression waves. The Icone Seismic system can be applied both onshore and offshore and has a -1,000 m water depth rating. In this paper the Icone Seismic system is described. Feedback from onshore fieldwork with the Icone Seismic will highlight the user experience with this approach.

Cone Penetration Testing 2022 – Gottardi & Tonni (eds)
© 2022 the Author(s), ISBN 978-1-032-35624-2

Tensiocone: A cone penetrometer with the facility to measure negative pore-water pressure

A. Tarantino & A. Capotosto
University of Strathclyde, Scotland, UK

F. Bottaro & M. Bellio
Tecnopenta S.r.l., Teolo (PD), Italy

D. Gallipoli
Università degli Studi di Genova, Italy

ABSTRACT: CPT sounding initially passes through the vadose zone, that is the zone above the phreatic surface where pore-water pressure is negative and degree of saturation is usually lower than unity. Negative pore-water pressure (suction) significantly affects tip resistance and sleeve friction and lack of knowledge of soil suction in this zone makes CPT data difficult if not impossible to interpret. For the case of Piezocone Cone Penetration Test (CPTU), crossing the vadose zone also exposes the cone tip porous filter to desaturation jeopardising the measurement of positive pore-water pressure once the cone penetrates the saturated zone below the phreatic surface. This paper presents the concept of the Tensiocone, a cone penetrometer with the facility to measure pore-water pressure in both negative and positive range. A first prototype was tested in the laboratory and in the field to investigate one of the major challenges in Tensiocone measurement, that is whether adequate contact can be established between the ground and the tensiometer porous filter during penetration.

Cone Penetration Testing 2022 – Gottardi & Tonni (eds)
© 2022 the Author(s), ISBN 978-1-032-35624-2

New portable pressiocone system for carrying out CPT+FDP tests

G. Vinco
Geotechnical Engineer, Technical Manager of URETEK SOLUTIONS Group

M. Sacchetto
Geotechnical Engineer, Freelance Consultant

ABSTRACT: Geotechnical investigations in the foundation soil of existing buildings have always been challenging due to limited space and difficult access. URETEK has developed a portable integrated system for simultaneously carrying out a CPT and a pressiometric test with Full Displacement Pressuremeter. The 30 kN thrust penetrometer to be used is very small. The reaction is given by two "microanchors". The cone is a standard 10 cm^2 digital memory cone (no cable), capable of measuring qc, fs, U every centimeter. Above the cone there is the FDP equipment with a rubber sheath covered by steel plates and connected, by a tube filled with water, to a device for creating pressure to inflate the sheath and measure pressure-volume curves as in a standard pressiometric test. The pressure-volume device and the depth transducer are connected to a microcomputer that is programmed to carry out CPT +FDP tests in an easy-to-use/user-friendly way.

Cone Penetration Testing 2022 – Gottardi & Tonni (eds)
© 2022 the Author(s), ISBN 978-1-032-35624-2

Characterisation of near-surface sediments using a blend of vertical and shallow rotational penetrometers

D.J. White
University of Southampton, UK

S.A. Stanier
University of Cambridge, UK

H. Mohr
University of Western Australia and MSMT Solutions, Perth, Australia

ABSTRACT: The mechanical properties of near-surface sediments – to a depth of approximately half a metre – are relevant to the design of cables, pipelines and other shallowly-embedded infrastructure, as well as benthic habitat characterisation. For this depth of interest, vertically-pushed penetrometers – such as the cone, T-bar or ball – can be supplemented by shallow rotational devices such as the toroid or hemiball. In this paper, we report vertical and shallow rotational penetrometer test procedures and show results obtained in project conditions on natural soil samples. By combining these different penetrometers, a wider and more reliable set of mechanical properties – spanning strength and consolidation behaviour – can be obtained, compared to conventional practice. The paper concludes with practical advice on testing protocols and interpretation methods to best characterise the shallow seafloor, including the use of novel shallow penetrometers.

Cone Penetration Testing 2022 – Gottardi & Tonni (eds)

Development of an enhanced CPT system for Dogger Bank

Tor Inge Yetginer-Tjelta
Equinor, Stavanger, Norway

Simon Bøtker-Rasmussen
Geo, Copenhagen, Denmark

Mike Rose
SSE Renewables, Perth, Scotland

Tom Lunne & Vaughan Meyer
NGI, Oslo, Norway

Callum Duffy
Logos Geoservices Ltd, Oxford, UK

ABSTRACT: An enhanced seafloor CPT system has been developed to support completion of the soil investigation campaign for Dogger Bank. This enhanced system has a demonstrable and significant performance increase over standard seafloor CPT systems; capable of pushing through dense sand layers with q_c>100 MPa and through tens of meters of very stiff clays. At Dogger Bank, this enhanced system has enabled CPT penetrations of more than 40 m below seafloor, in soils where standard systems could only average in the twenties. The system enhancement has been achieved through the application and adaption of techniques well known in the geotechnical industry (water lubrication and water injection), but which have never before been combined in an offshore seafloor CPT system. The performance of the enhanced CPT system has enabled a reliance on seafloor CPTs to acquire data to beyond monopile toe depths, therefore removing absolute reliance on boreholes to acquire data at turbine locations and facilitating the fast and efficient development of a geotechnical design basis.

Session 2: Interpretation

Cone Penetration Testing 2022 – Gottardi & Tonni (eds)

CSi – a joint industry project into CPTUs in silty soils

A.H. Augustesen
Ørsted, Denmark

P. Carotenuto, C. Bilici & T. Lunne
Norwegian Geotechnical Institute, Norway

R.C.J. Lindeboom & L. Krogh
Ørsted, Denmark

J. van den Bosch & R. Barth
Fugro, The Netherlands

C. Erbrich & S. Ingarfield
Fugro, Australia

D. Giretti
University of Bergamo, Italy

V. Fioravante
University of Ferrara, Italy

H. Dias
Equinor

M.-C. Sougle
Vattenfall

A. Barwise
RWE

S. de Wit
Shell

D. Burbury
Scottish Power Renewables

N. Adams
Carbon Trust

ABSTRACT: The CPTU constitutes the main *in situ* offshore investigation tool for geotechnical site characterisation and provision of soil input for the design of wind turbine foundations. CPTUs are typically performed at all foundation positions. Thus, all obtained results of supporting geotechnical *in situ*, model and laboratory testing need to be confidently correlated to the CPTU parameters. Most of the interpretation methodologies available for industry practice consider the soil behaviour around the cone either as fully drained or undrained, and acknowledged and well-proven correlations between CPTU parameters and classification and engineering properties exist for sand and clay. However, for transitional soils, e.g. silty soils, which may exhibit partial drainage during standardized cone penetration, such robust interpretation schemes do not exist. This paper presents the background, the objectives, setup and early field results of a joint industry project into CPTUs in silty soils for developing schemes for planning, execution and interpretation.

Cone Penetration Testing 2022 – Gottardi & Tonni (eds)
 ISBN 978-1-032-35624-2

Numerical investigation of piezocone dissipation tests in clay: Sensitivity of interpreted coefficient of consolidation to rigidity index selection

A. Barati Nia, D.M. Moug & A.P. Huffman
Department of Civil and Environmental Engineering, Portland State University, Portland, USA

J.T. DeJong
Department of Civil and Environmental Engineering, University of California, Davis, USA

ABSTRACT: Standard methods for interpreting the coefficient of consolidation (c_h) from CPTu dissipation tests require an estimation of the soil's rigidity index (I_r=G/s_u). For the Teh & Houlsby numerically derived dissipation solution, the values of shear modulus (G) and undrained shear strength (s_u) were straight forward to specify due to the use of the elastic-perfectly plastic Von-Mises soil constitutive model. However, estimation of appropriate G and s_u values is not as straight forward for field dissipation tests. The objective of this study is to examine the sensitivity of the interpreted c_h to various approaches for estimating I_r. The study is performed with an axisymmetric direct cone penetration model to simulate piezocone dissipation. Simulations are performed with the MIT-S1 constitutive model calibrated for Boston blue clay behavior. Analyses examine how the sensitivity changes for normally to lightly overconsolidated clay, and slightly to strongly anisotropic hydraulic conductivities. The results indicate that c_h interpretation is not highly sensitive to I_r estimation when I_r is within standard values for clay; however, the sensitivity appears to increase as the overconsolidation ratio increases.

Cone Penetration Testing 2022 – Gottardi & Tonni (eds)

Correlation of CPT measurements and VibroCore penetration speed for medium, calcareous sands: A case study of the cable route survey at the North Sea

K. Bartczak & G. De Vries
Marine Sampling Holland B.V., The Netherlands

ABSTRACT: The Cone Penetration Test (CPT) is one of the most popular in-situ testing methods used in qualitative and quantitative research of characteristics of the subsurface sediments, both onshore and offshore. In offshore cable route surveys, CPT tests are often accompanied by VibroCoring (VC) boreholes. The article endeavours to verify whether there is a clear correlation between the relative density of cohesion-less soils and the VC penetration characteristics. The research has been based on an offshore site investigation campaign in the North Sea consisting of over 100 CPTs and VCs, supplemented by an extensive laboratory testing program. After a strict selection of locations, around one-fifth of those tests have been utilised in the correlation studies. The ultimate goal of this study was to demonstrate if processing data recorded during VC testing can increase certainty in the prediction of soil's strength parameters.

Cone Penetration Testing 2022 – Gottardi & Tonni (eds)

CPT data interpretation for an improved characterization of the paleosol stratigraphy in the Po River Valley, Italy

I. Bertolini, M. Marchi, L. Tonni & G. Gottardi
DICAM Department, University of Bologna, Italy

L. Bruno
Department of Chemical and Geological Sciences, University of Modena, Italy

A. Amorosi
Department of Biological, Geological and Environmental Sciences, University of Bologna, Italy

ABSTRACT: Post-depositional phenomena can produce significant changes in the geotechnical properties of soils. Although the identification of the effect of such phenomena on the soil structure may contribute significantly to the interpretation of the soil mechanical behaviour, their characterization is not routinely carried out in practice. In this context, the novel application of paleosol features in the definition of geotechnical stratigraphic models can result in improved characterization of soil deposits in alluvial sites. Sedimentology uses paleosols (i.e. fossil soils with peculiar geologic properties) as markers for stratigraphic correlations in alluvial areas at large scale. This paper outlines the main properties of these geological objects and investigates the strategies for their identification in cone penetration tests (CPT). The analyses are discussed with reference to boreholes and CPTs data in 3 well documented sites in the Po River Valley (Italy). First, sedimentological identification of paleosols from borehole corings is introduced; then, geotechnical interpretation of CPT is carried out on such a basis, showing how poorly developed paleosols (Inceptisols) found in the investigated sites, are difficult to be identified through CPT logs and how they plot in the well-established Robertson's classification charts.

Cone Penetration Testing 2022 – Gottardi & Tonni (eds)
© 2022 the Author(s), ISBN 978-1-032-35624-2

Application of CPT to the evaluation of permeability in a Po river embankment prone to backward erosion piping

Ilaria Bertolini, Guido Gottardi, Michela Marchi & Laura Tonni
University of Bologna, DICAM, Italy

Agnese Bassi & Alessandro Rosso
Interregional Agency for the Po River, AIPO, Italy

ABSTRACT: Backward erosion piping is one of the most critical issues for the stability of many river embankments worldwide. The article presents results from an extensive geotechnical campaign carried out on a section of the Po river embankments (Italy), prone to backward erosion piping. The aim of the investigations was twofold: firstly, identifying the stratigraphic arrangement of the river system causing favourable conditions for piping occurrence, and secondly determining the permeability of the aquifer and the finer top stratum (blanket) in their undisturbed state, as well as that of the preexisting eroded zone in proximity of the sand boil (i.e. the volume around the exit hole). Indeed, analytical and numerical approaches available in the literature are particularly sensitive to the value of permeability of the aquifer, which should therefore be determined in undisturbed conditions. CPT tests turned out to be suitable and economical to this scope. This study presents an integrated analysis of in situ testing data, collected from boreholes, piezocone test and Lefranc test, aimed at estimating the saturated permeability of the different soil units using different experimental methods and interpretation approaches. A comparison between alternative CPT-based methods and field permeability tests is proposed. A good agreement between the different methods was found, thus suggesting the suitability of CPT for the hydraulic characterization of soils.

Cone Penetration Testing 2022 – Gottardi & Tonni (eds)
© 2022 the Author(s), ISBN 978-1-032-35624-2

Dissipation tests to evaluate the equilibrium pore pressure

F.A.B. Danziger, G.M.F. Jannuzzi & A.V.S. Pinheiro
Coppe and Polytechnic School, Federal University of Rio de Janeiro, Brazil

ABSTRACT: Dissipation tests are carried out with two main purposes: the estimation of the horizontal coefficient of consolidation, c_h, and the evaluation of the equilibrium pore pressure. Regarding the first case, even in soft clays the duration of the test doesn´t need to be more than 40 minutes, in most situations. However, in the latter case a long time is needed to reach equilibrium, even when probes with small diameters are used. The present paper presents results of long-term dissipation tests performed in a soft clay layer, aiming at the decision of executing or not an underpinning of shallow foundations of a damaged warehouse which suffered significant settlements.

Cone Penetration Testing 2022 – Gottardi & Tonni (eds)
© 2022 the Author(s), ISBN 978-1-032-35624-2

Dynamic characteristics of the soils by Cone Penetration Tests (CPT)

A. Cavallaro
CNR-ISPC, Catania, Italy

ABSTRACT: A geotechnical project requires a site-specific investigation to collect data regarding the subsurface conditions. Soil explorations must be made to determine the presence and identification of underlying strata, groundwater conditions, types of geomaterials, their depths and thicknesses, and the associated engineering parameters required for geotechnical design. The Cone Penetration Test (CPT) represents one of the most widely used on-site investigation methods in geotechnical engineering. This paper presents some empirical correlations available by Cone Penetration Tests (CPT) for determining the main geotechnical characteristics of soils. Moreover it aims to evaluate the small strain shear modulus by means of empirical correlations based on penetration tests results, CPT, Standard Penetration Tests (SPT) and Flat Dilatometer Marchetti Tests (DMT) or laboratory geotechnical investigations. It is aimed to achieve a better understanding of the obtained geotechnical parameters and the empirical correlations, justifying the investigation effort and enabling reliable input data for advanced dynamic analysis.

Cone Penetration Testing 2022 – Gottardi & Tonni (eds)
© 2022 the Author(s), ISBN 978-1-032-35624-2

Validating cone penetration test in partially drained conditions using a simplified numerical modelling method

R.W.L. Chia, Z.Z. Wang & S.H. Goh
Department of Civil and Environmental Engineering, National University of Singapore, Singapore

ABSTRACT: A simplified numerical modelling method, the Press-Replace Method (PRM), is applied to simulate the cone penetration process under conditions that range from fully undrained to fully drained. In this study, rigorous validations of the PRM technique for simulating the cone penetration test were performed. The results for a fully undrained and drained penetration are compared against published information using cone factor solutions. Coupled-consolidation analyses at different cone penetration rates were then carried out for a smooth cone, and the results processed to obtain the dimensionless backbone curves for comparison with published data. The consolidation analyses were further extended to a rough cone subjected to different penetration rates, for which there is little, if any, numerical simulation results reported in the literature. The results and computed backbone curves obtained using the smooth and rough soil-cone interfaces are presented and compared in this paper.

Cone Penetration Testing 2022 – Gottardi & Tonni (eds)
© 2022 the Author(s), ISBN 978-1-032-35624-2

Cone penetration testing to constrain the calibration process of a sand plasticity model for nonlinear deformation analysis

A. Chiaradonna
University of L'Aquila, L'Aquila, Italy

T.J. Carey
The University of British Columbia, Vancouver, Canada

K. Ziotopoulou & J.T. DeJong
University of California at Davis, Davis, USA

ABSTRACT: A reliable prediction of liquefaction-induced damage typically requires nonlinear deformation analyses with an advanced constitutive soil model calibrated to the site conditions. The calibration of constitutive models can be performed by relying primarily on a combination of commonly available properties and empirical or semi-empirical relationships, on laboratory tests on site-specific soils, on in-situ penetration tests, or a combination thereof. Chiaradonna et al. (2022) described a laboratory-based calibration approach of the PM4Sand constitutive model and evaluated the prediction accuracy against the response of a centrifuge experiment of a submerged slope. This paper addresses an alternate calibration approach in which the PM4Sand model is calibrated using centrifuge in-situ CPT data. The model performance for the resulting calibration is evaluated against the centrifuge experimental data and prior simulations from Chiaradonna et al. (2022). In this case, the CPT-based calibration resulted in more accurate estimations of the dynamic response and permanent displacements.

Cone Penetration Testing 2022 – Gottardi & Tonni (eds)

Probabilistic delineation of soil layers using Soil Behavior Type Index

S. Collico & M. Arroyo
Universidad Politécnica de Cataluña, Barcelona, Cataluña, Spain

M. DeVincenzi, A. Rodriguez & A. Deu
Igeotest s.l. Figueres, Cataluña, Spain

ABSTRACT: CPTu-based soil profiling has become a key component in the geotechnical design process. However, this is an interpretative process, affected by the inherent variability of soil properties, measurement noise and subjective heuristics. These are difficult to communicate to other interpreters or, even for the same interpreter, to transfer across profiles. A semi-automated tool for CPTu data interpretation is presented as an aid in this interpretation process. A probabilistic-based algorithm is employed to elicit the implicit heuristics in CPTu-based soil profiling and facilitate transference. Univariate normal distributions fit Soil Behavior Type Index data. Soil class boundaries, taken from a conventionally accepted chart, are sequentially activated with user-specified refinement. Thin layers under cone resolution are merged using well-established criteria. An application to CPTu records on finely interlayered deltaic deposits is illustrated, in which output delineations resulting from different analyst choices are compared among themselves and with one based on core description.

Cone Penetration Testing 2022 – Gottardi & Tonni (eds)

Clustering analysis to improve total unit weight prediction from CPTu

S. Collico & M. Arroyo
Department of Civil and Environmental Engineering (DECA), Universidad Politecnica de Cataluña, Barcelona, Spain

M. DeVincenzi, A. Rodriguez & A. Deu
Igeotest s.l. Figueres, Spain

ABSTRACT: Accurate estimates of soil unit weight are fundamental for correctly post process CPTu data and making use of Soil Behavior Type-based classification systems. Soil-specific and global regressions have been proposed for this purpose. However, soil-specific correlation might pose a problem of pertinence when applied at new sites. On the other hand, global correlations are easy to apply, but generally carry large systematic uncertainties. In this context, this work proposes a data clustering technique applied to geotechnical database aiming to identify hidden linear trends among dimensionless soil unit weight and normalized CPTu parameter according to some unobservable soil classes. Global correlations are then revisited according to such data subdivision aiming to improve accuracy of soil unit weight prediction while reducing transformation uncertainty. A new probabilistic criterion for soil unit weight prediction is also obtained. The potential benefits of the proposed procedure are illustrated with data from a Llobregat delta site (Spain).

Cone Penetration Testing 2022 – Gottardi & Tonni (eds)

Combining CPTU and UMASW to characterise Irish offshore deposits

M. Coughlan
School of Civil Engineering, University College Dublin, Belfield, Dublin, Ireland
SFI Research Centre for Applied Geosciences (iCRAG), Ireland
Gavin and Doherty Geoslutions Ltd., Dublin, Ireland

A. Trafford, M. Long & S. Donohue
School of Civil Engineering, University College Dublin, Belfield, Dublin, Ireland
SFI Research Centre for Applied Geosciences (iCRAG), Ireland

S. Corrales
APEX Geophysics, Gorey, Wexford, Ireland

ABSTRACT: The north Irish Sea is earmarked for the development of several offshore wind farms. The complex Quaternary history of this area has resulted in a legacy of geotechnically challenging deposits, including heterogeneous glacial deposits, soft sediments and gas-charged sediments. Multichannel Analysis of Surface Waves (MASW) is commonly used onshore to produce shear wave velocity (V_s) profiles and is increasingly being used offshore (UMASW). When used in tandem with *in-situ* Cone Penetration Testing (CPTU) profiles it is a powerful tool for characterising offshore deposits. UMASW profiles were obtained from sites across the north Irish Sea coincident with CPTU profiles. V_s profiles from UMASW show good agreement with CPTU parameters in a range of geological settings. The purpose of this paper is to present a characterisation of geological deposits using a combined approach of CPTU and UMASW. By doing so the aim is to further develop classification charts that can give reliable characterisations of offshore deposits by combing non-invasive geophysical techniques and traditional CPTU.

Cone Penetration Testing 2022 – Gottardi & Tonni (eds)
© 2022 the Author(s), ISBN 978-1-032-35624-2

Cone factor from CPTU tests in very soft clays at the east of Mexico's valley

J.M. De La Rosa R. & F.A. Flores López
Ingenieros Geotecnistas Mexicanos. S.C., México.

ABSTRACT: From the results obtained during an exploration campaign, the correlation between CPTU resistance and the undrained shear strength of soft clays at the valley of Mexico, is evaluated in this paper. This campaign consisted of 50 piezocone penetration tests, 50 soil sampling borings and laboratory tests carried out on more than 300 undisturbed soil samples recovered along 10 square kilometers. The samples were tested by means of unconfined compression and unconsolidated undrained triaxial compression tests. The correlation of the determined undrained shear strength and the experienced CPTU results was evaluated and the empirical cone factor (Nkt) was determined for all tests. The main innovation lies in the fact that for the eastern zone of the Valley of Mexico, the geotechnical characterization is not abundant and there are information gaps for this and other design parameters.

Keywords: CPTU, Nkt, Undrained shear strength, Valley of Mexico

Cone Penetration Testing 2022 – Gottardi & Tonni (eds)

Piezocone testing in Nordic soft clays: Comparison of high-quality databases

M. D'Ignazio & B. Di Buò
Tampere University, Tampere, Finland
Ramboll Finland Oy, Tampere, Finland

T. Länsivaara
Tampere University, Tampere, Finland

J.-S. L'Heureux & P. Paniagua
Norwegian Geotechnical Institute, Trondheim, Norway

J. Selänpää
Destia Oy, Tampere, Finland

ABSTRACT: Soft and sensitive clays are widespread in Scandinavia. Piezocone correlations for Norwegian clays have been previously proposed based on high-quality block samples from several sites. Recently, a large database of Finnish soft clays was compiled by Tampere University from piezocone measurements as well as high-quality laboratory tests on specimens from large tube samples. Finnish and Norwegian clays exhibit some differences in terms of basic properties. Norwegian clays show lower water content, lower organic content, higher silt content and lower plasticity than the clays from Finland. This may be linked to the source of the materials, their depositional and post-depositional processes that in turn impact on the mechanical behaviour.

This paper aims to compare piezocone Norwegian and Finnish data with focus on strength and stress history. The database trends are compared for relevant engineering parameters. The data and its variability are critically discussed considering differences in geological history, basic properties, sampling techniques and disturbance.

Cone Penetration Testing 2022 – Gottardi & Tonni (eds)
© 2022 the Author(s), ISBN 978-1-032-35624-2

Prediction of resilient modulus of cohesive subgrade soils from CPTU data using polynomial neural networks

Wei Duan
College of Civil Engineering, Taiyuan University of Technology, Taiyuan, China
School of Transportation, Southeast University, Nanjing, China

Zening Zhao
School of Transportation, Southeast University, Nanjing, China

Guojun Cai*
School of Transportation, Southeast University, Nanjing, China
School of Civil Engineering, Anhui Jianzhu University, Hefei, China

Anhui Wang
China Construction Industrial & Energy Engineering Group Co., Ltd., Nanjing, China

Ruifeng Chen
School of Transportation, Southeast University, Nanjing, China

Anand J. Puppala
Zachry Department of Civil and Environmental Engineering, Texas A&M University, USA

Songyu Liu
School of Transportation, Southeast University, Nanjing, China

Surya Sarat Chandra Congress
Zachry Department of Civil and Environmental Engineering, Texas A&M University, USA

ABSTRACT: Although various empirical models have been proposed to predict the resilient modulus (M_r) from laboratory test parameters, less effort has been made in developing reliable empirical models using piezocone penetration test (CPTU) data. Moreover, the prediction accuracy of the existing empirical models is not high enough. In the present study, a novel empirical model was proposed to predict the M_r from CPTU data based on the polynomial neural network. To this end, a comprehensive database comprising 16 different sites in Jiangsu province, China, was firstly compiled, which contains 124 sets of M_r, cone tip resistance (q_c), sleeve friction (f_s), pore water pressure (u_2), moisture (w), and dry density (γ_d) values at the in-situ stress condition. Taking the in-situ M_r values as reference values, three empirical models was developed using the group method of data handling (GMDH) neural network. The results show that proposed GMDH model 3 (GMDH method) with the input parameters of q_c, f_s, w, and γ_d can accurately predict the M_r. The obtained specific expressions for prediction of M_r further prove the reliability of the GMDH model. Overall, the new GMDH method can more accurately predict the M_r of subgrade soil and guide engineering practice.

*Corresponding author

Cone Penetration Testing 2022 – Gottardi & Tonni (eds)
 ISBN 978-1-032-35624-2

Effect of sand bio-cementation on cone tip resistance: A numerical study

M.El Kortbawi, K. Ziotopoulou & J.T. DeJong
University of California, Davis, CA, USA

D.M. Moug
Portland State University, Portland, OR, USA

ABSTRACT: Understanding the effect of soil cementation on cone measurements is important for the identification of naturally cemented soil deposits and for the verification of soil improvement achieved by various forms of artificial cementation, including bio-cementation. This paper presents the results of an effort to connect cone tip resistances with fundamental, constitutive level, bio-cementation-induced changes in soil behavior. To this end, a direct axisymmetric cone penetration model using the Mohr-Coulomb constitutive model and grid rezoning and remapping algorithms is used to model cone penetration in bio-cemented sands. By connecting the results of cone penetration simulations in a Mohr-Coulomb material to real data and established relationships, this work will guide selection of equivalent strength properties of these challenging materials. More specifically, the apparent cohesion, peak friction angle, dilation angle, and small-strain shear modulus within the Mohr-Coulomb constitutive model are varied parametrically across a reasonable range of parameter values informed by past laboratory, bench-scale, and centrifuge tests. Results show that cone penetration resistance in bio-cemented sands is mostly influenced by the interconnected apparent cohesion and the small-strain shear modulus, while other parameters play a secondary role.

Cone Penetration Testing 2022 – Gottardi & Tonni (eds)
© 2022 the Author(s), ISBN 978-1-032-35624-2

A data-driven approach to predict shear wave velocity from CPTu measurements

I. Entezari & J. Sharp
ConeTec Group, Burnaby, British Columbia, Canada

P.W. Mayne
Georgia Institute of Technology, Atlanta, Georgia, USA

ABSTRACT: The use of machine learning modelling to predict shear wave velocity (V_S) from piezocone penetration tests (CPTu) is presented. A large dataset of paired V_S-CPTu data (n = 104,054) compiled from seismic piezocone (SCPTu) soundings completed in a wide variety of soil types with various stress histories and geological environments was used to develop machine learning models to directly estimate V_S from CPTu data. The impact of soil microstructure on the results was investigated and separate models were developed to predict V_S in cemented and uncemented soils. The results of machine learning models outperformed the existing widely used CPT-based relationships to predict V_S.

Cone Penetration Testing 2022 – Gottardi & Tonni (eds)

Interpretation and comparison of CPT derived soil properties to static and cyclic laboratory tests on unique fine-grained soils in Western Washington and Oregon

B. Exley, A. Pyrch, J. Jacoby & B. Thunder
Haley & Aldrich, Inc., USA

ABSTRACT: CPT interpretation methods are evaluated against laboratory and in-situ methods to identify and evaluate the soil properties of multiple non-textbook soil deposits located in Western Washington and Oregon. These deposits include unique glacially overridden soils which are widespread in the Puget Sound region, as well as younger soils which vary widely in depositional history and stress history.

The glacially overridden deposits can be very thick and are heavily overconsolidated, often showing signs of fissures and slickensides when they are fine-grained, with SPT blowcounts frequently much larger than N=100 in coarse grained deposits. Based on the CPTs included in this paper, the fine-grained overconsolidated materials are typically identified by the CPT as sand-like and dilative, and generally have I_c values between 2.0 and 2.6 despite frequently having index properties which classify the soils as stiff to hard high plasticity clay. This paper documents the characteristics of these soils to facilitate identification of them by the CPT.

Additionally, the correlated strength, preconsolidation pressures, and cyclic resistance of the younger fine-grained soils are evaluated against laboratory testing methods. This includes documenting a site in Beaverton, Oregon with several CPTs paired with constant rate of strain consolidation laboratory tests and both monotonic and cyclic direct simple shear tests in the fine-grained Missoula Flood Deposits as part of a study on the susceptibility of the silt to liquefaction or cyclic softening.

Cone Penetration Testing 2022 – Gottardi & Tonni (eds)
© 2022 the Author(s), ISBN 978-1-032-35624-2

Evaluation of cyclic softening potential using CPTu and assessment with cyclic triaxial test results: A case study

K. Fakharian, M. Bahrami & M. Kashkooli
Department of Civil & Environmental Engineering, Amirkabir University of Technology, Tehran, Iran

H. Vaezian & T. Bahrami
Pars GeoEnviro Inc., Tehran, Iran

ABSTRACT: An extensive geotechnical investigation was performed on an oil terminal site located in a seismically active field. Combination of *in situ* and laboratory tests were performed including Piezocone Penetration Tests (CPTu), some of which were adjacent to the borehole locations, providing a detailed stratigraphic soil profile and different physical and mechanical parameters. In this paper, the application of the CPTu test for analyzing cyclic liquefaction is evaluated and discussed. Both "liquefaction" and "cyclic softening" potentials are evaluated depending on the magnitude of I_c factor. A clayey silt layer was identified having an I_c factor of about 2.5 between 12 and 19 m of depth. The average q_c of this layer was 6 MPa and friction ratio (RF) was 2.5. The induced pore pressure, u, was measured between 300 to 1800 kPa from 12 to 19 m of depth. The equivalent SPT counts from correlations with CPT were 15 while the real SPT counts were about 50 or higher. According to CPT data, the factor of safety of liquefaction was calculated having an average of 0.56 for a 0.49g earthquake with M=7. Undisturbed samples were taken from the layer onto which cyclic undrained liquefaction tests were carried out. The results, however, revealed that specimens were resisting the cyclic liquefaction with sufficiently high factor of safety. The soil classification in the laboratory showed that the soil was clay and not silty clay. Detailed analyses and interpretations are carried out to find out the reasons of the differences between CPTu correlations and real response of the specimens to cyclic loading. The results show that modifications might be required at times in interpretations of soil classification as well as liquefaction potential evaluation for soils at the borderline of low plasticity clay and silty clay.

Cone Penetration Testing 2022 – Gottardi & Tonni (eds)
 ISBN 978-1-032-35624-2

Application of integrated Game Theory-optimization subground stratification (-IGTOSS) model to Venetian Lagoon deposits

M.S. Farhadi & T. Länsivaara
Faculty of Built Environment, Department of Civil Engineering, TERRA Research Center
Tampere University, Tampere, Finland

L. Tonni
Department of Civil, Chemical, Environmental and Materials Engineering (DICAM)
University of Bologna, Bologna, Italy

ABSTRACT: The geographical and geological spatial variabilities raise challenges for geotechnicians to devise globally applicable subground stratification models working based on cone penetration testing (CPT). Recently, a novel CPT-based stratification and classification model was proposed in Tampere University, Finland. It combines the soil behavior type (SBT) classification chart proposed by Robertson (1990) with a novel integrated Game Theory-optimization subground stratification model (denoted herein as RIGTOSS). The model has already been verified based on few test sites results from Taiwan and the U.S. Therefore, in this paper, the RIGTOSS model is developed further, and it is evaluated based on the stratification profiles provided by CPT experts for the Venetian Lagoon deposits. The test site has been selected because of high variability of CPT measurements and the thin transient soil layers in the area. The results indicated comparable stratification profiles from the developed model(s) and the profiles by experts, derived based on field and laboratory tests.

Cone Penetration Testing 2022 – Gottardi & Tonni (eds)
© 2022 the Author(s), ISBN 978-1-032-35624-2

Application of two novel CPTu-based stratification models

M.S. Farhadi & T. Länsivaara
Faculty of Built Environment, Department of Civil Engineering, TERRA Research Center
Tampere University, Tampere, Finland

J.-S. L'Heureux & T. Lunne
Norwegian Geotechnical Institute, Geotechnics and Natural Hazards, Oslo, Norway

ABSTRACT: Two novel CPTu-based stratification models are applied to the measurements at four Norwegian sites, including a broad range of soil behavior types (SBTs). The two models, called RIGTOSS and SIGTOSS, are developed based on the recently proposed integrated Game Theory-optimization subground stratification (-IGTOSS) model. The RIGTOSS model includes two submodels classifying the SBTs based on either the F_r-Q_t or B_q-Q_t charts by Robertson (1990); and similarly, the SIGTOSS model includes two submodels classifying the SBTs based on either the F_r-Q_t or $(\Delta u_2/\sigma'_{v0})$ Q_t charts by Schneider et al. (2012). The resulting stratification profiles of the four submodels were compared with the ones provided by experts, derived from extensive field and experimental testing. Similarity was observed and the results are promising. Differences at this stage can be attributed to incompatibility of the classification charts with the Norwegian soils and more specifically soil heterogeneities at the Øysand test site.

Cone Penetration Testing 2022 – Gottardi & Tonni (eds)
 ISBN 978-1-032-35624-2

Numerical simulation of CPT in sands using DeltaSand and Hardening Soil models

M. Fetrati
MARUM–Center for Marine Environmental Sciences, University of Bremen, Bremen, Germany

V. Galavi
Deltares, Delft, The Netherlands

M. Goodarzi
MARUM–Center for Marine Environmental Sciences, University of Bremen, Bremen, Germany
COWI A/S, Hamburg Branch Office, Germany

S. Kreiter
MARUM–Center for Marine Environmental Sciences, University of Bremen, Bremen, Germany

T. Mörz
MARUM–Center for Marine Environmental Sciences, University of Bremen, Bremen, Germany
Geo-Engineering.org GmbH, Bremen, Germany

ABSTRACT: In this paper, Hardening Soil model with small-strain stiffness (HSsmall) and DeltaSand model are used as constitutive soil models to simulate the Cone Penetration Test (CPT) using the Material Point Method (MPM). Both models are formulated within the double hardening framework, in which independent yield surfaces represent mechanical behavior of soil under deviatoric and volumetric loadings. DeltaSand is a new advanced state-dependent constitutive model in which the relative density is incorporated in the formulation to represent the mechanical behavior of soil under deviatoric and volumetric loading in different stresses and relative densities. The numerical simulations are compared with CPTs in Cuxhaven Sand in a calibration chamber. Both constitutive soil models are compared with each other and DeltaSand is found to be capable to capture the soil behavior during quasi-static CPT.

Cone Penetration Testing 2022 – Gottardi & Tonni (eds)

CPT calibration in centrifuge: Effect of partial saturation on cone resistance

V. Fioravante
Department of Engineering, University of Ferrara, Italy

D. Giretti
Department of Engineering and Applied Science, University of Bergamo, Italy

E. Dodaro, C.G. Gragnano & G. Gottardi
Department of Civil, Chemical, Environmental and Materials Engineering, University of Bologna, Italy

ABSTRACT: When dealing with unsaturated soil conditions, the influence of matric suction on cone tip resistance of the soil above the ground water table is typically neglected in engineering practice, with consequent possible misinterpretation of soil features. In the last decades, various researchers have investigated the influence of suction on cone resistance for sands, whilst still little is known for silty materials, whose contribution can be significant and extended for many meters above the groundwater table. Such issue is especially relevant for compacted earth structures, like river embankments, typically made of a heterogeneous mixture of intermediate soils. With the aim of providing a contribution and stimulating its correct implementation into geotechnical practice, a set of miniature piezocone tests have been carried out in a centrifuge on both saturated and partially saturated silty sand models. The interpretation of CPT results is discussed, highlighting the effect of partial saturation on cone tip resistance.

Cone Penetration Testing 2022 – Gottardi & Tonni (eds)
 ISBN 978-1-032-35624-2

Calibration cone penetration testing in silty soils

V. Fioravante
University of Ferrara, Italy

D. Giretti
University of Bergamo, Italy

T. Lunne & P. Carotenuto
Norwegian Geotechnical Institute, Norway

A.H. Augustesen, R.C.J. Lindeboom & L. Krogh
Ørsted, Denmark

H. Dias
Equinor

M.-C. Sougle
Vattenfall

A. Barwise
RWE

S. de Wit
Shell

D. Burbury
Scottish Power Renewables

ABSTRACT: The Offshore Wind Industry's rapid expansion across the globe requires geotechnical modelling of sites that are often characterized by layers of silty sand and silt mixtures. The CPTU is the main *in situ* offshore investigation tool for defining the ground conditions and for establishing facility position and soil parameters for foundation design, but no simple and robust methodologies exist for characterizing transitional soils. This paper presents some results of CPTUs carried out in a large calibration chamber and in a centrifuge aimed at contributing to the development of guidelines for planning, specification, execution, and interpretation of CPTUs in transitional soils.

Cone Penetration Testing 2022 – Gottardi & Tonni (eds)

Comparison between coefficients of consolidation from CPTu and laboratory tests for Guaratiba's soft soil, Rio de Janeiro, Brazil

M.M. Freire, M.E.S. Marques & M.C. Tassi
Military Institute of Engineering, Rio de Janeiro, Brazil

L.A. Berbert
Prodec Consultoria, Rio de Janeiro, Brazil

ABSTRACT: This paper presents a discussion about vertical (c_v) and horizontal (c_h) coefficients of consolidation of Guaratiba's very soft clay, located at the lowlands of the West Zone of Rio de Janeiro city, Brazil. Due to the thickness of the deposit, high compressibility of the clay, and low undrained strength, the settlement stabilization, even with vertical drains, are known to take of about 30 months to occur at these lowlands. Thus it is essential to know c_v, c_h and c_h/c_v relationship in order to predict when settlement stabilization will occur. The variability of c_v and c_h values is very high, probably due to the wide range of organic matter, sand lenses and the sand content throughout the soft soil deposit. The aim of the paper is to compare these values of coefficients of consolidation obtained from the dissipation tests and from the consolidation tests and to discuss an approach to determine them for design considering their variability.

Cone Penetration Testing 2022 – Gottardi & Tonni (eds)
© 2022 the Author(s), ISBN 978-1-032-35624-2

Quantitative modelling of spatial variability of piezocone data from Venice lagoon silty soils

G. Gottardi, M. Ranalli & L. Tonni
Department of Civil, Chemical, Environmental and Materials Engineering, DICAM, Alma Mater Studiorum, University of Bologna, Italy

M. Uzielli
Department of Civil and Environmental Engineering, DICEA, University of Florence, Italy
Georisk Engineering S.r.l., Florence, Italy

ABSTRACT: As geotechnical research and design codes rely increasingly on probabilistic approaches, site characterization should also be conducted in the light of the explicit quantification of the uncertainty and spatial variability in soil properties. This paper provides a practical case-study application of spatial variability analysis of piezocone data obtained at a test site in the Lagoon surrounding the historic city of Venice, in North-Eastern Italy, where unusually dense and regularly spaced CPT test data were available. Empirical semivariograms are calculated for cone resistance, sleeve friction, and pore-water pressure, along with the soil behavior classification index obtained from these measurements at a set of reference measurement depths. A number of theoretical semivariogram models are fitted comparatively and best-fit models are selected based on objective criteria and subjective judgment. Characteristic semivariogram parameters, providing information on correlation distance and small-scale variability, are retrieved. Modeling options are explained and results are analyzed and assessed critically.

Cone Penetration Testing 2022 – Gottardi & Tonni (eds)
© 2022 the Author(s), ISBN 978-1-032-35624-2

Thin-layer detection from the cone resistance rate of change

H.B. Hammer
Dr.techn. Olav Olsen AS, Trondheim, Norway

S. Nordal
Norwegian University of Science and Technology (NTNU), Norway

J.-S. L'Heureux & H. Skrede
Norwegian Geotechnical Institute (NGI), Norway

ABSTRACT: True determination of soil parameters for thin clay layers in sand or silts from CPTU measurements is challenging. Cone resistance in thin layers is influenced by the surrounding layers resulting in unrepresentative measurements. Measurements should be corrected for this effect prior to interpretation of parameters. Such correction requires accurate information about the thickness of the thin layer. Previous studies have shown that the pore pressure parameter of the CPTU may not properly identify the layer interfaces for thin clay layers in sand. In this study a "cone resistance rate of change" parameter is suggested for thin-layer interface detection. Results from recently performed physical experiments at NTNU as well as from prior studies are used to evaluate the ability of this parameter to detect thin layers of clay in sand. The parameter appears to detect layer interfaces with good accuracy, even for layers with thickness as thin as the cone diameter. The results suggest that the approach may estimate layer thickness quite well and aid towards efficient correction of cone resistance to achieve more realistic soil parameters for thin clay layers in sand.

Cone Penetration Testing 2022 – Gottardi & Tonni (eds)
© 2022 the Author(s), ISBN 978-1-032-35624-2

Recalculation of in-situ CPTu in intermediate soils using G-PFEM

L. Hauser, S. Oberhollenzer, A. Gharehaghajlou, H.F. Schweiger & R. Marte
Institute of Soil Mechanics, Foundation Engineering and Computational Geotechnics, Graz University of Technology, Graz, Austria

C. Fabris
Geoconsult ZT GmbH, Puch bei Hallein, Austria

ABSTRACT: Piezocone penetration in intermediate soils is often associated with partially drained behaviour which has to be taken into account when interpretating the results. As part of the research project PITS, the effect of partial drainage on in-situ testing in postglacial silty deposits in Austria is investigated by means of in-situ testing campaigns, laboratory testing and numerical simulations. The present work presents a numerical recalculation of in-situ CPTu, carried out with different penetration velocities in a silty soil layer at the test site Rhesi near lake Constance, using the application G-PFEM. The results agree reasonably well with the in-situ measurements considering the given heterogeneity of the soil layer and suggest that partial drainage occurs during the penetration process.

Cone Penetration Testing 2022 – Gottardi & Tonni (eds)

Full-flow CPT tests in a nearshore organic clay

S. Hov
GeoMind, Stockholm, Sweden
Norwegian Geotechnical Institute, Trondheim, Norway

K. Borgström
GeoMind, Stockholm, Sweden

P. Paniagua
Norwegian Geotechnical Institute and Norwegian University of Science and Technology, Trondheim, Norway

ABSTRACT: The T-bar is a full-flow CPT with a larger probe surface area compared with the conventional CPT probe. The tip has the shape either of a ball or as an upside-down T, a so-called T-bar. The term 'full-flow' comes from the assumption that the earth 'flows' around the tip, which is a realistic assumption for soils with extremely low undrained shear strength. This paper presents a case study where a large number of T-bar tests have been performed in a nearshore organic clay with high water content and very low undrained shear strength. The test site is located in Stockholm, Sweden, where planned land reclamation and capping of contaminated top soils are challenging from a stability perspective due to the low strength of the soil. T-bar tests were thus performed to characterise the shear strength profile of the soil in detail, especially at shallow depths where sampling was difficult and the shear strength values were under 5 kPa. A N-factor relating the net cone resistance (q_{net}) and the undrained shear strength of the soil (c_u) was evaluated based on T-bar measured parameters and direct simple shear tests and undrained triaxial tests on samples taken at greater depths. This allowed to estimate shear strength profiles with depth. For all tests, both the penetration and extraction cone resistance were measured, and a good correlation was obtained between this ratio and the soil sensitivity measured in the laboratory. In addition, the sensitivity was correlated to the organic content of the clay. These correlations were found to be OCR-dependent. Further, cyclic tests were performed, and their results were correlated with the remoulded shear strength values. This facilitated mapping of the soil conditions across the site.

Cone Penetration Testing 2022 – Gottardi & Tonni (eds)
© 2022 the Author(s), ISBN 978-1-032-35624-2

Bayesian supervised learning of 2D subsurface soil stratigraphy using limited cone penetration tests with consideration of uncertainty

Yue Hu
Department of Civil and Environmental Engineering, National University of Singapore, Singapore

Yu Wang
Department of Architecture and Civil Engineering, City University of Hong Kong, Hong Kong, China

ABSTRACT: Cone penetration tests (CPT) have been widely used for soil stratification in geotechnical site investigation for decades. However, due to time and budget limits, the layout of CPT sounding at a specific project site is often sparse, leading to significant interpolation uncertainty in the development of subsurface soil 2D cross-section, particularly at locations without CPT measurements. Such development is often combined with empirical classification criteria, which further introduce model uncertainty to soil stratification. These uncertainties may pose great risks to the geotechnical engineering practice. A Bayesian supervised learning method is presented in this paper for probabilistic soil stratification in a 2D cross-section using limited CPT. The proposed method can not only automatically stratify soils in a 2D cross-section from limited CPT soundings, but also can properly quantify the associated uncertainties. Complete 2D CPT data cross-section is firstly learned from limited number of 1D CPT profiles using Bayesian supervised learning. The associated interpolation uncertainty is modelled numerically using non-parametric random field simulation based on the results of Bayesian supervised learning. Parametric autocorrelation function of CPT data along either vertical or horizontal direction is not needed. A probabilistic model is also developed to account for the model uncertainty of an empirical soil behavior type classification chart. The interpolation uncertainty and soil classification model uncertainty are then evaluated simultaneously in a Monte Carlo simulation framework. A simulated data example is used for illustration. The results suggest that the proposed method performs well.

Cone Penetration Testing 2022 – Gottardi & Tonni (eds)

Estimation of constrained modulus from CPT measurements in case of Holocene sands

Zsombor Illés, István Kádár, Gábor Nagy, András Mahler & László Nagy
Budapest University of Technology and Economics, Budapest, Hungary

ABSTRACT: In case of major projects there are numerous measurements which provide a sufficient set of data to determine characteristic values based on statistical methods. In the current paper Cone Penetration Test (CPT), oedometer and Flat Dilatometer Test (DMT) results are analyzed. Constrained or oedometric modulus is one of the most important deformation parameters, which has a key role in settlement estimation. There are various recommendations to derive E_{oed} from CPT tip resistance values. Three of these recommendations are compared with the results of the oedometer tests and the correlation used in case of DMT measurements.

Keywords: Constrained modulus, CPT, Tip-resistance, DMT, Normally-consolidated layers

Cone Penetration Testing 2022 – Gottardi & Tonni (eds)
© 2022 the Author(s), ISBN 978-1-032-35624-2

Evaluation of complex CPTu dissipation tests of B.E.S.T.

E. Imre
BGK, HBM EKIK, Óbuda University, Budapest, Hungary

M. Hegedűs
BME, Budapest, Hungary
Apostol Brc, Budapest, Hungary

L. Bates & S. Fityus
University of Newcastle, Newcastle, Australia

ABSTRACT: The u_2 complex pore water pressure dissipation tests made in NC-LOC sandy soils at the Bolivian test site were evaluated with the variants of a mathematically precise, automatic method in the function of the testing time. The identified *c* became near constant and its error became small after a so called minimum testing time $t_{min} < t_{50}$ depending on the variant. The traditional t_{50} method failed in some cases since the tests were with types I, II, III, V and t_{50} varied from 4s to a value being larger than 15 min (not measured).

Cone Penetration Testing 2022 – Gottardi & Tonni (eds)
© 2022 the Author(s), ISBN 978-1-032-35624-2

Simulation of CPT penetration in sensitive clay

J. Isaksson
Chalmers University of Technology, Sweden
NCC AB, Sweden

J. Yannie
NCC AB, Sweden

M. Karlsson & J. Dijkstra
Department of Architecture and Civil Engineering, Chalmers University of Technology, Sweden

ABSTRACT: This paper presents the results from numerical simulations of CPTu penetration in a natural clay combining the SCLAY1S constitutive model with a large deformation Finite Element framework including a coupled deformation and porewater pressure formulation. The hierarchical model formulation of SCLAY1S captures many features of a natural sensitive clay, such as the evolving anisotropic strength-stiffness response, as well as the degradation of the initial bonding. A sensitivity analysis is performed varying the overconsolidation ratio (*QCR*), bonding and anisotropy, also the hydraulic conductivity (hence, c_v) of the clay. The findings indicate that some soil properties (the c_v and *OCR*) impact both the normalised cone resistance Q_t and the generation of excess porewater pressures. In contrast the sensitivity S_t of soft soils primarily affects Q_t. In the current work it seems that the effects of the inherent and stress induced (from CPT penetration) anisotropy is not detected using these normalised plots.

Cone Penetration Testing 2022 – Gottardi & Tonni (eds)

A CPT-based method for estimation of undrained shear strength of sands and transitional soils

K. Kaltekis & J. Peuchen
Fugro, Nootdorp, The Netherlands

ABSTRACT: This paper presents a practical approach for developing a site-specific CPT-based method for monotonic undrained shear strength (s_u) in sands and transitional soils, using results of laboratory undrained triaxial compression (CU) tests on reconstituted and undisturbed specimens as reference. The methodology includes use of net cone resistance values normalised to vertical effective stress, a procedure for pairing of CPT data with CU test results, and definition of a practical failure criterion for deriving s_u from CU test data. The presented approach is particularly useful for application in offshore wind, where the economics of wind farm development favour performing only a single cone penetration test (CPT) per wind turbine location. This setting drives development of CPT-based methods for key geotechnical parameters for foundation design.

Cone Penetration Testing 2022 – Gottardi & Tonni (eds)
© 2022 the Author(s), ISBN 978-1-032-35624-2

Comparison of frozen soil strength characteristics by cone penetration and triaxial compression testing

Daniil Lagosha
Lomonosov Moscow State University, Russian Federation

Ivan Sokolov & N.G. Volkov
Fugro, Russian Federation

ABSTRACT: The paper provides comparison of field and laboratory tests of permafrost soil. Electrical resistivity and temperature piezocone penetration tests (RTCPTu) were carried out in the area of sporadic distribution of permafrost. RTCPTu detected islands of permafrost at depth more than 10 meters. Based on this information, undisturbed samples of warm permafrost soils with massive cryogenic structure and thick ice lenses (up to 10 cm) were collected using CPT equipment with a direct push soil sampler (MOSTAP) mounted instead of cone. The samples were kept in frozen condition and transported to the soil lab, where physical properties and triaxial compression tests were carried out. The CPT results included the characteristics of long-term frozen soil strength such as σ_c – the long-term cone resistance and σ_n – the net long-term cone resistance. The results of triaxial tests provided the characteristics of peak strength: σ_1 – the maximum vertical stress and σ_1-σ_3 – deviator stress at failure. The frozen soil samples were cleared from ice lenses since it is not possible to test them with thick ice lenses. The results of CPT and triaxial testing showed fairly similar values, when the cone measured resistance in frozen soil with massive cryogenic structure. When the cone was located near ice lenses, the net long-term cone resistance appeared to be much lower compared to deviator stress. This observation is in line with conventional theory on physics of ice which states that ice has near-zero long-term strength. The obtained results clearly show the value and advantage of CPT in permafrost (in-situ testing) compared with triaxial compression testing (laboratory). The properties of frozen soil in-situ may significantly differ from properties of frozen soil in a sample. This must be considered for designing civil structures on permafrost.

Cone Penetration Testing 2022 – Gottardi & Tonni (eds)
© 2022 the Author(s), ISBN 978-1-032-35624-2

Fincone: A study on the use of CPT in soft sensitive clays

T. Länsivaara
Tampere University, Tampere, Finland

B. Di Buò
Tampere University/Ramboll Finland Oy, Tampere, Finland

J. Selänpää
Tampere University/Destia, Tampere, Finland

M. Knuuti & M. Haikola
Tampere University, Tampere, Finland

ABSTRACT: The paper presents the results of a recent study on the application of CPTU on soft sensitive Finnish clays. An extensive field and laboratory investigation program was carried out, including 9 test sites of which 5 were studied in more detail. In each of the test sites, a minimum of 4 CPTU soundings, 2-3 field vane test with a new type of down hole vane, sampling with a newly developed large diameter tube sampler and an extensive laboratory investigation program were carried out. Transformation models were developed to estimate the undrained shear strength, preconsolidation stress, and constrained modulus for the over consolidated region. In addition, new information about anisotropy of Finnish clays were obtained. In general, the CPTU tests proved to be very reliable with very good repeatability. However, the measurement of sleeve friction proved to be somewhat problematic for the very soft and sensitive clays studied.

Cone Penetration Testing 2022 – Gottardi & Tonni (eds)

CPT-based unit weight estimation extended to soft organic clays and peat: An update

H.J. Lengkeek
Delft University of Technology, Delft, The Netherlands
Witteveen+Bos, Deventer, The Netherlands

R.B.J. Brinkgreve
Delft University of Technology, Delft, The Netherlands
Bentley Systems, Delft, The Netherlands

ABSTRACT: Various CPT-based correlations exist for the unit weight of natural soils. One such correlation includes organic soils Lengkeek et al. (2018). This correlation is presented as a framework where the coefficients can be optimized and is based on predominantly Class 2 CPT records. This publication uses an expanded database which includes additional pairs of predominantly Class 1 CPT records selected from Holocene deposits in the Netherlands, on mineral clays, organic clays and peats. This results in a more extensive database and an improved CPT-based unit weight correlation for the whole range of soil types, which is proposed to replace the existing correlation. In addition, a specific unit weight correlation for peats is presented.

Cone Penetration Testing 2022 – Gottardi & Tonni (eds)
© 2022 the Author(s), ISBN 978-1-032-35624-2

CPT-based classification of soft organic clays and peat

H.J. Lengkeek
Delft University of Technology, Delft, The Netherlands
Witteveen+Bos, Deventer, The Netherlands

R.B.J. Brinkgreve
Delft University of Technology, Delft, The Netherlands
Bentley Systems, Delft, The Netherlands

ABSTRACT: An updated CPT-based classification system of organic clays and peat is proposed based on an extensive pairwise established database of classification tests and CPT measurements. This new classification system is proposed to supplement the existing dimensionless q_t/p_a-Rf-chart of Robertson (2010). The Robertson (2010) dimensionless classification system is selected for refinement because it appears to perform better than normalized systems for peats with very low stresses (<20 kPa). A combination with Robertson (2009 and 2016) is possible in cases where a stress normalization cut-off is used.

Cone Penetration Testing 2022 – Gottardi & Tonni (eds)

Shear wave velocity – SCPTU correlations for sensitive marine clays

M. Long
School of Civil Engineering, University College Dublin (UCD), Ireland

J.-S. L'Heureux
Norwegian Geotechnical Institute (NGI), Trondheim, Norway

ABSTRACT: The purpose of this paper is to encourage the use of the seismic cone penetrometer (SCPTU) in soil characterisation studies. There has been an increase in use of shear wave velocity (V_s) data in geotechnical engineering. This has been prompted by improvements in measurement and analytical systems. A significant advantage, as is confirmed here, is that V_s can be measured easily and repeatedly by several different techniques in the sensitive marine clays under consideration here. Here the focus is on the derivation of preconsolidation stress (p_c') from V_s. A rational method of determining a V_s/p_c' relationship is outlined with resorting to empirical data analysis. The proposed relationship is shown to work well for Canadian sensitive clay data as has been shown previously for Norwegian and Swedish clays

Cone Penetration Testing 2022 – Gottardi & Tonni (eds)

A simplified method to incorporate the benefits of microstructure for cyclic liquefaction analyses using the SCPT

K. Lontzetidis
CMW Geosciences, New Zealand

P.K. Robertson
Gregg Drilling Inc., USA

D.J. Morton
CMW Geosciences, New Zealand

ABSTRACT: Cyclic liquefaction resistance of sand deposits can increase due to microstructure that results from several factors, such as aging and bonding. Available empirical correlations, using either SPT, CPT or shear wave velocity (V_S) data, were derived from cyclic liquefaction case histories that considered very young Holocene-age, essentially normally consolidated, unbonded, silica-based soils that may not apply to soils with significant microstructure. Seismic CPT (SCPT) data can be used to identify soils with significant microstructure, since both aging and bonding tend to increase the small-strain stiffness (reflected in the measured V_S) significantly more than they increase the large-strain strength of a soil (reflected in the CPT penetration resistance). Hence, for a given soil, both age and bonding tend to increase V_S more than the larger-strain cone resistance, all other factors (such as, in situ stress state, density, etc.) being constant. The normalized rigidity index (K_G) has been proposed (Robertson, 2016) as a parameter, that combines V_S and normalized CPT tip resistance (Q_{tn}), to detect and quantify the presence of microstructure. This paper presents and discusses a suggested method to quantify the increased resistance to cyclic liquefaction due to microstructure by utilizing K_G. Results will be presented from older soil deposits in New Zealand well as compare the results with existing correlations that account for "aging".

Cone Penetration Testing 2022 – Gottardi & Tonni (eds)
© 2022 the Author(s), ISBN 978-1-032-35624-2

Classification of Miocene deposits using CPT data

A. Makra
GR8 GEO, Athens, Greece

H. Kim
DL E&C Co. Ltd, Seoul, South Korea

ABSTRACT: Soil classification based on CPT data is commonly carried out using classification charts which link the type of soil behaviour to the measured tip resistance, sleeve friction and pore water pressure. These charts have been successful in interpreting soil behaviour in young, saturated, uncemented soil deposits but often diverge from observations in samples in older deposits. The geotechnical investigation conducted for the 1915 Çanakkale bridge included 20 offshore boreholes with 18 adjacent CPT soundings conducted downhole that penetrated several meters into Miocene-age sediments. This paper investigates those data sets and proposes a modified chart for the classification of the Miocene deposits accounting for their microstructure. The proposed modification is useful for classifying Miocene deposits in similar geological settings through the performance of CPTs.

Cone Penetration Testing 2022 – Gottardi & Tonni (eds)
© 2022 the Author(s), ISBN 978-1-032-35624-2

On the interpretation of piezocone dissipation testing data in clay

F.M. Mántaras
Geoforma Engenharia Ltda, Brazil

F.S. Pereira
University of Western Australia, Australia

E. Odebrecht
State University of Santa Catarina, Brazil
Geoforma Engenharia Ltda, Brazil

F. Schnaid
Federal University of Rio Grande do Sul, Brazil

ABSTRACT: Piezocone dissipation tests are routinely used to estimate the *in situ* coefficient of consolidation (c_h). This is achieved either by extracting a single point of the dissipation curve (usually, the time for 50% dissipation, t_{50}) or by adjusting a theoretical solution to the experimental data. Results from a piezocone investigation carried out in the Holocene clay deposit of Tubarão are reported and interpreted to evaluate the accuracy of existing procedures to estimate c_h. Explicit recommendations on how to curve fit dissipation curves and use the first and second derivates to estimate t_{50} are presented. Validation of the proposed method is provided from field tests reaching up to 75% dissipation. The proposed method does not require accurate measurements of equilibrium pressure and relies less on engineering judgment when compared to reference procedures. Furthermore, the method performed reasonably well with a shorter dissipation time (40% dissipation).

Cone Penetration Testing 2022 – Gottardi & Tonni (eds)
© 2022 the Author(s), ISBN 978-1-032-35624-2

Determination of fine-grained soil parameters using an automated system

I. Marzouk & F. Tschuchnigg
Graz University of Technology, Institute of Soil Mechanics, Foundation Engineering and Computational Geotechnics, Graz, Austria

F. Paduli, H.J. Lengkeek & R.B.J. Brinkgreve
Delft University of Technology, Delft, The Netherlands

ABSTRACT: Performing numerical analysis successfully depends on several factors. One of the most important factors is determining the constitutive model parameters correctly. It is often the case that these parameters are determined based on limited soil data. Using in-situ tests for determining these parameters has several advantages such as minimal disturbance of the soil and lower cost compared to laboratory tests. However, it is not possible to determine soil parameters directly from in-situ tests results. Thus, empirical correlations are required for interpreting soil parameters. Generally, several correlations exist for the same parameter, which will lead to calculating several values for the same parameter. An ongoing research project focuses on formulating an automated parameter determination (APD) framework that uses a graph-based approach to identify constitutive model parameters based on in-situ tests. This is achieved by using two spreadsheets as an input, one for parameters and the other for equations (correlations used to calculate parameters). Based on these two spreadsheets, the system generates paths between the parameters and calculates the value(s) for each individual parameter. So far, the research project focused on determining the parameters for coarse-grained soil based on cone penetration test (CPT) results. Due to the fact that the system was set up in a modular and adaptable way, it is possible to expand the system to accommodate more soil types and in-situ tests. It is the aim of the research project to increase the reliability of the parameters values (required to perform numerical analysis) determined from in-situ tests. This paper focuses on expanding the current framework to determine parameters for fine-grained soil. By using the two spreadsheets as an input, the system successfully calculates the value(s) for fine-grained parameters. Further validation, dealing with several values for each parameter, determining the accuracy of derived parameters and expanding the system to accommodate other in-situ tests and types of soils are part of ongoing research.

Cone Penetration Testing 2022 – Gottardi & Tonni (eds)

Undrained shear strength of clays from piezocone tests: A database approach

P.W. Mayne
Georgia Institute of Technology, Atlanta, GA, USA

J. Peuchen
Fugro, Nootdorp, The Netherlands

ABSTRACT: Deriving undrained shear strength of clays from piezocone tests (CPTU) suits the use of a database approach, particularly because good correlations are expected on the basis of theoretical correspondence. Benefits of a database approach include a minimal environmental footprint and field schedule shortening, primarily because of reduced borehole sampling and laboratory testing focused on verification rather than development of comprehensive site-specific data sets. This paper expands on the classic expression $s_u = q_{net}/N_{kt}$ where s_u is a reference undrained shear strength obtained by laboratory testing, q_{net} is the net cone resistance and N_{kt} is a cone bearing factor that is noted to decrease with increasing values of CPTU pore pressure ratio B_q. The database includes CPTU results and high-quality laboratory triaxial compression tests from 70 different clay deposits, of which 8 represent new case studies. The clays are allocated to 5 main categories: (a) soft-firm offshore; (b) soft-firm onshore; (c) soft sensitive; (d) stiff overconsolidated intact; and (e) stiff fissured clays. Organic clays and cemented clays are excluded.

Cone Penetration Testing 2022 – Gottardi & Tonni (eds)
© 2022 the Author(s), ISBN 978-1-032-35624-2

Evaluating geoparameters of Maine sensitive clay by CPTU

P.W. Mayne
Georgia Institute of Technology, Atlanta, GA USA

P. Paniagua
Norwegian Geotechnical Institute and Norwegian University of Science & Technology, Trondheim, Norway

B. Di Buò
Tampere University, Tampere, Finland

S.S. Agaiby
Cairo University, Giza, Egypt

ABSTRACT: Using two sets of analytical solutions for CPTU in clays, a suite of theoretically-consistent geoparameters is shown to be in good agreement with independent laboratory and field test results obtained on sensitive Presumpscot clay in Portland, Maine, USA. Fall cone tests indicate a mean sensitivity of $S_t \approx 37$. Values of undrained rigidity index (I_R), undrained shear strength (s_u), and yield stress ratio (YSR) are provided by a modified spherical cavity expansion-critical state hybrid model while an effective stress limit plasticity solution is utilized to assess the effective friction angle of the sensitive clay at both peak strength [ϕ' at q_{max}] and also at maximum obliquity [ϕ' at $(\sigma_1'/\sigma_3')_{max}$]. A CPTU screening method that uses three simplified equations for YSR helps to identify that the clay is sensitive.

Cone Penetration Testing 2022 – Gottardi & Tonni (eds)

CPTU evaluations in Appalachian Piedmont residual sandy silts

P.W. Mayne
Georgia Institute of Technology, Atlanta, GA, USA

E. Cargill
ConeTec Group, Richmond, VA, USA

ABSTRACT: Results from recent piezocone testing are compared with a statistical set of 22 prior CPTU soundings at the Opelika National Geotechnical Experimentation Site in Alabama. The site is underlain by residua comprised of fine sandy silts of the Appalachian Piedmont geologic province in the eastern USA. The cone resistance and sleeve friction from a 2016 CPTU compare well with the mean values from previous soundings while the porewater pressure readings are similar yet different because of a drop in the groundwater table from 2 m to 10 m over the period of study. Evaluations of the effective friction angle and yield stress from CPTU in residual soils are presented and compared with benchmark values obtained from laboratory triaxial and consolidation test results. Most interestingly, as the soil is intermediate with about 50-50 silt and sand, both undrained and drained penetration give more or less the same results.

Cone Penetration Testing 2022 – Gottardi & Tonni (eds)
© 2022 the Author(s), ISBN 978-1-032-35624-2

Soil unit weight prediction from CPTs for soils and mining tailings

T. Menegaz & E. Odebrecht
University of Santa Catarina State, Joinville, Santa Catarina, Brazil

H.P. Nierwinski
Federal University of Santa Catarina, Joinville, Santa Catarina, Brazil

F. Schnaid
Federal University of Rio Grande do Sul, Porto Alegre, Rio Grande do Sul, Brazil

ABSTRACT: An accurate estimation of the total unit weight (γ_t) of soils and mining tailings is a requirement for several geotechnical engineering applications. In the interpretation of CPT testing data, the value of γ_t is needed for the evaluation of both the total and effective stresses, as well as in the accurate determination of subsequent parameters. Current methods for determining γ_t based CPT data have been developed for a wide range of natural soils where the specific gravity of solids (G) generally ranges from 2.5 to 2.7. However, a large gap exists for soils or mining tailings where G is often higher, such as in the case of bauxite and zinc. This paper presents a compilation of results of CPT tests, and laboratory measurements of specific gravity of solids and total unit weight on bauxite and zinc mine tailings. The application of the literature methodologies was evaluated for this database and the statistical coefficient of determination (r) was presented, which is determined between the measured value and the value estimated by the correlations

Cone Penetration Testing 2022 – Gottardi & Tonni (eds)
© 2022 the Author(s), ISBN 978-1-032-35624-2

Use of DMT and CPTU to assess the G_0 profile in the subsoil

Z. Młynarek
University of Life Sciences, Poznań, Poland

J. Wierzbicki
Institute of Geology, Adam Mickiewicz University, Poznań, Poland

P. Monaco
University of L'Aquila, L'Aquila, Italy

ABSTRACT: The subject of this article is the analysis of the relationship between G_0/M_{DMT} and K_D, where G_0 is the small strain shear modulus, while M_{DMT} and K_D are respectively the constrained modulus and the horizontal stress index determined from DMT tests. This relationship allows to determine a profile with depth of G_0 from standard DMT test results, useful when data from non-seismic DMT investigations are available. The analysis was based on a large amount of data for a wide range of soils of different origins in Poland. The dataset included OC and NC loams, silts, medium sands, silty sands and fine sands. The overconsolidation ratio (*OCR*) was estimated from CPTU and DMT tests. The obtained empirical G_0/M_{DMT} vs. K_D relationships were compared with the correlations established by Marchetti et al. (2008) for different soil types. To account for the significant influence of overconsolidation, an original empirical relationship between G_0/σ'_p and K_D, where σ'_p is the preconsolidation stress, was determined based on data from all investigated fine-grained soils.

Cone Penetration Testing 2022 – Gottardi & Tonni (eds)
© 2022 the Author(s), ISBN 978-1-032-35624-2

Physical and numerical modelling of T-CPT for mechanisms of penetration and heat transfer

P.Q. Mo & L. Gao
China University of Mining and Technology, Xuzhou, China

H.S. Yu
University of Leeds, Leeds, UK

X.L. Tao
Jiangsu Vocational Institute of Architectural Technology, Xuzhou, China

Q.Z. Ma
CCCC-FHDI Engineering Co., Ltd. GuangZhou, China

ABSTRACT: Thermal-Cone Penetration Test (T-CPT) is proposed as a new type of in-situ soil testing method for estimations of both mechanical and thermal properties of soil layers. Physical modelling of T-CPT is conducted in this study to obtained the effects of soil density, penetration depth and heating duration on the penetration resistance and thermal responses. Discrete Element Method (DEM) is then adopted to simulate the processes of penetration, heating and cooling, for investigations of penetration induced stress and temperature distributions in the surrounding soil. The measurements of thermal responses are analyzed to validate the simulation and to evaluate the interpretation method for thermal properties. The results could be used as a benchmark for further analyses of mechanisms of penetration and heat transfer. The interpretation method for T-CPT data is to be developed to improve the performance, with implications to utilization of geothermal energy.

Cone Penetration Testing 2022 – Gottardi & Tonni (eds)

Soil stratigraphy from seismic piezocone data and multivariate clustering in alluvial soil deposits: Experience in the Lower Tagus Valley region

F. Molina-Gómez, D. Cordeiro, C. Ferreira & A. Viana da Fonseca
CONSTRUCT-GEO, Faculdade de Engenharia da Universidade do Porto (FEUP), University of Porto, Portugal

ABSTRACT: The identification of soil stratigraphy at a given site is crucial for geotechnical analysis and design. This study addresses the delineation of soil stratigraphy from seismic piezocone penetration test (SCPTu) measurements. For this purpose, SCPTu data from an experimental site in the Lower Tagus Valley region (close to Lisbon) were compiled. Soil stratigraphy was obtained by applying a multivariate clustering approach using the direct measurements reported by SCPTu. The hierarchical clustering results were compared against the soil behaviour index profile. Moreover, the differences/similitudes obtained from multivariate clustering were validated by contrasting statistical results against the visual description of samples collected in the experimental site at Lower Tagus Valley using advanced sampling techniques; namely, Gel-Push sampler. The main findings showed that the clustering approach implemented herein detects and groups soil layers with similar soil behaviour types, allowing delineating the soil stratigraphy in multilayer alluvial soil deposits.

Cone Penetration Testing 2022 – Gottardi & Tonni (eds)
© 2022 the Author(s), ISBN 978-1-032-35624-2

Undrained strength from CPTu in brittle soils: A numerical perspective

L. Monforte, M. Arroyo & A. Gens
CIMNE-UPC, Barcelona, Spain

ABSTRACT: Static liquefaction of soils that have a brittle undrained response (hydraulic fills, mine tailings or sensitive clays) may lead to sudden failures of large consequence. Given the importance of undrained failure, obtaining precise estimates of peak and residual yield strength is important. The CPTu plays a major role in the geotechnical characterization of these geomaterials and so do CPTu-based estimates of undrained strength. Most of the methods available for CPTu-based estimation of undrained strength are empirical, based on correlation with other laboratory or field tests. When such correlations are established difficulties appear due to variable disturbance affecting the reference laboratory samples and parasitic effects, such as unaccounted for partial drainage during penetration or unknown side friction, affecting the cone results. Such difficulties are not present when using numerical simulation. The paper builds upon a series of CPTu simulations using a model able to represent brittle undrained failure. Confounding factors such as partial drainage and cone side friction are systematically varied to examine their effect on the results. The results are then employed to examine the performance of several empirical methods frequently employed to obtain peak and residual strength from CPTu.

Cone Penetration Testing 2022 – Gottardi & Tonni (eds)

Assessment of deltaic soil behavior classification using AUT: GMD database regarding CPTu records

Mohammad H. Naghibi, Abolfazl Eslami & Sara Heidarie Golafzani
Department of Civil and Environmental Engineering, Amirkabir University of Technology (AUT), Tehran, Iran

ABSTRACT: Cone penetration tests (CPT and CPTu) are unique geotechnical in-situ tools for soil behavior classification (SBC). Identification and classification of subsurface marine layers help geotechnical engineers analyze, design, and monitor. Thus, a significant database of 398 cases was collected from 58 sites in 18 countries with CPTu soundings, and soil profiling obtained by direct boring and laboratory testing, namely AUT: CPTu&GMD (Amirkabir University of Technology): Geo-Marine CPTu Database. Furthermore, diverse essential parameters such as normalized cone resistance q_t, friction ratio R_f, soil behavior type index I_{SBT} are available at 0.1 m intervals. CPTu-based classification methods were subdivided into three generations, and several methods for marine deposits were examined utilizing 57 case studies from the AUT: CPTu&GMD. The results indicated that directly incorporating all three parameters of q_c, f_s, and u_2 is recognized to be more accurate than utilizing mathematical relationships, which reduces uncertainties and increases accuracy in identifying subsurface layers.

Cone Penetration Testing 2022 – Gottardi & Tonni (eds)
© 2022 the Author(s), ISBN 978-1-032-35624-2

A review of methods for estimating undrained brittleness index from the CPT

Y. Narainsamy & S.W. Jacobsz
University of Pretoria, Pretoria, South Africa

ABSTRACT: The Cone Penetration Test (CPT) is a useful tool for soil profiling due to its near continuous data measurements, low cost and repeatability. When conducting designs or safety evaluations of slopes, it is often of interest to understand the undrained response of the soil at large strains. This behaviour can be defined in terms of the Undrained Brittleness Index (I_B) which relates the yield undrained shear strength to the steady state undrained shear strength. Over time, a number of field case histories where strain softening during undrained shear was deemed to have occurred have been assessed, and a relationship between corrected tip resistance and I_B was proposed (Sadrekarimi, 2014). A quick clay test site and a silt test site in Norway included in a recently published open access geotechnical database were assessed. It was found that the method proposed by Sadrekarimi (2014) underestimated the I_B for both the quick clay and silt test sites. It was also noted that there are some limitations with a popular screening method for identifying soils susceptible to strength loss during undrained shear.

Cone Penetration Testing 2022 – Gottardi & Tonni (eds)

A comparative study on CPTu-based soil classification methods: Case studies

T.D. Nguyen & P.S. Khin
VNU Vietnam Japan University, Hanoi, Vietnam

Q.N. Pham
Vietnam Petroleum Institute, Hanoi, Vietnam

A.T. Vu
Le Quy Don Technical University, Hanoi, Vietnam

ABSTRACT: This paper presents a comparative study on the applicability of three CTPu-based soil classification charts, namely, Robertson's normalized (SBTn) chart (Robertson 2009), Eslami-Fellenius (EF)'s non-normalized chart (Eslami and Fellenius 1997) and Robertson's non-normalized (SBT) chart (Robertson 2010). A well-monitored database of CPTu data at six study sites of different geological conditions was used for the analyses. It is found from the study that when the effective stress (σ'_{v0}) is small, typically smaller than 150 kPa, the EF's and Robertson's SBTn charts results in similar soil types obtained from visual classification procedure (VCP) whereas the Robertson's SBT chart results in similar or slightly coarser soil types. When the σ'_{v0} becomes larger, the Robertson's SBTn chart still results in soil types well matched with those obtained from the VCP but both the non-normalized charts tend to result in coarser soil types. This feature of the non-normalized charts is more pronounced in stiff clayey or dense sandy soils.

Cone Penetration Testing 2022 – Gottardi & Tonni (eds)
© 2022 the Author(s), ISBN 978-1-032-35624-2

Characterization of young sediments using CPTu and Medusa SDMT

S. Oberhollenzer, L. Hauser, F. Brand, R. Marte & H.F. Schweiger
Institute of Soil Mechanics, Foundation Engineering & Computational Geotechnics, Graz University of Technology, Graz, Austria

D. Marchetti
Studio Prof. Marchetti, Rome, Italy

S. Pfeifer
illwerke vkw AG, Bregenz, Austria

ABSTRACT: Piezocone penetration tests (CPTu) and seismic flat dilatometer tests (SDMT) present cost- and time-efficient insitu investigation techniques for onshore as well as offshore projects. Since soil sampling is often related to a strong soil disturbance in fine-grained sediments, parameter identification is frequently based on insitu measurements in combination with correlations. As shown in previous studies, correlations are more difficult to develop in silt-dominated sediments. To overcome this problem, the research project PITS (parameter identification using insitu tests in silts), was launched by Graz University of Technology in cooperation with the Federal Chamber of Architects and Chartered Engineering Consultants. To investigate the influence of time effects (age) and microstructure on the load-settlement behaviour of normallyto slightly underconsolidated sediments, deposits younger than 50 years have been investigated at the water storage reservoir Raggal (Austria) using CPTu as well as Medusa SDMT. Both probes were pushed first through the water and subsequently into the sediments by means of a stand-alone pushing device, where the testing setup was installed on a floating pontoon. In order to prevent buckling of the penetration rods, additional casing tubes (along the water) were used. In a last step, soil sampling was executed using the CPT-Ranger system by Geomil. To characterize and quantify the sediments microstructure, shear wave velocities determined insitu by means of SDMT ($V_{S,SDMT}$) are compared with measurements on reconstituted soil samples using bender elements ($V_{S,BE}$). Ratios $V_{S,BE}/V_{S,SDMT} \approx 1$ indicate the presence of no or moderate soil microstructure. On the other hand, postglacial deposits of similar grain size distribution are characterized by smaller $V_{S,BE}/V_{S,SDMT}$ ratios, indicating a higher microstructure.

Cone Penetration Testing 2022 – Gottardi & Tonni (eds)
© 2022 the Author(s), ISBN 978-1-032-35624-2

PFEM modeling of CPTu tests in saturated structured soils

Kateryna Oliynyk
School of Science and Engineering, University of Dundee, Dundee, UK
Department of Civil and Environmental Engineering, University of Perugia, Perugia, Italy

Matteo O. Ciantia
School of Science and Engineering, University of Dundee, Dundee, UK

Claudio Tamagnini
Department of Civil and Environmental Engineering, University of Perugia, Perugia, Italy

ABSTRACT: The conventional interpretation of CPTu tests is typically based on empirical and semi-empirical correlations based on very crude descriptions of soil behavior, such as the total stress approach coupled with the Terzaghi-Rendulic pseudo-3d consolidation theory for modeling the time evolution of excess pore water pressure. The aim of this work is to show that a more rational interpretation of the coupled deformation and flow processes occurring in the soil during a CPTu test is possible by resorting to the numerical solution of the relevant governing equations, incorporating a realistic constitutive model for the soil. In order to deal with the large displacements and deformations induced by the cone penetration, the Particle Finite Element Method code G-PFEM, recently developed for geomechanical applications, has been used for this purpose. A key feature of the present work is the use of a finite deformation version of a non-associative isotropic hardening plasticity model for structured geomaterials - the FD_Milan model. The model is equipped with a *structure-related* internal variable which provide a macroscopic description of the effects of structure in natural, fine-grained soils. In order to deal with strain localization, typically observed in structured geomaterials upon yielding, the model has been equipped with a non-local version of the hardening laws, which has demonstrated capable of regularizing the pathological mesh dependence of classical FE solutions in the post-localization regime. A number of PFEM simulations of CPTu tests on a soft structured natural clay has been performed in order to assess the effects of the initial bond strength and permeability on the predicted results of the test, as well as on the spatial distributions of accumulated plastic strains, internal variables and excess pore water pressures. The results obtained represent a promising step towards a more rational interpretation of the CPTu tests in structured geomaterials and for their use in the calibration of advanced soil models.

Cone Penetration Testing 2022 – Gottardi & Tonni (eds)

Effect of the scatter between CPTU measured parameters in soil classification

P. Paniagua
Norwegian Geotechnical Institute & Norwegian University of Science & Technology, Trondheim, Norway

J.-S. L'Heureux
Norwegian Geotechnical Institute, Trondheim, Norway

ABSTRACT: Using piezocones from different manufacturers may yield different results even if the equipment complies with international standards. This causes problems when soil investigation contractors, using different cones, operate in the same area, and especially on the same project. Studies done in soft clay, sand, silt, and quick clays from different Norwegian sites show that repeatability for the cone resistance measurements and penetration pore pressure is good and that it has improved from one cone type to another. However, the scatter in the measured sleeve friction, and hence the friction ratio, is still very significant. Here, an attempt is made to qualitatively describe the impact of the scatter in the soil classification based on CPTU parameters. The most common soil classification charts are used to illustrate this scatter.

Evaluation of shear wave velocity profiles in alluvial and deltaic soils using a CPT database

J. Paredes & F. Illingworth
Subterra, Ecuador

ABSTRACT: A seismic piezocone SCPTu database from Guayaquil (EC), consisting of Holocene alluvial and estuarine deltaic soil deposits, is evaluated in this document through V_s mapping when comparing more than 600 CPTu profiles and 45 seismic downhole testing locations. Shear wave velocity downhole measurements were used to adjust CPT-based V_s estimates and compared with cone resistance values. A suitable fit is observed between Robertson (2009a) calculation and downhole measurements, but it tends to over-estimate by 20-30% in soft clayey soils. An age correction factor as a function of normalized s_u/σ'_v ratio has been determined for clayey soils while a V_s-q_t correlation is provided for clays and all soils. A V_{s30} map has been developed based on proposed shear-wave velocity calibration.

Cone Penetration Testing 2022 – Gottardi & Tonni (eds)
© 2022 the Author(s), ISBN 978-1-032-35624-2

Upscaling 1 500 000 synthetic CPTs to voxel CPT models of offshore sites

J. Peuchen & W. van Kesteren
Fugro, Nootdorp, The Netherlands

V. Vandeweijer & S. Carpentier
TNO, Utrecht, The Netherlands

F. van Erp
RVO – Netherlands Enterprise Agency, Utrecht, The Netherlands

ABSTRACT: In 2020, the Dutch government published 1 500 000 synthetic CPT profiles for use in development of the Hollandse Kust (west) Wind Farm Zone, offshore Netherlands. The scale of this approach was novel at that time and possibly first-ever. The synthetic CPT profiles were derived from a training data set of 122 actual CPTs and ultra-high-resolution (UHR) seismic reflection traces, using machine learning by a convolutional neural network. The synthetic CPT profiles were limited to positions along the 162 UHR survey track lines (2D) and were limited to cone resistance to a depth of 50 m below seafloor. The UHR track lines were spaced at about 400 m. This paper explores upscaling the synthetic CPT approach to voxel (3D) models and adding (S)CPT-based parameters such as shear modulus at small strain G_{max}. Future added-value is expected from post-2020 improvements seen in seismic reflection data resolution, attribute extraction and neural network architecture.

Cone Penetration Testing 2022 – Gottardi & Tonni (eds)

Automated CPT interpretation with a Convolutional Neural Network

M. Pippi, R. Vink & J. Haasnoot
CEMS Crux Engineering MicroServices BV, Amsterdam, The Netherlands

S. Bersan
CRUX Engineering, Amsterdam, The Netherlands

ABSTRACT: Cone Penetration Tests are widely used in the Netherlands, due to their ease of execution in the Dutch delta deposits and their relatively low cost. As the amount of performed CPTs increases, an automated soil interpretation becomes more and more relevant. Attempts to automate soil classification have been done in the past, but the empirical formulas commonly used do not always provide a satisfactory interpretation for engineering purposes. Besides that the soil type is often not interpreted correctly, there is also the problem that the classification is provided for each measurement (every 2 centimeters) and no strategy is provided to aggregate those tiny layers. This paper shows how a data driven approach can yield better results than the traditional empirical methods. A machine learning model is presented which is based on a Convolutional Neural Network. The Neural Network has been trained on 1800 pairs of CPTs and boreholes that met the condition of being less than 6 meters apart. An algorithm based on the theory of signals is applied to the classification given by the model to group measurements into soil layers. The paper explains the theory behind the model, shows a comparison with the soil classification given by the Robertson correlation and shows how the model can be used in the geotechnical practice.

Cone Penetration Testing 2022 – Gottardi & Tonni (eds)

Watch out for the use of global correlations and "black box" interpretation of CPTU data

J.J.M. Powell
Geolabs Limited, Watford, UK

L. Dhimitri
In Situ Site Investigations Ltd, UK

ABSTRACT: Many people interpret soil properties from CPTU measurements based on correlations embedded in software packages without ever questioning the validity of those correlations. This could be termed the 'black box' approach! This paper aims to show how dangerous this can be but also the power of the CPTU in helping to show the variations in soil properties within profiles. Based on correlations properties can be both over and under-estimated, which can of course result in both unsafe design and over design. Too often one correlation must be used to derive a soil property required in another correlation, further compounding the potential for errors. A range of sites will be examined with a range of soils varying from very soft clays and silts to stiff clays, sands and soft rocks.

Cone Penetration Testing 2022 – Gottardi & Tonni (eds)

New methods for assessing Plasticity Index and Low-strain Shear Modulus in fine-grained offshore soils

N. Ramsey
Fugro Australia Marine Pty. Ltd., Perth, Western Australia, Australia

K.K. Tho
Fugro Singapore Marine Pte. Ltd., Singapore

ABSTRACT: This paper introduces a new method for predicting Plasticity Index, PI, directly from CPT data. Plasticity Index values predicted using the new method have been designated "PI_{CPT}" to emphasise their CPT-based origin. The new PI_{CPT} method has, in turn, enabled a new method of assessing Low Strain Shear Modulus, G_{max}, to be proposed, based on a well-established equation published by Viggiani & Atkinson (1995). As an additional benefit, the predicted PI_{CPT} profiles have proven very useful as a simple means of identifying soil-type variations and layer boundaries. This paper presents details of the new formulae that have been developed to estimate PI_{CPT} and G_{max}, the database that has been used to calibrate the formulae, and the results that have been obtained. Some limitations of the method are presented and discussed, as well as some of the benefits.

Cone Penetration Testing 2022 – Gottardi & Tonni (eds)
© 2022 the Author(s), ISBN 978-1-032-35624-2

Cone penetration testing and interpretation in the holds of two ore-carrying vessels

N. Ramsey
Fugro, Australia

ABSTRACT: This paper provides details of two in-situ testing programmes performed to assess ore properties in the holds of ore carrying vessels. The first case comprised CPTs in the hold of an iron-ore carrying vessel - at both the ports of loading and discharge – enabling assessment of the ore properties after loading and the changes caused by the voyage. The second case comprised two CPTs and one Seismic Dilatometer Test (SDMT) in the hold of a coal carrying vessel shortly after loading – enabling assessment of the coal properties after loading, and direct comparison of CPT and SDMT interpretations.

Cone Penetration Testing 2022 – Gottardi & Tonni (eds)

Suction influence on CPT and DMT for some Brazilian tropical soils

B.P. Rocha
Federal Institute of São Paulo (IFSP), Advanced Campus of Ilha Solteira, Brazil

R.A. Rodrigues & H.L. Giacheti
Department of Civil and Environmental Engineering, São Paulo State University (Unesp), Brazil

ABSTRACT: Suction plays an important role in geotechnical engineering practice since it influences the mechanical behavior of unsaturated soils. Interpretation of in situ test data at unsaturated soil sites must consider both spatial and seasonal variability, especially in strength and stiffness parameters. This paper presents and discusses CPT and DMT performed at different seasons at two sites with tropical sandy soils. CPT and DMT data were interpreted considering the influence of soil suction. Suction was estimated from water content profiles and soil water retention curves (SWRC). Failure to consider the effect of soil suction in the site investigation of unsaturated soils can lead to inappropriate soil classification as well as significant and unconservative overestimates of geotechnical parameters.

Cone Penetration Testing 2022 – Gottardi & Tonni (eds)

Study of SPT-CPT and DP-CPT correlations for sandy soils

M.D. Santos
Department of Civil Engineering, Federal University of Espirito Santo, Vitória, Brazil
Department of Civil Engineering, COPPE, Federal University of Rio de Janeiro, Rio de Janeiro, Brazil

K.V. Bicalho
Department of Civil Engineering, Federal University of Espirito Santo, Vitória, Brazil

ABSTRACT: Field penetration tests are widely used in geotechnical engineering for site investigation. The most common in-situ penetration tests are: the Standard Penetration Test (SPT), the Cone Penetration Test (CPT), and the Dynamic Probing (DP). Since there are limitations on their test methodologies and equipment, equations that can describe the relationship between their data are important in geotechnical engineering. This study presents a review of SPT-CPT and CPT-DP correlations for sandy soils, and discusses the uncertainties involved in some previously published correlations in literature. It is also recommended adjustments accounting for variations in soil properties input. Linear regression and residual analysis were carried out in several experimental data sets from different sandy soils and locations. The observed sandy soils' heterogeneity implies in the need of correlations obtained at a regional level. Thus, this study does not aim to exhaust the subject about in situ penetration tests correlations, but rather to recommend and stimulate the critical evaluation and use of previously published correlations and those that will be formulated in the future

Cone Penetration Testing 2022 – Gottardi & Tonni (eds)
© 2022 the Author(s), ISBN 978-1-032-35624-2

Correlation between SPT and CPT tests in liquefiable deposits

Rashid Shahgholian, Cristiana Ferreira & António Viana da Fonseca
CONSTRUCT-GEO, Faculty of Engineering of University of Porto, Porto, Portugal

ABSTRACT: Cone penetration test (CPT) started being used for soil investigation almost four decades after the standard penetration test (SPT). Despite their differences, both tests estimate the resistance of the soil against penetration of a probe. Although it is expected that the results of these tests are comparable and directly correlated, different driving mechanisms (static versus dynamic), sensitivity, type and frequency of measured parameters, operational errors, among others, compromise such correlations. In addition, the diverse nature of soils in terms of grain size, packing and fabric, shear strength, permeability, and other soil characteristics also limit the derivation of a unique correlation between CPT and SPT. Many authors have studied these in situ tests, recognizing that such correlation would be valuable, as it would enable the use of design methods and other libraries of correlations available for each individual test. A variety of formulas and charts has been proposed in the literature to describe this relationship, which demonstrates that this is a complex, not a global and simply predictable correlation. In this research, some of the most usual CPT-SPT relations are evaluated for a large dataset of field measurements from a pilot site on liquefiable soils, near Lisbon in Portugal. The extensive analysis of these data showed that the application of previously established, tested, and published correlations, available in the literature for specific sites, requires great care. One of the requirements is the verification of applicability, by comparison between the type of soils under study and those in the literature. Specific correlations for each geological condition must be defined before making engineering estimations based on correlated parameters.

Cone Penetration Testing 2022 – Gottardi & Tonni (eds)

CPTU-detection of thin clay layers in sand: Results from calibration chamber tests

H. Skrede
Norwegian Geotechnical Institute (NGI), Norwegian University of Science and Technology (NTNU), Norway

H.B. Hammer
Dr.techn. Olav Olsen AS, Norway

S. Nordal
Norwegian University of Science and Technology (NTNU), Norway

J.-S. L'Heureux
Norwegian Geotechnical Institute (NGI), Norway

ABSTRACT: The detection of thin clay layers (i.e. ≤ 20 cm) is challenging for all conventional geotechnical field investigations techniques, including high quality CPTU tests. During the last two years, a research program has been carried out in the geotechnical laboratory at NTNU in Trondheim. The work aims to identify possibilities and limitations in detecting thin clay layers and assess their properties using the CPTU tool. Tests were run in a pressurized chamber where thin horizontal clay layers of both pottery clay and natural, sensitive clay were embedded in a homogenous, medium dense sand. Both a standard piezocone (10 cm^2) and a mini-piezocone (5 cm^2) have been utilized. The results show to what degree the CPTU response in thin layers is influenced by the surrounding sand, and how this influence in practice may lead to serious overestimation of shear strength in thin layers. In addition, the effect of depth-offset of measurements in connection with soil type characterization was evaluated.

Cone Penetration Testing 2022 – Gottardi & Tonni (eds)
© 2022 the Author(s), ISBN 978-1-032-35624-2

Determination of hydraulic conductivity using HPT & CPTu

M. Slowiok, S. Oberhollenzer & R. Marte
Institute of Soil Mechanics, Foundation Engineering & Computational Geotechnics, Graz University of Technology, Graz, Austria

T. Freudenthaler
mjp ZT GmbH, Saalfelden, Austria

ABSTRACT: In geotechnical engineering insitu tests are becoming increasingly popular to determine the hydraulic conductivity of soils. Direct push methods such as piezocone penetration test CPTu or the hydraulic profiling tool HPT allow characterizations along a linear profile. The determination of hydraulic conductivity with HPT is an alternative to CPTu for drained and partially drained conditions. Thereby, water is continuously injected from a small screened port at the probe into the soil. Based on the required pressure to inject the water into the soil and the constant flow rate, the hydraulic permeability can be determined for a certain range of drainage conditions. The present article compares HPT and CPTu results with slug tests, executed at different test sites in a wide range of grain size distributions. Based on this comparison, a new correlation was elaborated for HPT which enables an improved characterization of hydraulic conductivity.

Cone Penetration Testing 2022 – Gottardi & Tonni (eds)
© 2022 the Author(s), ISBN 978-1-032-35624-2

DEM-FDM coupling simulation of cone penetration tests in a virtual calibration chamber

Y. Song, X.Q. Gu & J. Hu
Tongji University, Shanghai, China

ABSTRACT: We performed numerical simulation of the cone penetrating process with discrete element method (DEM) coupled with finite difference method (FDM) to investigate the mechanics of granular soil interacted with a cone penetrometer. The efficacy of particle size amplification method to estimate soil physical indexes has yet to be established and it was validated in this work. The effects of cone-particle chamber size, confining stress and sand modulus on cone tip resistance are analyzed. The results show that the effect of confining stress is consistent with the empirical relationship and the cone tip resistance correlates positively with the number of particles contacting with the cone penetrometer. Furthermore, the distribution of radial stress agrees well with the cavity expansion theory that a nearly linear relation between cone tip resistance and radial stress near the cone penetrometer can be observed.

Cone Penetration Testing 2022 – Gottardi & Tonni (eds)

Data-driven soil profile characterization using statistical methods and artificial intelligence algorithms

R.L. Spacagna
CNR IGAG Area della Ricerca Roma 1, Montelibretti (Roma), Italy

A. Baris, L. Paolella & G. Modoni
University of Cassino and Southern Lazio, Cassino, Italy

ABSTRACT: CPT soil profile interpretation represents a fundamental aspect for subsoil stratigraphic reconstruction of complex geological contexts. In some situations, the soil profile may not exhibit evident boundary changes, making the interpretation more difficult. This crucial aspect plays a key role in the layers boundaries discontinuities identification and the construction of bi-dimensional and three-dimensional geotechnical models. In this paper, CPT and boreholes are used to calibrate and validate a massive and automated site characterization by combining statistical tools and artificial intelligence algorithms (AI). The procedure is applied in the complex stratigraphic context of Terre del Reno (Italy). The proposed data-driven analysis allows to combine the geological and geotechnical knowledge of the subsoil in an efficient and automatic way based on site-specific data, obtaining reliable and indispensable results for the construction of a robust and coherent geotechnical model of the subsoil.

Cone Penetration Testing 2022 – Gottardi & Tonni (eds)

Capability of seismic CPTu and DMT in assessing propagation velocity of body waves: A comparative study

S. Stacul, D. Lo Presti & N. Nenci
University of Pisa, Pisa, Italy

F. Fiera & M. Perini
Geo-Energizers Snc, Cascina (Pisa), Italy

D. Marchetti
Studio Prof. Marchetti s.r.l., Rome, Italy

E. Pagani & M. Siviero
Pagani Geotechnical Equipment s.r.l., Calendasco (Piacenza), Italy

ABSTRACT: The paper shows the geotechnical characterization of a soft clay test site by means of laboratory testing. Such a test site was established in the campus of the Department of Biology of the University of Pisa at about 1.6 km from the Leaning Tower square. The main (but not exhaustive) scope of the present research activity was to compare the propagation velocity of body waves as obtained from various seismic techniques. A preliminary site characterization, based on both laboratory and in situ tests, was carried out. In situ testing consisted of a preliminary CPTu, three boreholes with four high quality Osterberg-type undisturbed sampling, cross-hole, surface seismic reflection test named Multichannel Analysis of Reflection Waves (MARW). Laboratory tests consisted of soil classification (Atterberg Limits and grain size distribution), incremental loading oedometer tests (OE), direct shear tests (DS) and resonant column tests (RC). Moreover, Pagani seismic piezocone and Marchetti seismic dilatometer were used for the assessment of the propagation velocity of body waves.

Cone Penetration Testing 2022 – Gottardi & Tonni (eds)
© 2022 the Author(s), ISBN 978-1-032-35624-2

Bayesian estimation of small-strain shear modulus from offshore CPT tests in the North Sea

B. Stuyts & C. Sastre Jurado
OWI-Lab, Vrije Universiteit Brussel, Brussels, Belgium
UGent, Ghent, Belgium

D. Gomez Bautista & A. Kheffache
OWI-Lab, UGent, Ghent, Belgium

ABSTRACT: The rapid expansion of offshore wind energy requires accurate and cost-effective site characterisation. The dynamic response of offshore wind turbine structures installed on monopile foundations strongly depends on the stiffness of the subsoil where the small-strain shear modulus governs the behaviour during operational loading. Direct measurement of the small-strain shear modulus with the seismic CPT (S-PCPT) is increasingly performed offshore. However, the additional cost associated with this test leads to incomplete data coverage. In the past, correlations between cone tip resistance and small-strain shear modulus have been developed for onshore conditions. The increasing availability of offshore site data has allowed a unique dataset of over 2000 small-strain shear modulus measurements from S-PCPT to be created for North Sea soil units. The comparison of the small-strain shear modulus inferred from the S-PCPT data to the corresponding cone tip resistance allows recalibration of the existing correlations. This paper proposes a Bayesian approach to update the correlation model for sands proposed in the literature based on the developed dataset for offshore conditions in the North Sea.

Cone Penetration Testing 2022 – Gottardi & Tonni (eds)
© 2022 the Author(s), ISBN 978-1-032-35624-2

Factors influencing CPT_U N_{kt} for marine clay in Singapore reclaimed land

C. Tanaka, M. Angeles & J.Y. Wong
Arup Singapore Pte. Ltd., Singapore

ABSTRACT: Cone penetration tests with porewater pressure measurements (CPT_U) are extensively adopted in Singapore due to the presence of soft soils from the young deposits of the Kallang Formation. The most prominent member of this formation is the soft Singapore marine clay, which can be present in thicknesses of up to 50m. Singapore is a relatively small island, and as such, over the past few decades has embarked on numerous reclamation projects to increase land to support their growing economy and population. Therefore, significant data in reclaimed land overlying soft deposits is available, especially in the east of Singapore. This has presented the opportunity to review the factors that influences N_{kt} in the marine clay located within this area. N_{kt} plays a significant role in the evaluation of the undrained shear strength (c_u), which is often derived from site-specific calibrations. This paper summarizes the CPT_U results carried out in reclaimed land to the east of Singapore. It discusses the calibration of N_{kt} with other site data, and proposes a range of N_{kt} to be adopted in this location.

Cone Penetration Testing 2022 – Gottardi & Tonni (eds)
© 2022 the Author(s), ISBN 978-1-032-35624-2

Parameters affecting the CPT resistance of reconstituted sands

Y. Tian & B.M. Lehane
The University of Western Australia, Perth, WA, Australia

ABSTRACT: Despite the popularity of the Cone Penetration Test, uncertainty remains regarding the stress level dependence of q_c in a sand of constant relative density, the effect of the mineralogy and the effect of saturation. This paper provides greater clarity on these effects by presenting results from a large series of drained cone penetration tests performed under controlled conditions in a laboratory pressure chamber. The experiments involve dry and fully saturated silica and carbonate sands placed at a number of relative densities (D_r) and tested at different stress levels and overconsolidation ratios. The observations are compared with existing relationships proposed between q_c and D_r and highlight the approximate nature of such relationships.

Cone Penetration Testing 2022 – Gottardi & Tonni (eds)

Practical experience with cone penetration in frozen soils

N.G. Volkov & I.S. Sokolov
GEOINGSERVICE (Fugro Group), Moscow, Russia

R.A. Jewell
Fugro GeoConsulting, Brussels, Belgium

ABSTRACT: This paper summarizes recent practical experience with CPT investigations at several permafrost sites in Russia. Some of the investigations were performed in unconventional conditions such as with ice cover, in a crawl space, or from a jack-up platform. All CPT measurements used a cone equipped with a temperature sensor to confirm the subzero temperature of frozen soil. Stress relaxation tests were performed in ice-rich permafrost soils to investigate long-term soil strength in both compression and shear. Sampling of frozen soil using direct push techniques was achieved at two sites.

Cone Penetration Testing 2022 – Gottardi & Tonni (eds)
© 2022 the Author(s), ISBN 978-1-032-35624-2

A site-specific relationship between CPT data and fines content for fine grained soil in the context of liquefaction analyses

C. Vrettos
Technical University of Kaiserslautern, Germany

ABSTRACT: Fines content is a crucial factor affecting the liquefaction resistance of soils within the frame of the CPT-based cyclic stress ratio approach. In contrast to highly susceptible sands and silty sands, on one hand, and non-liquefiable clays, on the other hand, the proper identification and classification of silts is still an unresolved issue. Several empirical equations have been proposed to correlate fines content with common CPT soil behavior indices. In the frame of a site-specific study comprising pairs of CPT soundings and adjacent exploration borings, gradation characteristics of retrieved soil samples are combined with the respective values of common soil behavior indices inferred from the CPT logs to examine available predictive equations and calibrate a new hyperbolic equation. The investigation confirms the large scatter in the data and the inherent difficulty in establishing a robust correlation.

Cone Penetration Testing 2022 – Gottardi & Tonni (eds)
 ISBN 978-1-032-35624-2

Numerical modelling of cone penetration tests in spatially variable clays

Ze Zhou Wang, Siang Huat Goh & Xiangxiang Zheng
Department of Civil and Environmental Engineering, National University of Singapore, Singapore

ABSTRACT: In soft clays, a cone factor N_{kt} is typically adopted to correlate net cone tip resistance q_t with undrained shear strength S_u. Many existing numerical studies on the estimation of N_{kt} are performed using analyses in homogeneous soil medium. However, soil properties exhibit spatial variability under real field conditions. In this study, the effects of spatial variability on the interpretation of CPT data were investigated using a finite element analysis technique, namely the Press-Replace method (PRM). This method, which can capture installation effects arising from the penetration of a piezocone, was implemented with the soil domain modelled as anisotropic random fields. Using a pre-determined N_{kt}, statistics of S_u were back-calculated using the q_t profiles obtained from different random field realizations. Comparisons of the back-calculated statistics with the true statistics used to generate the random fields revealed that the true statistics may not be fully recovered from the CPT data.

Cone Penetration Testing 2022 – Gottardi & Tonni (eds)
© 2022 the Author(s), ISBN 978-1-032-35624-2

Some aspects of in situ testing of clay-glacial till mixture redeposited as man-made fills

J. Wierzbicki, K. Stefaniak & S. Wilczyński
Institute of Geology, Adam Mickiewicz University, Poznań , Poland

B. Brzeziński
Labortest Sp. z o.o. Sp. k., Poznań, Poland

ABSTRACT: The paper presents the results of CPTU and FVT tests carried out in soils, which filled the former brown coal excavation. In their natural state, these lands were pre-consolidated glacial till and clay. Detachment of soil from the original deposit, transport and low-energy redeposition led to the creation of multi-meter-thick non-consolidated embankments, which are now used as a construction subsoil. These embankments consolidate under their own weight, however, unlike typical geological formations, they are characterized by a decreasing influence of consolidation on the strength properties along with depth.

Cone Penetration Testing 2022 – Gottardi & Tonni (eds)
© 2022 the Author(s), ISBN 978-1-032-35624-2

Cone penetration in a thin medium dense sand layer sandwiched by different clay layers – LDFE analysis

Q. Xie & Y.X. Hu
University of Western Australia, Perth, Australia

M.J. Cassidy
University of Melbourne, Melbourne, Australia

M. Zhou
South China University of Technology, Guangzhou, China

ABSTRACT: The cone penetration test (CPT) has been the most common site investigation practice for decades. This paper presents the results of large deformation finite element (LDFE) analysis that modeled a CPT into soil comprising a thin medium dense sand layer sandwiched between two clay layers of different undrained shear strengths. An extended critical state Mohr-Coulomb model is utilized to capture the stress-dependent drained behavior of sand, while Tresca failure criterion is adopted for the undrained behavior of clay. Because the sand layer and the surrounding clay layers work as a system that defines the mobilized soil stiffness and soil interface deformation, two soil layer profiles are analyzed to study cone penetration resistance responses, including soft clay on top and stiff clay at bottom, and vice versa. It is found that a peak resistance is registered in the thin sand layer, through the top and bottom clay layers show impact to the peak as well. The impact on the peak resistance from the clay layer ahead (i.e. bottom layer) weights more than the clay layer behind (i.e. top layer). The evolving soil flow mechanisms are revealed to explain the findings.

Cone Penetration Testing 2022 – Gottardi & Tonni (eds)
© 2022 the Author(s), ISBN 978-1-032-35624-2

Study on SPT N-values and relative density through various soundings in full-scale chamber test ground

H. Yabe, K. Harada, T. Ito & E. Watanabe
Geotechnical Division, Fudo Tetra Corporation, Chuo-ku, Tokyo, Japan

ABSTRACT: In Japan, the evaluation of ground strength, such as the liquefaction resistance, is often carried out through the *N*-value of the standard penetration test (SPT), and it is used in various designs. Other soundings are also used, including the cone penetration test (CPT) and the Swedish weight sounding (SWS) that is comparatively simpler when compared with SPT, and conversion formulae to *N*-value have also been proposed by various researchers. In this paper, the various soundings described above (SPT, CPT, and SWS) were carried out in a sandy ground, where the density was controlled and measured by Radio Isotope (RI) method. Then, the correlation between the SPT *N*-value and the relative density *Dr* was examined. Based on the results, the *N-Dr* correlation was examined using the results of various soundings in a full-scale chamber test where the ground density was controlled. The results confirmed the applicability of the correlation formulae, with good correlation confirmed not only between *Dr* and SPT *N*-value, but also with the *N*-value estimated from each sounding.

Cone Penetration Testing 2022 – Gottardi & Tonni (eds)
© 2022 the Author(s), ISBN 978-1-032-35624-2

Correcting measured CPT tip resistance for multiple thin-layer effects

K.M. Yost, J. Cooper, R.A. Green, E.R. Martin & A. Yerro
Virginia Tech, Blacksburg, VA, USA

ABSTRACT: Multiple interbedded fine-grained layers in a sand deposit have a “smoothing” effect on the measured tip resistance (q_c) from the cone penetrometer test (CPT). This can result in an underestimation of the predicted liquefaction resistance of the sand layers. Herein, the efficacies of two multiple-thin-layer correction procedures are evaluated using published calibration chamber test data. The results highlight limitations of the assessed procedures for profiles with layers less than 40 mm thick. A new approach to estimate the “true” q_c (i.e., values that would be measured in a stratum absent of multiple thin-layer effects) from measured q_c is explored. The proposed numerical optimization algorithm searches for “true” soil profiles with a finite number of layers. We compare two versions of the algorithm that numerically optimize different functions, one of which uses a logarithm to refine fine-scale details, but which requires longer calculation times to yield improved corrected q_c profiles.

Cone Penetration Testing 2022 – Gottardi & Tonni (eds)

Spatial interpolation of consolidation property of clays from limited CPTU dissipation data

Zening Zhao
School of Transportation, Southeast University, Nanjing, China

Wei Duan
College of Civil Engineering, Taiyuan University of Technology, Taiyuan, China
School of Transportation, Southeast University, Nanjing, China

Guojun Cai*
School of Transportation, Southeast University, Nanjing, China
School of Civil Engineering, Anhui Jianzhu University, Hefei, China

Meng Wu
School of Transportation, Southeast University, Nanjing, China

Anand J. Puppala
Zachry Department of Civil and Environmental Engineering, Texas A&M University, USA

Songyu Liu
School of Transportation, Southeast University, Nanjing, China

Surya Sarat Chandra Congress
Zachry Department of Civil and Environmental Engineering, Texas A&M University, USA

ABSTRACT: The coefficient of consolidation (c_h), an important soil property, is often estimated from the piezocone penetration test (CPTU) data. However, the number of interpreted c_h data is usually limited especially in the horizontal direction because the CPTU is usually performed vertically and the dissipation test is time-consuming. In the present study, a novel method is proposed for the interpolation of c_h data in a 2D vertical cross-section. Firstly, the original c_h is interpolated by CPTU dissipation data. Then, the c_h values are interpreted based on the dissipation data at every location including the untested locations using random field theory. Finally, real case is illustrated to prove the applicability of the proposed method. It is shown that the proposed method can reasonably interpolate the c_h at untested locations. Overall, the new method can spatially interpolate the c_h from limited CPTU dissipation data and significantly save the test time.

*Corresponding author

Session 3: Applications

Cone Penetration Testing 2022 – Gottardi & Tonni (eds)

Evaluation of the geotechnical behavior of mining tailings through CPTU tests in the soil improvement process for the decharacterization of upstream heightened dams

J.L. Albino
Universidade Federal de São Carlos – UFSCAR, Brazil

T.A.T. Souza J. & L.S. Machado
Universidade Federal de Viçosa – UFV, Brazil

ABSTRACT: In order to assess the pore pressure conditions, to verify the drained behavior or not of a reservoir of a tailings dam raised upstream, and to observe the evolution of the resistence of the material, CPTU surveys were carried out before and after the execution of works for the implementation of vertical geodrains in the reservoir of this dam. After the comparative tests, it was possible to indicate the safety gains of the structures resulting from the execution of the works.

Keywords: CPTU test, upstream tailing dam, pore pressure dissipation, geodrains, liquefaction

Cone Penetration Testing 2022 – Gottardi & Tonni (eds)
© 2022 the Author(s), ISBN 978-1-032-35624-2

VCPT: An in-situ soil investigation method to validate vibratory pile-soil interaction models

D. Al-Sammarraie, S. Kreiter & T. Mörz
MARUM–Center for Marine Environmental Sciences, University of Bremen, Germany

M.O. Kluger
School of Science/Te Aka Mātuatua, University of Waikato, Hamilton, New Zealand

M. Goodarzi
COWI – Hamburg, Germany
MARUM–Center for Marine Environmental Sciences, University of Bremen, Germany

ABSTRACT: Until now the vibratory pile driving method is not widely used because there is no reliable approach for drivability analysis. One of the main issues with the current drivability analysis methods are the inability to accurately predict the cyclic soil behavior during vibratory pile driving. In this research, the cyclic soil behavior was evaluated using a new in-situ soil investigation method, the vibratory cone penetration test (VCPT). VCPT penetrates the ground, while inducing controlled vertical cyclic strains, and measuring cone resistance, sleeve friction, and pore water pressure. Nine static CPTs and 15 VCPTs were performed at a constant frequency with three different amplitudes. The resulted cone resistance-displacement cycles were compared with the state-of-the-art soil-pile interaction models. The reduction in soil strength against vibratory loading was found to be amplitude-dependent; and it was observed that a cavity formed between the cone and soil during the upward movement of the cone for high displacement amplitudes.

Cone Penetration Testing 2022 – Gottardi & Tonni (eds)
© 2022 the Author(s), ISBN 978-1-032-35624-2

Monitoring ground improvement by Rammed Aggregate Piers using a combined CPTU and SDMT approach at a silty sand liquefaction-prone site in Emilia-Romagna

S. Amoroso
University of Chieti-Pescara, Pescara, Italy
Istituto Nazionale di Geofisica e Vulcanologia, L'Aquila, Italy

M.F. García Martínez, L. Tonni & G. Gottardi
University of Bologna, Bologna, Italy

P. Monaco
University of L'Aquila, L'Aquila, Italy

K.M. Rollins
Brigham Young University, Provo, Utah, USA

L. Minarelli
Istituto Nazionale di Geofisica e Vulcanologia, L'Aquila, Italy

D. Marchetti
Studio Prof. Marchetti, Rome, Italy

K.J. Wissmann
Geopier Foundation Company, Davidson, North Carolina, USA

ABSTRACT: Following the 2012 Emilia-Romagna earthquake, widespread liquefaction of silty sands was observed, providing the opportunity to gain a better understanding of the influence of fines content on liquefaction hazard and mitigation works. This paper presents the results of a thorough geotechnical investigation performed as part of a full-scale liquefaction experiment involving controlled blast tests in Bondeno, a small village that suffered liquefaction in 2012. Piezocone (CPTU) and seismic dilatometer (SDMT) tests were performed in natural and improved soils after Rammed Aggregate Pier® (RAP) treatment to provide accurate soil characterization and to evaluate the effectiveness of liquefaction mitigation. CPTU and SDMT results revealed a good agreement in the geotechnical characterization of the site, detecting homogenous soil properties in both the natural and treated soils and estimating CPTU-DMT coupled parameters in sandy layers (e.g. overconsolidation ratio, at-rest earth pressure coefficient), that are usually not determinable by the use of a single type of in situ test. In particular, the combined use of CPTU-DMT data provided verification of the increase in the lateral stress produced by the RAP installation. Data analyses revealed that the RAPs were an effective ground improvement technique despite the high percent of fines (≈ 25-35%).

Cone Penetration Testing 2022 – Gottardi & Tonni (eds)

A CPT-based method for monotonic loading of large diameter monopiles in sand

S. Bascunan
Ramboll, Hamburg, Germany

K. Kaltekis
Fugro, Nootdorp, The Netherlands

B. van Dijk
Arcadis, Amersfoort, The Netherlands

K.G. Gavin
Delft University of Technology, Delft, The Netherlands

ABSTRACT: A joint academia-industry project, the Pile Soil Analysis (PISA) project, resulted in an empirical method for assessing the monotonic lateral loading response of large diameter monopiles. The method predicts four soil reactions, namely the distributed load and the distributed moment along the pile shaft, the pile base shear and the pile base moment. The method considers pile load test data and 3D numerical modelling. A 1D framework allows prediction of the four soil reactions. In this paper, a CPT-based approach is proposed to derive the four soil reaction components for use in a 1D model for conceptual design of monopiles in sand subject to monotonic lateral loading. The approach relies on results from 3D finite element analyses that were performed considering soil conditions for a sand site used in the PISA project (Dunkirk site). The results are compared to pile load test data from the PISA project, showing good agreement, particularly for load levels related to the serviceability limit state.

Cone Penetration Testing 2022 – Gottardi & Tonni (eds)
© 2022 the Author(s), ISBN 978-1-032-35624-2

Automatic interpretation and statistical evaluation of soil conditions for preliminary design of offshore foundations using the cone penetration test

L. Berenguer Todo Bom & M. Kanitz
Ramboll, Hamburg, Germany

ABSTRACT: The increasing need for renewable energy has led to a substantial growth of the offshore wind industry. To meet the need of the demanding timelines of the industry, while maintaining a thorough and robust design of the offshore foundations, automation is key in supporting rapid and accurate geotechnical screening of any investigated site. The preliminary design of foundations for offshore wind turbines is a crucial step to determine the feasibility of the planned structures to build an offshore wind park at the site. Therefore, an initial estimate of the ground conditions, in terms of both soil stratigraphy as well as soil strength and stiffness properties are necessary. This contribution presents a MATLAB-based tool which performs an automatic interpretation and statistical evaluation of soil conditions based on solely cone penetration testing (CPT) data. The tool derives a preliminary ground model by assembly of similar soil type and strength and stiffness parameters are determined through available CPT correlations. A depth-dependent statistical evaluation of the strength and stiffness parameters for each soil layer in the respective soil profile is computed. The output profiles can be adopted directly for the preliminary design of offshore foundations such as jacket pin-piles, suction bucket or monopiles using the PISA rule-based method.

Cone Penetration Testing 2022 – Gottardi & Tonni (eds)
© 2022 the Author(s), ISBN 978-1-032-35624-2

Effect of dynamic pile driving parameters on vibratory penetration

Anchal Bhaskar, Stefan Kreiter & Dina Al-Sammarraie
MARUM–Center for Marine Environmental Sciences, University of Bremen, Germany

Tobias Mörz
MARUM–Center for Marine Environmental Sciences, University of Bremen, Germany
GeoEngineering.org

ABSTRACT: To reduce the harmful noise for marine life, generated during impact driving of offshore foundations, vibratory driving was presented as an alternative. Vibratory driving produces less noise and allows faster installation in sandy soil. This technique is cost-efficient for offshore pile installations and causes less pile fatigue than impact driving. One of the main issues with vibratory driving is the lack of a reliable driveability analysis. This is due to uncertainties in understanding of cyclic soil response during vibratory driving and the effect of different vibratory driving parameters on the soil response. In this study Vibratory Cone Penetration Test (VCPT) is used to assess the cyclic soil response during vibratory penetration. VCPT penetrates the soil while inducing cyclic loads and measures cone resistance. VCPTs were conducted in a calibration chamber to investigate the reduction of the cyclic cone resistance due to variation in frequency for the same penetration path. The VCPT results are discussed in terms of acceleration, energy, reduction ratio, which are considered as primary parameters in the different driveability analyses methods. The effect of cavitation during upward movement of the cone is investigated and discussed.

Cone Penetration Testing 2022 – Gottardi & Tonni (eds)
© 2022 the Author(s), ISBN 978-1-032-35624-2

Application of a new q_c averaging approach for end bearing of driven piles in sand

E.J. Bittar, Y. Tian & B.M. Lehane
The University of Western Australia, Perth, Australia

ABSTRACT: CPT-based methods to estimate axial pile capacity generally relate the pile base resistance with a q_c value averaged in the vicinity of the pile tip ($q_{c,avg}$) and hence implicitly acknowledge the greater zone of influence of a full scale pile compared to a cone. To account for layered deposits (e.g. loose over dense sand), most common designs methods average the cone tip resistance over 1.5 diameters (D) above and below the cone tip or use the Dutch methodology where the averaging is conducted over a zone of 0.7D to 4D below the pile tip and 6D to 8D above the pile tip. This paper examines an alternative averaging approach based on an algorithm which allows estimation of the steady state end bearing resistance of penetrometer with the same diameter as a pile (q_p). End bearing stresses determined at a base displacement of 10% of the pile diameter ($q_{b0.1}$) in a database of instrumented static load tests on driven piles are compared with the corresponding q_p values. It is shown that the $q_{b0.1}/q_p$ ratio varies with the effective area ratio of driven piles and is independent of the pile diameter. The best-fit equation of the database of end bearing measurements provide a rational and improved means of determining end bearing of driven piles.

Cone Penetration Testing 2022 – Gottardi & Tonni (eds)
© 2022 the Author(s), ISBN 978-1-032-35624-2

A review of a CPT based axial capacity prediction of screw piles in sand

E.J. Bittar, B.M. Lehane & S. Mahdavi
The University of Western Australia, Australia

A.P. Blake, D.J. Richards & D.J. White
The University of Southampton, UK

ABSTRACT: Screw piles are widely used in onshore engineering applications and have recently been considered as an alternative foundation solution for offshore wind turbines (OWTs) supported on jacket structures. The high loads required to support such structures demand a considerable up-scaling of the screw pile geometry typically used onshore. Driven piles followed a similar upscaling process for their use in offshore structures and the CPT tip resistance (q_c) value is now commonly used directly in design methods for both onshore and offshore driven piles. This paper evaluates the performance of a new CPT-based design method (UWA-SP-21) to predict the axial capacity of screw piles in tension and compression. To achieve this, a database of single helix load tested screw piles sand is collated and used for assessment of the method. The development of this CPT design approach reflects the process that was following for driven piles. UWA-SP21 has been developed through studies of onshore piles. The findings from this study can in the future be extrapolated appropriately to the larger pile sizes required offshore, in the same way that driven pile design has evolved.

Cone Penetration Testing 2022 – Gottardi & Tonni (eds)
© 2022 the Author(s), ISBN 978-1-032-35624-2

CPT-based liquefaction ejecta evaluation procedure

Jonathan D. Bray & Daniel Hutabarat
University of California, Berkeley

ABSTRACT: A CPT-based procedure to estimate the severity of liquefaction ejecta is presented. It employs a liquefaction ejecta demand parameter (L_D) that captures the amount of upward seepage pressure that can produce artesian flow due to elevated excess hydraulic head and a crust layer resistance parameter (C_R) that captures the strength and thickness of the nonliquefiable crust layer. L_D tends to increase systematically as ejecta severity increases at the thick, clean sand sites, and low L_D values are estimated at stratified soil sites that did not produce ejecta. C_R captures the differing performances of sites with and without a competent crust layer overlying a thick liquefiable layer with a high L_D value. The proposed $L_D - C_R$ liquefaction ejecta severity chart separates cases with severe or extreme ejecta, which have high L_D and low C_R values, from cases with minor or no ejecta, which have low L_D and high C_R values.

Cone Penetration Testing 2022 – Gottardi & Tonni (eds)
© 2022 the Author(s), ISBN 978-1-032-35624-2

Settlement estimations for buildings founded on saturated silty sands from CPT and DMT results

Maxwell Cáceres, Javier Fumeron & Felipe A. Villalobos
Department of Civil Engineering, Universidad Católica de la Santísima Concepción, Chile

Ricardo Moffat
LMMG Geotechnics & Faculty of Engineering and Sciences, Universidad Adolfo Ibáñez, Chile

ABSTRACT: The estimation of building settlements founded in saturated sandy and silty soils is a relevant part of foundation design. Settlement calculation methodologies are still based on SPT and plate-load tests. SPT results can have significant dispersion and soil stiffness estimations are obtained from correlations without a physical meaning and valid only for particular soils and geology conditions. Moreover, plate-load tests are normally limited to shallow depths. CPT and DMT can obtain reliable results in engineering units. These tests are operator independent and the equipment is truly standard worldwide. Results obtained using CPT and DMT equipment for an urban renovation project with buildings founded on saturated sands and silts in Concepción city in Chile, are presented. These results allow estimations of static and liquefaction-induced settlements obtained from calculation methods developed for CPT and DMT.

Cone Penetration Testing 2022 – Gottardi & Tonni (eds)

Numerical modeling of static load test in drilled shaft using CPTu results

M.A. Camacho
Independent Engineer, Cochabamba, Bolivia

C.B. Camacho
Universidad Mayor de San Simón, Cochabamba, Bolivia

V.H. Miranda
Independent Engineer, Cochabamba, Bolivia

ABSTRACT: The projects with deep foundations use information from field tests (e.g. CPTu and static load test) to predict an ultimate capacity. The static load test verifies foundation design parameters and installation method. However, this test establishes the real behavior of the drilled shaft under specific conditions and the possibility of experimenting in different conditions becomes extremely expensive and unfeasible. Therefore, the numerical modeling of the static load test becomes important because the initial soil profile and dimensions of the drilled shaft could be changed. This article presents the comparison of the load–displacement curves by numerical modeling in fine soils with that load–displacement curve through a static load test in a drilled shaft built in Cochabamba, Bolivia.

Cone Penetration Testing 2022 – Gottardi & Tonni (eds)

Characterization of geotechnical spatial variability in river embankments from spatially adjacent SCPT

F. Ceccato
DICEA-University of Padua, Italy

M. Uzielli
DICEA – University of Florence, Italy
Georisk Engineering S.r.l., Florence, Italy

P. Simonini
DICEA-University of Padua, Italy

ABSTRACT: characterization of the spatial variability of geotechnical properties of river embankment soils is important for the enhanced modelling and assessment of embankment stability. In practice, available data is usually limited. This study presents the results of a quantitative statistical analysis of the spatial variability of cone tip resistance from the results of 16 closely spaced SCPT carried out near a recent breach in a levee on the Panaro river in northern Italy. Two geotechnical homogeneous soil units are preliminary identified in the levee to ensure the meaningfulness of the analysis in terms of soil type. For each unit, the horizontal and vertical spatial variability and spatial correlation structures of cone tip resistance are investigated by a two-step procedure involving the calculation of empirical semivariograms and the subsequent fitting of semivariogram models. Horizontal and vertical scales of fluctuation are estimated based on fitted semivariogram model parameters.

Cone Penetration Testing 2022 – Gottardi & Tonni (eds)
© 2022 the Author(s), ISBN 978-1-032-35624-2

Coupling site wide CPT profiles and genetic algorithms for whole-site offshore windfarm layout optimization

J.A. Charles & S.M. Gourvenec
University of Southampton, Southampton, UK

M.E. Vardy
SAND Geophysics, Southampton, UK

ABSTRACT: Offshore windfarm layout is driven by optimizing wind potential at the selected site subject to minimum spacing between turbines. Within these constraints, this paper explores the optimization of turbine layout within a selected site based on geotechnical site conditions. Utilizing an irregular but site-wide grid of CPT profiles, a simple pile design method and a cubic interpolation of pile dimensions, heatmaps of pile length and diameter at every possible coordinate within a site can be generated, providing the opportunity for optimization of layout. This paper demonstrates the ability to optimize the location of a prescribed number of wind turbine foundations, in conjunction with other constraints such as minimum spacing, to minimize total steel usage (a simple proxy for foundation cost and embodied carbon) in the foundation system using a genetic algorithm (GA)-based approach. Existing work on windfarm layout optimization uses such techniques to maximize wind energy generation but negligible work exists on extending the methodologies to minimize foundation costs. This paper demonstrates the viability of using a set of CPT profiles with a GA-based approach for a geotechnically-informed windfarm layout and explores the effects of various meta parameters using publicly available datasets. The work demonstrated in this paper is directly relevant to ongoing advances in geophysics and machine learning that would allow for the generation of a synthetic CPT profile at any point on the site, eliminating the drawbacks of interpolation between actual CPT profiles for design parameters.

Cone Penetration Testing 2022 – Gottardi & Tonni (eds)
© 2022 the Author(s), ISBN 978-1-032-35624-2

CPT-based model calibration for effective stress analysis of layered soil deposits

A. Chiaradonna
University of L'Aquila, L'Aquila, Italy

N. Ntritsos & M. Cubrinovski
University of Canterbury, Christchurch, New Zealand

ABSTRACT: Recent research recognized the advantages of effective stress dynamic analysis in estimating the seismic response of layered soil profiles. One of the key challenges in performing effective stress analysis is the calibration of constitutive models able to simulate complex soil behavior under seismic loading. To overcome this problem, calibration procedures have been developed to guide the definition of model parameters on the results of in-situ tests. In this study, a reference layered soil profile from Christchurch in New Zealand, which experienced liquefaction during the 2010-2011 Canterbury earthquake sequence, was used to verify the goodness of the CPT-based calibration procedure of a simplified pore water pressure model. Comparisons with the seismic response predicted using an advanced state concept-based constitutive model were also reported. Despite the simplifications adopted in the calibration process, the analysis results replicate the occurrence of liquefaction and encourage CPT-based model parameter definition for routine effective stress dynamic analyses.

Cone Penetration Testing 2022 – Gottardi & Tonni (eds)
© 2022 the Author(s), ISBN 978-1-032-35624-2

Stress increase induced by impact precast pile driving

V. Colella
Geofondazioni Ingegneria e Lavori s.r.l., Martellago, Italy

G. Cortellazzo
University of Padua, Padua, Italy

A. Dei Svaldi
Desam ingegneria e ambiente s.r.l., Italy

S. Amoroso & L. Minarelli
University of Chieti-Pescara, Istituto Nazionale di Geofisica e Vulcanologia, Italy

K.M. Rollins
Brigham Young University, Utah, USA

ABSTRACT: The paper describes the experimental activities carried out in a test site, set up to evaluate the increase of soil stiffness and horizontal effective stress, mainly in saturated low-medium density sandy layers and in silty sand, after the driving of tapered precast piles.

The experimentation consisted in driving some prefabricated tapered piles with different energies and spacing between them in an area where some soil layers had a high liquefaction potential. To evaluate the pile driving effects on the stress state around them, preliminary CPTu and DMT tests were carried out and repeated after the driving activity.

In particular, the change of the CPTu sleeve resistances was compared with that of the DMT K_D data, to evaluate the increase in horizontal stress using different methods, a phenomenon influencing the soil susceptibility to liquefaction and the pile bearing capacity.

Cone Penetration Testing 2022 – Gottardi & Tonni (eds)

CPT based liquefaction potential of flood defences in The Netherlands

T. de Gast & K.G. Gavin
Section of Geo-Engineering, Faculty of Civil Engineering and Geosciences, Delft University of Technology, Delft, The Netherlands

P.D. Notenboom & R. Abraimi
Afdeling Kennis & Advies, Waterschap Hollandse Delta, Ridderkerk, The Netherlands

C. Reale
Department of Architecture & Civil Engineering, Centre for Infrastructure, Geotechnical and Water Engineering Research (IGWE), University of Bath, Bath, UK

ABSTRACT: The paper describes a study on the liquefaction potential of flood defences along the rivers running through the delta area of the Netherlands. The study concentrates on an area south of Rotterdam. The dykes used as primary flood defences protect an urban, rural, and industrial area of 102,400 ha. In this paper the data from more than 4200 Cone Penetration Tests, CPT traces are used to assess more than 200 km of dykes. The pore pressure, u2 data is analysed, then used to separate the material response into contractive and dilative zones. Using the separation of liquefaction susceptible soils, and geometry of the riverbed a regional hazard map is generated. The choices for the data visualisation and their effect on the generated map are discussed and presented. The final liquefaction susceptibility map is used by the water governing authority Waterschap Hollandse Delta as a decision-making tool to improve the efficacy of liquefaction hazard assessment such as the location and return period of bathymetry measurements, and the scale of site- and laboratory investigation.

Cone Penetration Testing 2022 – Gottardi & Tonni (eds)
© 2022 the Author(s), ISBN 978-1-032-35624-2

Application of the CPT for liquefaction assessment of gravelly reclamations at the port of Wellington

R. Dhakal & M. Cubrinovski
University of Canterbury, Christchurch, New Zealand

J.D. Bray
University of California, Berkeley, USA

ABSTRACT: Widespread liquefaction occurred in the reclamations at CentrePort, Wellington (New Zealand) during the 2016 Kaikōura earthquake (M_w7.8) which produced a relatively moderate seismic demand (peak horizontal ground acceleration of 0.2-0.3g at the site). Most of the damage occurred in the end-dumped fills (i.e., the Thorndon reclamation) which are comprised of 60-80% gravels and 20-40% sand-silt mixtures. This study presents results from CPT-based liquefaction assessment on a representative profile from the Thorndon reclamation using both a semi-empirical simplified method and advanced dynamic effective stress analysis. The simplified assessment results are generally consistent with observations since the CPT is able to capture the response of the finer sand-silt fraction in the matrix, which likely controlled the liquefaction behavior during the Kaikōura earthquake. The preliminary effective stress analysis provides insights on the timing and onset of liquefaction, and processes involving dynamic interactions within the deposit, thus illustrating important response characteristics which are beyond the capability of simplified methods. This paper illustrates the use of the CPT in performing conventional simplified and dynamic liquefaction analysis when applied to non-conventional or problematic soils for the assessment with reference to a well-documented case study.

Cone Penetration Testing 2022 – Gottardi & Tonni (eds)

Concept design of a new CPT module for direct in situ measurement of p-y soil responses

A. Diambra, J. Creasey, J. Leonet, A. Conn, E. Ibraim & G. Mylonakis
University of Bristol, UK

D. White, B. Cerfontaine & S.M. Gourvenec
University of Southampton, UK

D. Igoe
Trinity College Dublin, Ireland

ABSTRACT: Due to the rapid expansion of the offshore energy market, driven by the installation of wind turbines founded on single- or multi-piled foundations, the design and optimisation of laterally loaded piles has attracted enormous interest in the last decade. Current industrial design practice, for offshore and onshore piles, typically employs lateral load-displacement springs (*p-y*) to model the soil response under serviceability, ultimate or fatigue limit state scenarios of this foundation type, supplemented by moment and axial springs. Correlations based on CPT data are commonly used to determine soil properties and, in turn, the stiffness and resistance of the *p-y* reaction curves, in advance of detailed laboratory testing that follows later in the project schedule. To extend the potential for in situ testing to support lateral pile design, this paper presents the novel idea of complementing a CPT device with a new module capable of probing the soil in such manner that the monotonic and cyclic *p-y* soil response (including its evolution during the foundation design life) can be directly measured in the field. The stress and strain fields induced by such a module resemble those of a miniature laterally loaded pile element and differ from circular or flat cavity expansions induced by pressuremeters or dilatometers. The new device will be developed in a collaborative research project 'ROBOCONE' financed by the UK and Irish research councils. The background thinking and initial conceptual design of the device, including the review and selection of appropriate motion mechanisms and instrumentation, is presented in this manuscript.

Cone Penetration Testing 2022 – Gottardi & Tonni (eds)
© 2022 the Author(s), ISBN 978-1-032-35624-2

The use of dynamic probing tests and cone penetration tests to verify the effectiveness of expanding polyurethane resin injections for ground improvement

A. Dominijanni
Politecnico di Torino, Italy

M. Gabassi
Uretek Italia Spa, Italy

A. Minardi
Geotechnical Engineering Ltd, Italy

S. Pavan
Uretek Italia Spa, Italy

ABSTRACT: Injection of expanding polyurethane resins is a popular method to improve both the stiffness and the shear strength of the ground below existing foundations. The effect of the polyurethane resin expansion is to increase the soil confining stress and density around the injection holes. An estimation of the horizontal stress and volumetric strain changes that are induced within the ground is derived from the theory of cavity expansion in elasto-plastic materials. A series of case-histories is presented to document the feasibility of different *in*-situ tests to evaluate the achieved ground improvement. The tests have been performed before and after the injection of polyurethane resins and the obtained results have been compared with theoretical predictions. The considered investigation methods include the dynamic probing tests and the cone penetration tests. The preliminary results that have been achieved using an experimental miniature cone penetration test are also illustrated. The advantages and limitations of different test methods are discussed and practical indications for conducting such verifications of polyurethane resin injection effectiveness are provided.

Cone Penetration Testing 2022 – Gottardi & Tonni (eds)

Evaluation of flow liquefaction susceptibility of a sandy-silt tailings using the CPTu

M.P. dos Santos Junior & R. César Gomes
Universidade Federal de Ouro Preto, Brazil

S.G. Silva Ribeiro
GeoFast – Geotechnical Modelling, Brazil

B.G. Delgado
CONSTRUCT-GEO, Faculdade de Engenharia da Universidade do Porto, Portugal

ABSTRACT: Flow liquefaction is a behavior observed in saturated or nearly saturated geomaterials that show a strain softening response during undrained shear, most common in very loose sands and silts as well as very sensitive clays. In the mining industry, flow liquefaction is a subject of high relevance due to the geotechnical characteristics of the tailings. Many methodologies have been developed to evaluate the susceptibility to flow liquefaction using in-situ tests, especially the CPTu test, since it provides high accuracy and good repeatability. This paper compares four methodologies based on the CPTu test to evaluate the susceptibility to flow liquefaction: i) Plewes et al. (1992), ii) Olson (2001), iii) Shuttle & Cunning (2008) and iv) Robertson (2016). The results obtained highlight the need to correct the original contour suggested by Olson (2001) for medium and high compressibility geomaterials. Furthermore, all other methods evaluated presented similar results in the overall classification.

Cone Penetration Testing 2022 – Gottardi & Tonni (eds)
© 2022 the Author(s), ISBN 978-1-032-35624-2

Prospects on data mining approach for pile geotechnical design utilizing CPT and CPTu records: Case study: AUT database

A. Eslami & S. Heidarie Golafzani
Amirkabir University of Technology, Tehran, Iran

S. Moshfeghi
University of British Columbia (UBC), Vancouver, Canada

ABSTRACT: A comprehensive database has been compiled including CPT soundings performed adjacent to pile load test and related geotechnical information, namely AUT (Amirkabir University of Technology): Geo-CPT&Pile Database. In this paper, after a brief review of existing CPT and pile databases, the specifications of the updated AUT: Geo-CPT&Pile Database as well as different categories of data are presented. Subsequently, several procedures developed using this database from different geotechnical aspects are reviewed and introduced. These implementations are extended for appraisal of currently used CPT-based methods for pile geotechnical design by focusing on methods screening, uncertainty- reliability measures, and Performance-Based Design (PBD) approach regarding resistance factors. After all, an algorithm is presented including a formulated procedure of pile geotechnical design by means of smart database collections, reproduction of CPT profile, assortment of competitive methods, integration of geotechnical aspects i.e., capacity, load-displacement performance and resistance distribution. Correspondingly, via realizing prospects on implementation of major aspects leads towards optimum pile performance-based design upon a data mining approach.

Cone Penetration Testing 2022 – Gottardi & Tonni (eds)

Evaluating mitigation of kinematic moments of precast driven piles in liquefiable layers using pre- and post-CPTu tests

K. Fakharian, D. Mohtashamamiri, K. Behroozian & M. Bahrami
Department of Civil & Environmental Engineering, Amirkabir University of Technology, Tehran, Iran

T. Bahrami & I.H. Attar
Pars GeoEnviro Inc., Tehran, Iran

ABSTRACT: Precast piles are driven to 23 m depth to improve the capacity of foundations located on saturated liquefiable sands on top underlain by a thick soft to medium stiff clay. Analyses show excessive kinematic moment at borderline of the two layers during liquefaction. It was decided to drive additional shorter piles (12 m) to mitigate liquefaction hoping that kinematic moments shall be reduced within allowable limits. CPTu tests were carried out before and after pile driving. The results show that q_c did not change considerably in the top saturated silty sand layer as well as the clayey soil underneath. The friction sleeve, however, increased more significantly due to radial displacement and compaction/consolidation attributed to cavity expansion during pile driving. Analyses showed that pile arrangement was not sufficient to overcome liquefaction during earthquake. However, safety factor increased sufficiently to reduce kinematic moment within the allowable limits. The details of the site operations, CPTu data before and after pile driving, analyses and interpretations are presented and discussed.

Keywords: Pizeocone Penetration Test (CPTu), Liquefaction, Precast pile, Kinematic moment

Cone Penetration Testing 2022 – Gottardi & Tonni (eds)
© 2022 the Author(s), ISBN 978-1-032-35624-2

Verification of 3D FEM analysis of ground improvement works using CPT test results

G.A. Faour & K.N. Khouri
NSCC International Ltd, Abu Dhabi, United Arab Emirates

ABSTRACT: The cone penetration test (CPT) has been widely used as one of the most well-known and reliable performance tests for ground improvement works. A comprehensive CPT testing campaign was conducted for the proposed ground improvement works of a highly technical project in Oman. The ground improvement works were evaluated based on the results of pre and post improvement CPT testing. CPT was the basis for choosing rigid inclusions (RI) and Vibro-compaction (VC) as the most suitable and economical ground improvement techniques for this project. The acceptance criteria based on the Post improvement CPT results incorporated the liquefaction potential and both allowable immediate and long-term settlements. The degree of ground improvement was evaluated by verifying that the final test results in the field met the project design criteria. It was concluded that CPT testing can be accurately correlated with multiple ground improvement acceptance criteria including liquefaction and settlement.

Cone Penetration Testing 2022 – Gottardi & Tonni (eds)

A comprehensive design procedure for pile groups in liquefiable soils

M. Franceschini, F. Fiorelli & E. Bandiera
Teleios Srl – Società di Ingegneria, Castel Maggiore, Bologna, Italy

ABSTRACT: Among the various strategies to mitigate the effects of soil liquefaction one of the most effective is the design of deep foundations. The designs of piles in liquefiable soils are too often concerned with the only axial bearing capacity, addressed by simply neglecting pile resistance in the liquefiable layer. This approach is inadequate to properly face the complexity of the problem. In the present paper we intend to examine this theme throughout its multiple aspects. We have synthesized a design procedure of analysis that has been already applied in some projects in different areas of Emilia-Romagna region characterized by high risk of liquefaction. The proposed design procedure is based on the most updated theories and design references concerning piles in liquefiable soil, such as those of Cubrinovsky, Olson & Stark, Rollins, Bhattacharya, Madabhushi and others, primarily referring to CPTu.

Cone Penetration Testing 2022 – Gottardi & Tonni (eds)
© 2022 the Author(s), ISBN 978-1-032-35624-2

Deep foundations of the new Pavilion 37 – fair quarter Bologna

M. Franceschini, F. Fiorelli & E. Bandiera
Teleios Srl – Società di Ingegneria, Castel Maggiore, Bologna, Italy

Vincenzo Colella
Geofondazioni Ingegneria e Lavori Srl, Martellago, Venezia, Italy

Giampaolo Cortellazzo
Università di Padova, Padova, Italy

Andrea Dei Svaldi
Desam Ingegneria e Ambiente s.r.l., Mogliano Veneto, Treviso, Italy

ABSTRACT: The present paper concerns the design of driven precast tapered piles, that have been used within the project of the new Pavillion 37 of the fair quarter of Bologna. A particular focus will be reserved to the evaluation of the axial bearing capacity of these elements and to the prevision of their load-settlement curve. The geotechnical characterization, as well as the prediction of the axial resistance of the piles, has been mainly based on Cone Penetration Test with piezocone CPTu. The previsions will be compared with the results of a series of static load tests performed during the construction.

Cone Penetration Testing 2022 – Gottardi & Tonni (eds)
© 2022 the Author(s), ISBN 978-1-032-35624-2

Consolidation settlement of coastal areas of the Emilia-Romagna region from cone penetration tests

D. Giretti
University of Bergamo, Bergamo, Italy

V. Fioravante
University of Ferrara, Ferrara, Italy

L. Perini & L. Calabrese
Geological, seismic and soil survey of the Emilia-Romagna Region, Bologna, Italy

ABSTRACT: This paper describes an attempt to develop and calibrate a simple tool to forecast the natural subsidence of coastal Holocene deposits. An empirical correlation has been calibrated between the penetration resistance measured by a static cone penetrometric tests and i) the normal compression line NCL of fine grained, coastal deposits ii) the current void ratio and iii) the void ratio on the NCL at the site effective stress. The comparison between the site void ratio and the normally consolidated void ratio can be used to roughly estimate if the soil is overconsolidated, normally consolidated or under-consolidated/structured and, in the latter case, if the fine layer is in the condition of further developing consolidation settlement to reach the self-weight equilibrium.

Cone Penetration Testing 2022 – Gottardi & Tonni (eds)

A comparative study on liquefaction assessment of Rajarhat area of Kolkata by using different approaches

Abhipriya Halder
Department of Civil Engineering, Narula Institute of Technology, India

Kaustav Das, Saptarshi Nandi & Kaushik Bandyopadhyay
Department of Construction Engineering, Jadavpur University, India

ABSTRACT: Liquefaction potential assessment is of particular significance to geotechnical engineers in order to evaluate the risks for construction of high rise buildings, railway embankments, canal, road embankments, earth dams etc. in areas susceptible to liquefaction. Investigations were made at Rajarhat area of Kolkata, India by means of CPT, DMT, SPT and SDMT measurements. The main purpose of carrying out these tests is to grab a relatively rare opportunity to compare the results of all these methods for arriving at a conclusive evidence in predicting liquefaction. It was found that the results of all the in-situ techniques were relatively consistent in predicting liquefaction within a depth range of 0 to 17.0 m below the ground surface. CPT, DMT, SPT and SDMT are basically field-based in-situ tests for measurements of important geotechnical parameters and also provide reliable data for assessment of liquefaction potential of a site. In recent times Finite Element Method based (FEM), a powerful viable alternative tool is available to the geotechnical engineers for application in various fields. In the present investigation, an attempt has been made to study the assessment of liquefaction based on the numerical simulation using Finite element software in addition to all the field techniques mentioned before. The main objective of the present study is to evaluate the correlation between factor of safety against liquefaction using different in-situ test approaches as well as the FEM tool. Finally, the paper presents the correlation between the depth and factor of safety for this particular Rajarhat area of Kolkata for understanding the vulnerability of this area to liquefaction.

Keywords: Dilatometer and seismic dilatometer test, cone penetration test, standard penetration test, Finite Element Method, liquefaction potential

Cone Penetration Testing 2022 – Gottardi & Tonni (eds)

Kriging analysis on CPTU data from offshore wind farm

Rongjie He & Jinhui Li*
Department of Civil and Environmental Engineering, Harbin Institute of Technology (Shenzhen), China

Shaoli Yang
Norwegian Geotechnical Institute (NGI), Oslo, Norway

Ben He
Key Laboratory for Far-shore Wind Power Technology of Zhejiang Province; PowerChina Huadong Engineering Corporation Limited, (HDEC), Hangzhou, China

ABSTRACT: In recent years, offshore wind power has been widely developed. Because of the complexity of marine geology and the large spatial variability of soils, it is necessary to obtain CPTU data before the foundation design. However, how to reasonably interpret the CPTU data will greatly affect the reliability and safety of the offshore windfarm foundation. In addition, there is a need to predict soil conditions at the locations without CPTU data. In this study, based on the available CPTU data of offshore wind farm in east China sea, kriging method is used to interpret the CPTU data, and the soil conditions in a large area can be predicted. This method can be used to develop ground model for a large offshore wind farm, and estimate the relevant soil parameters based on CPTU data. The analysis methods can provide reference for other projects of offshore wind farm.

Keywords: offshore wind farm, CPTU, kriging method, marine geology

Cone Penetration Testing 2022 – Gottardi & Tonni (eds)

Evaluation of CPT-based design method for offshore pile

B. Huang
School of Architecture and Civil Engineering, Huizhou University, China

E.J. Bittar
The University of Western Australia, Australia

Y. Zhang
The University of Newcastle, Australia

X. Fu
School of Civil Engineering, Wuhan University, China

ABSTRACT: In offshore engineering, it's difficult to obtain undisturbed samples. CPT-based design method of pile has become recommending method. In this paper focusing on offshore pile foundation engineering, the typical CPT-based design methods are introduced. Various design methods are evaluated in respect of friction fatigue, pile loading direction, and the plug ratio of open-ended pile in capacity contribution, and the determination of design parameters. CPT-based design methods in the clay are compared with API method. For the typical stratum in the China East Sea, the monopiles of an offshore wind farm are analysed. The calculated capacities of monopiles with different methods are compared with the measured value, and the reliability of methods are evaluated. At last according to the vertical loading condition of the offshore monopile, reasonable suggestions for the current design methods and parameters determination are given.

Cone Penetration Testing 2022 – Gottardi & Tonni (eds)
© 2022 the Author(s), ISBN 978-1-032-35624-2

Application of CPT based 3DFE approach for estimating monopile damping in sand

D. Igoe & M.B. Mohammed
Trinity College Dublin, Ireland

ABSTRACT: Monopiles are the most popular foundation type supporting offshore wind turbines. The use of 3DFE modelling to analyse the pile response has grown significantly in recent years, supported by the introduction of the PISA framework and the development of easy to use commercial software such as PLAXIS Monopile Designer. In sands, the small-strain hardening soil (HSS) model is widely adopted for monopile design, partially due to its relatively simple set of soil input parameters. However to date there is no widely used approach for determining the required soil parameters and approaches often differ across the industry. Igoe and Jalilvand (2020) developed an approach to derive the HSS soil model parameters directly from CPT data. This procedure has been validated by the authors for monotonic loading against a database of large-scale field tests including PISA field tests in the Dunkirk marine SAND. This paper examines the use of the 3DFE approach under cyclic loading to determine soil damping for monopiles. The results from the 3DFE analysis are compared to field test data and show a good match with the damping estimated under two-way cyclic loading.

Cone Penetration Testing 2022 – Gottardi & Tonni (eds)
© 2022 the Author(s), ISBN 978-1-032-35624-2

The use of CPTU for driven piles designed in a backfilled opencast 'marl hole' in an important post-industrial revolution area within the UK

D. Illingworth & C. Burton
Piledesigns Limited, UK

L. Dhimitri & D. Ward
In Situ Site Investigation, UK

P. Shelton
Phil Shelton Geotechnical Consultancy Limited, UK

ABSTRACT: The Etruria Formation has historically been an important resource for the pottery and brick-making industry in the Stoke-on-Trent area and was mainly exploited using opencast methods – or 'marl holes' as they were commonly known. These 'marl holes' some of which have been reported to be more than 100m deep, have been subsequently backfilled generally with non-organic pottery, brickmaking and domestic waste. This paper explores the redevelopment of land over a backfilled 'marl hole' at a site in Hanley, Stoke-on-Trent, Staffordshire, UK, for residential properties. The development posed a number of challenges for the design of the foundations due to the historical legacy of the site. Driven steel tubular piles were chosen as a suitable system as there is precedent with this foundation solution, which has been frequently adopted for buildings within the area. However, some of the key challenges for this site included the presence of 'high walls' along the edges of the opencast works where driven piles could be deviated off the sides along with the uncertainty regarding socket lengths of piles driven into the underlying solid strata. To establish a 3D ground model for the 'marl pit', underlying strata, piezocone tests (CPTU) were seen as the most cost-effective and practicable method of site characterisation. Using this data, piles were installed, test/production piles were installed, driven to a set/penetration into the bedrock criteria and then compared to this ground model.

Cone Penetration Testing 2022 – Gottardi & Tonni (eds)
© 2022 the Author(s), ISBN 978-1-032-35624-2

Consolidation settlement prediction using cone penetration testing

M. Kermani
Golder Associates Ltd., Montreal, Canada

F. Esford
Golder Associates Ltd., Vancouver, Canada

ABSTRACT: A reduction in the local groundwater table surrounding a mine, in the Abitibi-Témiscamingue region of Quebec, Canada, has led to dissipation of porewater pressure and displacement of the ground surface. The mechanical behavior of the site's fine-grained soil has been studied in the laboratory (consolidation tests) and in-situ tests, particularly cone penetration tests (CPT). Additionally, displacements of the ground surface have been monitored for over 10 years. The consolidation settlements due to the groundwater table drawdown were calculated based on tip resistance values measured during the CPT. By comparing the calculated magnitudes and trends of vertical displacement with measured values, the consolidation calculation method using CPT data was validated. At some locations, there was a good match between the calculated and measured trends. At other survey monuments, the measured displacements were greater than calculated values. It was concluded that additional mechanisms were contributing to the displacements. The possible influence of these mechanisms varies spatially within the studied site.

Cone Penetration Testing 2022 – Gottardi & Tonni (eds)
 ISBN 978-1-032-35624-2

3D FE derivation of CPT based soil reaction curves for monopile lateral static design in sand

Louis-Marin Lapastoure
Department of Civil, Structural & Environmental Engineering, Trinity College Dublin, Dublin, Ireland
Gavin and Doherty Geosolutions Ltd, Dublin, Ireland

David Igoe
Department of Civil, Structural & Environmental Engineering, Trinity College Dublin, Dublin, Ireland

ABSTRACT: This paper aims to offer novel correlation between all 4 components of PISA soil reaction curves for monopile modelling and Cone Penetration Testing (CPT) data. This approach requires minimum user inputs and offers a consistent approach to develop site-specific soil reaction curves when CPT data are made available. The procedure developed by Igoe and Jalilvand was used to derive small-strain hardening (HS-small) soil model parameters from a range of CPT profile. HS-small parameters are then used to model the monopile response to static lateral loading in sand using Plaxis 3D. This procedure has been validated by the authors against a database of large-scale field tests including PISA tests in the Dunkirk marine sand. Soil reaction curves were directly extracted from the Plaxis models then applied into Timoshenko beam elements to match the 3D FE response and validated the extraction process. It is demonstrated that current CPT based correlations do not match the shape nor the ultimate reaction of 3D FE extracted soil reaction curves. It is also shown that all four components of soil reactions are required to accurately model monopiles which are short and rigid.

Cone Penetration Testing 2022 – Gottardi & Tonni (eds)
© 2022 the Author(s), ISBN 978-1-032-35624-2

CPT-based assessment of densification induced by stone column installation

F. Marchi, E. Zambianchi, A. Boschi, A. Mastrangelo & G. Marchi
ENSER Srl, Faenza, Italy

G. Gottardi & L. Tonni
UNIBO, Bologna, Italy

ABSTRACT: The use of CPTs for liquefaction susceptibility assessment has become increasingly common in design practice. In the past, the design of densification soil improvement methods against liquefaction has been commonly based on experience, large scale field tests and, possibly, validation after the intervention. Nevertheless, a design method to assess quantitatively the densification effect induced by stone column installation can be developed following a workflow based on CPTs, in order to estimate the efficiency of the soil improvement and to check the compliance with safety requirements before accessing the site. As by common practice, the assessment of liquefaction susceptibility is performed by means of CPT data and related empirical methods. The design method proposed in this paper is based on the estimation of modifications induced by the soil densification technique, easily detectable by the CPT profile, in order to carry out a liquefaction susceptibility analysis. The approach is applied in particular to a well-documented case study in Northern Italy (Bondeno, Ferrara province), providing a consistent dataset of piezocone tests carried out before and after soil treatment through stone columns, as part of a blast-induced liquefaction experiment. In this way, a clear comparison between pre- and post- intervention conditions along the CPT profiles with depth can be performed, thus allowing a validation of the design methodology.

Cone Penetration Testing 2022 – Gottardi & Tonni (eds)

Geopier Impact technology for liquefaction risk mitigation based on CPTu investigations

G. Martinez & K.J. Wissmann
Geopier Foundation Company, Davidson, USA

M. Franceschini, E. Bandiera & F. Fiorelli
Teleios Srl-Società di Ingegneria, Castel Maggiore, Bologna, Italy

ABSTRACT: The present paper is focused on the Impact Rammed Aggregate Pier system, a ground improvement technology developed by Geopier Foundation Company. From extensive in-situ tests also including blast tests, Impact elements have proven to be very effective at densifying in-situ granular soils and increasing the factor of safety against liquefaction. The intent of the paper is to verify the methods that are used to evaluate liquefaction mitigation with Impact piers. One method to evaluate the mitigation of liquefaction risk for design of Impact piers is to estimate the increase in CPT tip resistance q_c expected after installation of the Impact elements. In the paper we will consider CPTu tests performed before and after Impact installations in various sites worldwide to attempt to verify the accuracy of formulas from literature and to try to understand which aspects can be refined in future studies. The future aim is to possibly modify the formulas to account for local soil conditions in the Emilia-Romagna region in Italy.

Cone Penetration Testing 2022 – Gottardi & Tonni (eds)
© 2022 the Author(s), ISBN 978-1-032-35624-2

Estimating bearing capacity of polar snow using the Cone Penetration Test (CPT)

A.B. McCallum & G. White
School of Science, Technology and Engineering, University of the Sunshine Coast, Sippy Downs, Australia

ABSTRACT: Increased access to the polar regions requires increased infrastructure. Design and construction of this infrastructure can only be achieved by accurately estimating surface bearing capacity. We reviewed contemporary methods for determining bearing capacity in soils and applied these to data obtained from almost 100 cone penetration tests conducted in Antarctica. Numerous direct and indirect methods exist to enable estimation of surface bearing capacity in polar snow, but the preferred method is the rate-controllable, friction-sleeve equipped, cone penetration test (CPT). Application of this and similar techniques is essential to ensure the correct design and construction of infrastructure in ever-more trafficked polar regions.

Cone Penetration Testing 2022 – Gottardi & Tonni (eds)

The use of CPT based metamodels to predict the performance of offshore anchor piles

A Mentani & L. Govoni
University of Bologna, Bologna, Italy

F. Bourrier
French National Research Institute for Agriculture, Food and Environment, Grenoble, France

ABSTRACT: The paper presents the development of metamodels for the prediction of the load-displacement response of steel piles driven in sand subjected to pull-out. Two metamodels are created for the evaluation of the tensile capacity and initial stiffness of the pile. They were developed based on the outcomes of a finite element testing campaign, employing models of parameters derived from the tip resistance of cone penetration tests. Two hundreds finite element simulations, which included various soil-pile configurations, were required to calibrate accurate metamodels. Assessment of the procedure was carried out with reference to available data on a model pile and related cone penetration test results. The approach relies on particularly simplified finite element models, but it can be extended to accommodate modelling features of higher complexity. The results find application to the design of off-shore piles used as anchors for floating structures.

Cone Penetration Testing 2022 – Gottardi & Tonni (eds)
© 2022 the Author(s), ISBN 978-1-032-35624-2

Spatial interpolation of sparse PCPT data to optimise infrastructure design

M.P. O'Neill, M.F. Bransby, J.P. Doherty & P. Watson
ARC Research Hub for Transforming energy Infrastructure through Digital Engineering (TIDE), Oceans Graduate School, The University of Western Australia, Australia

ABSTRACT: In an offshore setting the geotechnical data available to infrastructure designers is usually sparse, and judgement is required in using information from sampled locations to estimate design parameters at unsampled locations. Recent interest in data-centric methods has seen advances in the interpolation of sparse data via statistical and analytical approaches. This paper demonstrates the implementation of one such approach, applying Bayesian Compressive Sensing and Markov Chain Monte Carlo techniques to sparse two-dimensional PCPT data. Through a simplified case study, the paper highlights how the method incorporates estimation uncertainty and its associated impact on the geotechnical design of a representative foundation.

Cone Penetration Testing 2022 – Gottardi & Tonni (eds)
© 2022 the Author(s), ISBN 978-1-032-35624-2

Use of CPT as a soil mixing verification tool: Some practical observations

A. O'Brien & I. Murray
GE Solutions Consulting Ltd., Whitburn, UK

ABSTRACT: Mass soil mixing is a versatile ground improvement technology for marginal and brownfield sites. Dry soil mixing is relatively common in the UK for the improvement of ground with very wet and/or organic materials. Wet mixing is less commonplace and involves introduction of a fluid grout with simultaneous rotavating of the soil with a mixing tool. This paper presents the results of CPT investigations into wet mixed soils with a view to use the results as a means to verify the strength and consistency of the mixed materials. A high volume of data has been extracted from two active sites. CPT soundings are found to be purposeful in terms of demonstrating the overall improvement effect and integrity of the mixed soil volume spatially and with depth. Existing CPT testing of mass mixed soil is very limited and this testing has allowed the mixed soil to be compared to existing CPT soil behaviour indexes to give an insight into their usefulness for characterising mixed material and its behaviour.

Cone Penetration Testing 2022 – Gottardi & Tonni (eds)

Liquefaction damage assessment using Bayesian belief networks

L. Paolella, A. Baris & G. Modoni
University of Cassino and Southern Latium, Italy

R.L. Spacagna & S. Fabozzi
CNR IGAG, Area della Ricerca Roma 1, Italy

ABSTRACT: The seismic and liquefaction risk assessment implies introducing methods based on different hypotheses and dealing with different levels of uncertainty affecting the whole process from triggering to surficial manifestation. In this context, soft computing methods, like Bayesian Belief Networks (BBN) and artificial intelligence algorithms, provide the logic framework for cause-effect relationships and the statistical statement to manage uncertainties. Taking advantage of the significant amount of geotechnical data and post-earthquake surveys, an application of BBN versus the forecasting of liquefaction-induced ground damage is proposed considering three main shocks of the 2010 – 2011 Christchurch (New Zealand) Earthquakes Sequence. The BBN algorithms are firstly employed to identify significant variables and learn the relationships among them, then a direct and graphical link between input and target data is created. The quantitative validation of the built architecture enables to advantageously queried the net to predict the result of new datasets.

Cone Penetration Testing 2022 – Gottardi & Tonni (eds)
© 2022 the Author(s), ISBN 978-1-032-35624-2

Geotechnical zoning of deltaic and alluvial soils of Guayaquil (Ecuador) using CPT and N_{kt} calibration based on FVT

J. Paredes & F. Illingworth
Subterra, Ecuador

R. Luque
Geosísmica, Ecuador

ABSTRACT: This document summarizes the results obtained from geotechnical zoning analysis in the city of Guayaquil and its surroundings, which stratigraphy consists mostly of alluvial and estuarine deltaic soils, by compiling over 600 CPTs and 800 boreholes. Products generated include a map of rigid stratum depth with high impedance contrast as well as spatial distribution of liquefaction susceptibility using a simplified methodology. Furthermore, shear strength s_u peak and remolded have been evaluated by comparing CPT-borehole pairs and 26 sites with CPT and electric FVT. It has been found that s_u peak from laboratory tests adjusts well to the FVT s_u remolded and f_s, suggesting sample disturbance. It has been verified that N_{kt} values measured with FVT are lower than average reported values in previews studies, possibly due to diatoms identified in scanning electron microscope (SEM). F_r based estimates appear to yield an upper limit of N_{kt} factor in Guayaquil soft clays, while s_u maps for 5 m soil segments are provided from CPT-based N_{kt} values.

Cone Penetration Testing 2022 – Gottardi & Tonni (eds)
© 2022 the Author(s), ISBN 978-1-032-35624-2

Use of piezocone with dissipation tests CPTu_Δu, in tailings dams in Mexico: Case history

J.L. Rangel-Núñez
Universidad Autónoma Metropolitana, Plantel Azcapotzalco, Mexico

E. Ibarra-Razo & R. Flores-Eslava
inGeum Ingeniería, Mexico

ABSTRACT: This paper describes and analyzes several aspects observed during the exploration of mine tailings deposits in Mexico using the CPTu_Δu, test: high ground temperatures, interstratification, chemical reaction, dissipation curves with negative pore pressure, and S-wave profile. It was observed that the use of sensors with temperature compensation for the q_T and f_s measurements produces satisfactory results if temperature changes are not abrupt. For the case of the dynamic pore pressure sensor u_2, where no temperature compensation was made, the measurements are within the range mentioned in the standard except in the case when the temperature increases to values of 50°C or higher. The presence of high interstratification between thin layers of hard and soft soils, with different degrees of saturation, leads to errors in the measurement of the tip resistance and can produce deviations from the verticality of the cone, making it necessary to carry out corrections. The different types of dissipation curves observed in the tailings dams studied, where the soils are partially saturated, are generally the same as those described by Sully *et al.* (1999), with the particularity that in some cases, the equilibrium pressure reaches negative values. Finally, the S-wave profile obtained with correlations reproduces the general trend of the measured S-wave profile. Nevertheless, it was not possible to capture the variations in this type of deposit. In the case history described in this paper, the maximum values were not reproduced. Therefore, in this type of deposit, it is advisable the direct measurement of the shear wave and to use the smaller intervals of measurement according to the technique used.

Cone Penetration Testing 2022 – Gottardi & Tonni (eds)
© 2022 the Author(s), ISBN 978-1-032-35624-2

Assessment of the spatial variability of a Croatian flood embankment using the cone penetration test

C. Reale
University of Bath, UK

M.S. Kovačević & M. Bacic
University of Zagreb, Croatia

K.G. Gavin
Delft University of Technology, The Netherlands

ABSTRACT: Understanding how soil varies spatially is necessary in order to accurately quantify the reliability of geotechnical infrastructure. For long linear infrastructure such as flood embankments, incorporating vertical and horizontal scales of fluctuation can have a significant impact on stability assessments. This paper presents preliminary results and discussion from a field test designed to determine the vertical and horizontal scales of fluctuation of a Croatian flood embankment. A series of 15 CPTUs were carried out over a 200m length of the embankment with a Multi-channel Analysis of Surface Waves (MASW) survey done on the same section. CPT spacing was designed specifically to determine horizontal variation with multiple CPTs carried out in close proximity to each other. There was significant variation in soil stratigraphy over the embankment section with pockets of increased strength and stiffness showing up in the MASW and CPT results. This paper discusses dealing with horizontal correlation in challenging deposits and presents initial findings from the underlying sand layer.

Cone Penetration Testing 2022 – Gottardi & Tonni (eds)
© 2022 the Author(s), ISBN 978-1-032-35624-2

Prediction of bearing capacity and settlement using penetrometer design method for shallow foundation and load transfer curves

P. Reiffsteck & F. Szymkiewicz
GERS-SRO, Université Gustave Eiffel, IFSTTAR, Marne la Vallée, France

M.A. Benz Navarrete & T.A. Luong
Research and Development, Sol-Solution, Riom Cedex, France

ABSTRACT: This paper presents a simple, economic, and sound design method that could be easily implemented worldwide, based on a new generation version of Dynamic Cone Penetrometer Test (DCPT) devices giving access to the whole cone resistance versus penetration depth of the cone curve, to predict the bearing capacity, as well as to estimate the load – settlement curve of a shallow foundation which is partly inspired from the French Penetrometer design method. A methodology is then proposed in this study for the conception of shallow foundation based on load transfer curves obtained from DCPT.

Cone Penetration Testing 2022 – Gottardi & Tonni (eds)
© 2022 the Author(s), ISBN 978-1-032-35624-2

Probabilistic analysis of gyttja undrained strength from CPTU data for slope stability analysis

S. Rios, L. Sousa & A. Viana da Fonseca
CONSTRUCT-GEO, Faculty of Engineering, University of Porto, Porto, Portugal

P. Milheiro-Oliveira
CMUP, Faculty of Engineering, University of Porto, Porto, Portugal

O. Hededal
COWI A/S, Kongens Lyngby, Denmark

ABSTRACT: The design of an artificial peninsula near Copenhagen led to an extensive experimental campaign where a significant number of CPTU tests were performed. The CPTU data presented a significant dispersion due to material heterogeneity, which motivated a probabilistic analysis of the soft soil undrained strength. For that purpose, a probabilistic density function of the undrained strength ratio was estimated and introduced as input in a limit equilibrium slope stability analysis. This enables the evaluation of the probabilistic density function of the factor of safety, the probability of failure and the reliability index instead of a single deterministic factor of safety. This information can be used by the designer or project owner to choose between different solutions based on the required level of risk.

Cone Penetration Testing 2022 – Gottardi & Tonni (eds)
© 2022 the Author(s), ISBN 978-1-032-35624-2

Monitoring ground improvement using in situ tests in Guayaquil, Ecuador

F. Ripalda, D. Falquez & D. Besenzon
ESPOL Polytechnic University, Escuela Superior Politécnica del Litoral, ESPOL, Department of Earth Sciences, Guayaquil, Ecuador

R. Luque
Geosísmica, Guayaquil, Ecuador

F. Illingworth
Subterra, Guayaquil, Ecuador

S. Amoroso
University of Chieti Pescara, Pescara, Italy
Istituto Nazionale di Geofisica e Vulcanologia, Italy

ABSTRACT: The present work describes the use of the seismic dilatometer test (SDMT) and the piezocone test (CPTu), to assess the effects of ground improvement at a wastewater treatment plant in Guayaquil, Ecuador. The ground improvement consisted of 15 m-long, 0.55 m-diameter and 2 m-spacing stone columns built with vibro-replacement technique. The tests were carried out both in natural and in treated soils to compare the variation of the geotechnical parameters in the analyzed deposits. The results show specific sensitivity of the DMT over the CPTu tests to the ground improvement into the layer composed of sands and sandy silts, while VS values show a limited increase in the treated area.

Cone Penetration Testing 2022 – Gottardi & Tonni (eds)
© 2022 the Author(s), ISBN 978-1-032-35624-2

Plate anchor capacity estimation through CPT tip resistance in sand

A. Roy & S.H. Chow
Department of Infrastructure Engineering, the University of Melbourne, Australia

ABSTRACT: Reliable estimation of plate anchor uplift capacity in sand through analytical and empirical equations is often complicated due to uncertainties in estimation of soil properties required in the equations. In order to address this uncertainty, this study proposes a correlation to estimate plate anchor vertical uplift capacity in sand based on cone tip resistance measured from cone penetrometer tests (CPT). The correlation was established using a database of reported centrifuge experiments on circular, rectangular and strip anchors in loose and dense silica sand at various embedment depths and *g*-levels, along with the corresponding centrifuge CPTs performed in the same testing boxes. The centrifuge cone tip resistances were also depth-corrected to remove the effect of shallow embedment. Through regression analyses, the correlation between plate anchor capacity and cone tip resistance in dimensionless form was developed, with different coefficients fitted for circular, rectangular and strip anchors respectively.

Cone Penetration Testing 2022 – Gottardi & Tonni (eds)

CPT results and installation parameters for CFA piles in pyroclastic soils

G. Russo & M. Ramondini
Department of Civil and Environmental Engineering, University of Napoli Federico II, Italy

A. Vecchietti
Department of Engineering, University of Perugia, Italy

G. Russo
Department of Earth Science, Environment and Resources, University of Napoli Federico II, Italy

ABSTRACT: A large number of bored CFA piles was installed during the construction of a big treatment plant in Poggiomarino (Naples), where the subsoil is formed by sub-horizontal strata of alluvial soils of pyroclastic origin interbedded with organic silt layers, overlying the pyroclastic base formation. The installation parameters of CFA piles (rate of revolution, rate of penetration, torque, etc.) have been recorded during construction. In the paper the statistical correlation between CPT results and monitoring data of CFA piles installation was analysed. The statistical correlation allowed evaluating the bearing capacity of each CFA pile based on the monitoring of installation data. The correlation was developed by considering different pile groups located around the CPT verticals, in order to analyse the sensitivity to the spatial variability of geotechnical subsoil properties. The statistical based prediction of bearing capacity was validated by considering two load tests on pilot CFA piles at the study site.

Cone Penetration Testing 2022 – Gottardi & Tonni (eds)
© 2022 the Author(s), ISBN 978-1-032-35624-2

Validation of CPT-based methods for estimation of footing settlement in sand

V.A. Sakleshpur, M. Prezzi & R. Salgado
Lyles School of Civil Engineering, Purdue University, West Lafayette, IN, USA

ABSTRACT: A variety of design methods have been proposed to estimate the settlement of footings in sand using cone penetration test (CPT) results. Yet, due to the limited number of well-documented, high-quality footing load tests, efforts still need to be made to assess the performance of these methods. This paper presents two case histories of full-scale, instrumented load tests performed on axially loaded, square footings in silica sand reported in the literature. The performance of the traditional Schmertmann method and more modern CPT-based methods are evaluated through a detailed comparison between the predicted footing load-settlement curves and those obtained from footing load test measurements. For the case histories considered, these modern methods are shown to generally produce more accurate, reliable, and consistent predictions of footing response in sand than the traditional Schmertmann method.

Cone Penetration Testing 2022 – Gottardi & Tonni (eds)
© 2022 the Author(s), ISBN 978-1-032-35624-2

Derivation of SRD for driven piles from CPT data

J.A. Schneider
U.S. Army Corps of Engineers, St. Paul, MN, USA

M.F. Randolph & J.P. Doherty
The University of Western Australia, Australia

ABSTRACT: Measurements from the cone penetration test are analogous to parameters used in assessing the static component of the soils pile driving resistance. This paper presents comparisons of pile driving data from 610 mm open ended pipe piles driven through lake clays into underlying glacial till, with estimates of driving resistance using the web-based software, IMPACT. IMPACT uses a continuum-based soil model that explicitly separates internal and external shaft friction in the soil-pile model. CPT based expressions for evaluating the static resistance to driving include factors that account for end condition, friction fatigue and differences in the diameter of the CPT and the pile. Evaluation of static tension load test results requires additional estimates of setup and relative contribution of driving resistance from the soil plug.

Cone Penetration Testing 2022 – Gottardi & Tonni (eds)
© 2022 the Author(s), ISBN 978-1-032-35624-2

New method for assessing soil liquefaction resistance using a cyclic cone penetrometer

A. Sharma, P. Rapanakis, F. Incardona, C. Dano, L. Sibille & B. Chareyre
University of Grenoble Alpes, CNRS, Grenoble INP, 3SR, Grenoble, France

H.H. Sadrabadi
Equatech R&D, Equaterre, Annecy, France

ABSTRACT: This paper introduces a new methodology and machinery to extract soil properties and assess its liquefaction susceptibility. A mechanical Gouda tip cone penetrometer is used to apply a cyclic load on the soils. This test aims a load closer to real world liquefaction events, more robustness and a cheaper cost than a conventional CPTu. Since the cone tip used for the test does not have a pore pressure sensor, nor a friction sleeve and electrical or optical cables, it is of low maintenance, easy to operate and can be used to drill through gravel or any dense layer to reach layers of interest without digging a borehole, saving time and money. The liquefaction resistance is found using the response of soils to cyclic loading on the cone tip. The methodology also involves measuring the change of tip resistance before and after applying a cyclic load and correlates it to the volume changes that took place during it. For this, an experimental campaign was carried out in a new calibration chamber using a mini cone with stress and pore pressure changes measured independently in the soil during the cyclic load. The different response to cyclic load of clean Fontainebleu GA39 sand under different vertical stresses, saturation conditions and created by different sample preparation methods is studied. In turn, the pore pressure increase and the corresponding number of cycles during the cyclic cone penetration tests are compared with similar results from cyclic compressive triaxial tests. This allows us to find a relation between the loading applied in the two methods to cause a similar increase of pore pressure. Hence, a method is suggested to compare the loading applied by the cone penetrometer to the Cyclic Stress Ratio(CSR) expected in a liquefaction event. Eventually, using the CSR and CRR defined, the liquefaction susceptibility of the soils can be calculated.

Cone Penetration Testing 2022 – Gottardi & Tonni (eds)
© 2022 the Author(s), ISBN 978-1-032-35624-2

Evaluation of deformation modulus during Cone Loading Tests (CLT) and settlement of shallow foundations

A. Teyssier, M. Rispal & C. Jacquard
Fondasol, Avignon, France

P. Reiffsteck
GERS-SRO, University Gustave Eiffel, IFSTTAR, Marne-la-Vallée, France

ABSTRACT: The static loading of a piezocone is one of the only in-situ tests leading to a measure of compression of soil under the cone, shear stress on the shaft and pore pressure between these two points. These three load-deformations curves give valuable information usually impossible to reach with another ground investigation technique. The incremental loading of the cone up to failure, called Cone Loading Test (CLT), provides the vertical loading/unloading soils curve. From this curve, an estimation of deformation modulus, which is a key parameter in structure design such as foundations, is possible. CLT's deformation modulus will be compared to modulus obtained from traditional investigation methods (Menard pressuremeter tests). An experimental survey, including pressuremeter tests (PMT), Cone penetration test (CPT) and shallow foundation loadings tests, was undertaken to compare the results with CLT moduli measurements. A calculation procedure, based on CLT's deformation modulus, is proposed to design shallow foundations settlement. The article presents prediction of long-term foundation settlements (i.e. effect of time) both with the conventionnel methods and the new CLT method.

Cone Penetration Testing 2022 – Gottardi & Tonni (eds)
© 2022 the Author(s), ISBN 978-1-032-35624-2

Assessment of the potential for liquefaction using CPTu in the tailings dam I – Feijão

M. Tincopa
Universidad de Ingeniería y Tecnología - UTEC, Peru

G. Carnero-Guzman
Golder Associates Pty. Ltd., Australia

ABSTRACT: In Brazil in 2019, the failure of the tailings dam I - Feijão costed a total of 7 billion dollars in civil and environmental reparation. The report of the expert panel that evaluated the failure of dam I concluded that the failure was due to static liquefaction, which is one of the most common dam failures in the world. However, the identification of the liquefaction potential has challenges such as identifying the behavior of the tailings (contractive and dilative behavior) and extracting undisturbed samples for sophisticated laboratory tests. CPTu is a fast and reliable field test, and, through empirical relationships, the liquefaction potential of soil can be estimated using the state parameter (ψ). The objective of this article is to present the results of piezocone (CPTu) in the different in situ geotechnical campaigns in Feijão Dam with focus on the state parameter (ψ) data used to determine the liquefaction potential. The methodology used herein is that $\psi > -0.05$ means contractive behavior or high change of potential for liquefaction. The results show that the CPTu tests in dam I - Feijão showed that there was high liquefaction potential during the geotechnical campaigns (2005, 2016, and 2018) by using this procedure.

Cone Penetration Testing 2022 – Gottardi & Tonni (eds)

Is CPT a suitable in situ test for characterizing gravely sands?

G. Togliani
Geologist, Massagno, Switzerland

ABSTRACT: A recent study (Han et al, 2019) showed that CPTs performed in sandy soils with a variable but sometimes significant gravel content, despite special operational measures (alternating drilling and CPT), frequently provide unrealistic q_c values which adversely affect the piles capacity predictions even using the most representative CPT-based design methods. However a series of SPT verticals are also available and the conversion of N_{30} into q_c values has allowed, via a number of steps, the development of four virtual CPTu, two of which were then used for the pile capacity predictions This procedure has proven to be effective, so that, even employing far less authoritative methods, were obtained sufficiently approximate shaft and toe resistances.

Cone Penetration Testing 2022 – Gottardi & Tonni (eds)
© 2022 the Author(s), ISBN 978-1-032-35624-2

A CPT-based diameter-dependent m-θ spring model for lateral pile analysis

J. Tott-Buswell & L.J. Prendergast
Department of Civil Engineering, University of Nottingham, Nottingham, UK

ABSTRACT: Offshore Wind Turbines (OWT) are a successful renewable energy solution; however, emerging turbine sizes require pile geometries beyond the calibration range of existing design standards. This necessitates soil sampling and rigorous finite element modelling, which is problematic for quick design estimates. Cone Penetration Test (CPT)-based *p-y* methods can provide preliminary deflection estimates, although their applicability becomes increasingly uncertain as pile slenderness ratios, length normalized by diameter (*L/D*), reduce. This is due to the increase in diameter incurring additional resistances that *p-y* models alone cannot account for. To incorporate the additional resistance, this paper defines a CPT-based moment-rotation (m-θ) model by rescaling empirically derived axial capacity functions (known as τ-w curves). Various monopile dimensions are simulated and pile-head displacements are compared for CPT-based *p-y* models with and without m-θ springs. The net effect of incorporating m-θ springs increases as monopile diameters (and rigidity) increase and diminishes as piles increase in slenderness.

Cone Penetration Testing 2022 – Gottardi & Tonni (eds)

An investigation into the use of the Vibdrive and β-methods for calculating the SRV of offshore piled foundations

M.P. Trubshaw & T. Joseph
Ramboll, London, UK

G. Giuliani
COWI, London, UK

ABSTRACT: Offshore piled foundations are typically installed using an impact hammer. Increasingly, more attention is being given to use of vibratory hammers due to added benefits of prolonged pile life, reduced installation noise, and reduced risk of observing pile-run. This paper considers a case study in the North Sea, consisting of six positions where piles are installed in layered sands and clays. The Cone Penetration Test (CPT) data for the positions have been used in calculating the Soil Resistance to Vibratory Driving (SRV) as part of the Vibdrive method, from Holeyman (2002). This method has been compared with the SRV calculated using the non-CPT based β-method, from Jonker (1986). By varying the SRV used in driveability predictions, the results can be compared with the available driving records. It was found that the CPT based Vibdrive method consistently produced more representative results while still maintaining a conservative prediction.

Cone Penetration Testing 2022 – Gottardi & Tonni (eds)
© 2022 the Author(s), ISBN 978-1-032-35624-2

Potential of the Cone Pressuremeter Test for obtaining stiffness degradation for offshore wind turbine monopile foundations

G.W. Tucker, C.T. Leth, L. Krogh & P. Ladefoged
Ørsted

T. Lunne
Norwegian Geotechnical Institute

M. Taylor
In Situ Site Investigation

ABSTRACT: The piezocone (CPTU) offers a quick and repeatable investigation tool with instantly available high-resolution data. For these reasons it is both technically and commercially attractive to further instrument the CPTU to acquire additional measurements with more sensors either within the cone itself or as add-on modules behind the cone. Many ideas for further instrumentation of the CPTU have been conceptualised or tested but one that could offer direct benefit for offshore wind turbine foundation design is the Cone Pressuremeter (CPM). This in situ tool includes a pressuremeter module behind the standard cone that can measure ground displacement or expansion as a function of pressure applied during the loading, unloading and reloading of the surrounding soil when penetration is paused. From these measurements a number of soil parameters may be interpreted including the stiffness degradation (G/G_0) of the soil when combined with reliable in situ or laboratory measured G_0 values. With increasingly larger wind turbines being used and deeper water sites selected for construction, soil stiffness is an increasingly critical input for the design of many offshore wind turbine foundations. CPM testing has been trialed in the field onshore at a glacial till site and a medium to very dense sand site. The results of the CPM are compared in this paper to site specific results for the self-boring pressuremeter and advanced laboratory-based alternatives. The practical considerations of using the CPM offshore in seafloor mode are also reviewed.

Cone Penetration Testing 2022 – Gottardi & Tonni (eds)

Centrifuge study on the CPT based *p-y* models for the monopiles

Huan Wang & Daan Veldhuijzen van Zanten
Delft University of Technology, The Netherlands

D.A. de Lange
Deltares, The Netherlands

Federico Pisanò, K.G. Gavin & Amin Askarinejad
Delft University of Technology, The Netherlands

ABSTRACT: The *p-y* load transfer method is the most widely used for the design of monopile foundations. Recently, many CPT based *p-y* models were proposed experimentally or numerically to overcome the uncertainty in deriving stiffness and strength parameters of soil from in-situ testing. This paper aims to evaluate the performance of existing CPT based *p-y* models for predicting the monopile behavior, by conducting a series of centrifuge tests in dense Geba sand. Four different diameter (D = 0.9 m, 1.8 m, 1.25 m and 2.5 m) monopiles with two aspect ratios (L/D) of 3.6 and 5 were studied. The lateral load was applied at two different loading eccentricities of $0.67L$ and L above the ground surface. It was found that the model of Novello (1999) model gave the best overall prediction (at both small and large deflection) for the tests in this study, while the model of Dyson and Randolph (2001) model always underestimate the pile response. Compared with the Novello (1999), the model of Suryasentana and Lehane (2014) and Li et al. (2014) will overestimate the pile response at large deflection, but, can well predict the pile response at small deflection. In all cases, the API *p-y* model significantly overestimates the response with a factor more than 2. The experimental observations in this study proves the feasibility and reliability of developing CPT based *p-y* model for monopile.

Cone Penetration Testing 2022 – Gottardi & Tonni (eds)
© 2022 the Author(s), ISBN 978-1-032-35624-2

Using near-surface CPT data to predict foundation skirt embedment in partially drained carbonate sands

H.M. Wroth, M.F. Bransby & C.D. O'Loughlin
Centre for Offshore Foundation Systems, Oceans Graduate School, The University of Western Australia, Australia

M.F. Silva, M. Cocjin, N. Levy & H.E. Low
Fugro Australia Marine, Australia

ABSTRACT: A series of CPT and skirt penetration tests have been conducted in carbonate silty sand in a range of drainage conditions ranging from fully drained to fully undrained. The drainage transition response for each has been quantified in order to provide guidance in how to use the results from field CPTs to predict soil resistance during the installation of skirted foundations.

Cone Penetration Testing 2022 – Gottardi & Tonni (eds)

The role of cone penetration testing in the Dogger Bank offshore wind farm

Tor Inge Yetginer-Tjelta, Jérôme De Sordi & Leonardo Caferri
Equinor, Stavanger, Norway

Mike Rose
SSE Renewables, Perth, Scotland

Callum Duffy
Logos Geoservices Ltd, Oxford, UK

Tom Lunne, Øyvind Blaker, Stein Strandvik & Vaughan Meyer
NGI, Oslo, Norway

ABSTRACT: The cone penetration test is the preferred tool for measuring the in-situ soil conditions for offshore wind farm developments. This preference stems from the consistent and reliable data quality, low relative cost, and speed of data acquisition which enables developers to characterise large areas with numerous structure locations to support foundation concept feasibility studies, installation risk assessments and geotechnical engineering analyses to optimise the foundation design by reducing the uncertainty in soil input parameters. However, the interpretation of CPT data is largely dependent on site-specific or published correlations, which can present challenges when relying on the CPT as the primary tool for stratigraphic profiling, interpretation of soil behaviour and for defining geotechnical parameters for engineering design. This paper presents recent developments and processes that enabled the delivery of an optimised CPT based geotechnical design basis for the Dogger Bank Wind Farm. The processes and tools that enabled this approach comprised a novel SI strategy, focussed CPT interpretation and CPT tool development. The strategic approach for the novel SI strategy is presented and the individual elements described. Challenges and limitations associated with the interpretation of CPT data are discussed, including size and scale effects, resolution, calibration with laboratory and large-scale field test data and rate effects. Finally, the challenges with using statistical methods to derive design values from CPT records are briefly discussed.

Cone Penetration Testing 2022 – Gottardi & Tonni (eds)

Procedures to evaluate seismic settlement in dry sand based on CPT data – an update

Fred Yi
Mid Pacific Engineering Inc., California, USA

ABSTRACT: The seismic settlement of dry sands becomes one of the major geological hazards in areas where loose sandy deposits exist, and groundwater is relatively deep as human activities and urban development moving towards these areas. Research related to the evaluation of seismic settlement of dry sand seems paused for a period since pioneer work of Silver and Seed (1971) in the early 1970's. Later research related to the topic was found by Tokimatsu and Seed (1987) and Pradel (1998). Both papers were based on the use of Standard Penetration Test (SPT). Yi (2010) was the first attempting to evaluate the seismic settlement of dry sand using Cone Penetration Test (CPT) data. In this paper, the procedures to evaluate seismic settlement in dry sand based on CPT data was summarized and the relationship between volumetric strain and cyclic shear strain was updated for CPT data by considering various effects such as fine contents and overburden pressure based on the most recent research findings.

Author Index